ALGEBRA

ALGEBRA

G. Santhanam

Alpha Science International Ltd.
Oxford, U.K.

Algebra
320 Pages.

G. Santhanam
Department of Mathematics & Statistics
Indian Institute of Technology
Kanpur

ALPHA SCIENCE INTERNATIONAL LTD.

7200 The Quorum, Oxford Business Park North
Garsington Road, Oxford OX4 2JZ, U.K.

www.alphasci.com

ISBN 978-1-78332-302-9

Printed from the camera-ready copy provided by the Author.

To
All my Teachers

Preface

This book has evolved from nearly 15 years of teaching the courses

- Algebra to students at Indian Institute of Technology, Kanpur (IIT, Kanpur),
- Groups and Symmetry to level I participants of Mathematics Training and Talent Search Programme, India (MTTS Programme) and
- Elementary Number theory (Two weeks course) to level O participants of Mathematics Training and Talent Search Programme, India.

At IIT Kanpur there were students from other sciences and Engineering as well. For some of these students it was of interest to learn some advanced mathematics beyond Calculus and elementary linear algebra while for some of them it was foundation for future advanced courses. I felt that these students should be taught some of the main themes in abstract algebra (in one semester). At the same time I also decided that concrete example should precede abstract definitions. I believe that concrete examples would interest the students and it would make it easy for the students to understand the abstract definitions and also appreciate the need for the abstraction. This has worked well for me in IIT, Kanpur and MTTS Programme and this approach was appreciated by the students as being somewhat different from the existing books.

The book contains a wealth of examples and exercises with varying levels of difficulty- quite a few of them drawn from other branches of mathematics. Further concrete examples are worked out illustrating a concept or result. For some of the exercises, I have included a solution right after the exercise as a separate paragraph of less width. I would like the students to try and find a possibly different solution than the one I have given, before taking the easy route!

I have also departed from the tradition of listing exercises at the end of each chapter. I find that, when exercises are given at regular intervals after each section, the students are able to see the results and conepts that are needed to solve the problems. This helps them in understanding and appreciating the results and concepts they have learned and also builds their confidence!

Now I turn to the contents of the book. My general approach is holistic and I hope the readers will appreciate this as did the students of my course.

In Chapter 1 I discuss some basic concepts in Elementary Number Theory. While discussing divisibility properties, irreducibility and primality of an integer is defined separately and shown to be equivalent. When the readers encounter irreducible and prime elements in Ring Theory they are able to appreciate these conecpts. Then I discuss Linear Diophantine Equations and Linear Congruences and end this chapter with a proof of Chinese Remainder Theorem. I find that the students who learn this first are able to appreciate Group Theory and Ring Theory better. This chapter also serves as a gentle introduction to the art of writing proofs and abstraction. This helps the sudents immensely.

Chapter 2 discusses Elementary Group Theory. The first two sections deal with simple properties of groups and subgroups. It also provides some practice in writing proofs of elementary results. One of the first and most important result in finite groups, namely Lagrange's Theorem (Theorem 2.3.1) is proved in Section 2.3. To bring out the significance of this beautiful result, I have proved few results- Theorems 2.3.9, 2.3.11- which are typically proved using Sylow's type argument.

Groups were invented to study the symmetries of field extensions and since symmetry is common in all branches of sciences it is still one of the ways Group Theory is applied. In section 2.8, we study symmetric groups or permutation groups and Dihedral Groups as Group of symmetries of regular n-sided polygon.

In the next Chapter, Ring theory is studied. Starting with the divisibility of integer in school arithmetic to finding roots of polynomials and factorizing a polynomial, the idea of factorization has been the recurring theme in theory of rings. To bring out this aspect I have emphasised on factorization in Rings. To do this I have introduced divisibility in rings very early.

Proving Euclidean domains are Principal ideal domains and Principal ideal domains are Unique factorization domians have been the standard practice to show the unique factorization in Euclidean domains and Principal Ideal domians. In this book I have proved this directly. The treatment of factorization of polynomials also somewhat different in this book. (See sections 3.9.3 and 3.9.4).

In Chapter 4 I do Galois Theory. Historically Abel proved the impossibility of solving the general quintic equation in radicals before Galois. Further the works of Abel and Galois laid the foundation for modern Galois Theory and Group Theory. Hence my aim in this chapter is to bring this out and prove these two outstanding results efficiently (Theorems 4.2.2 and 4.2.4). I believe this is one of the high points of this book. I also find that through this approach the students are able to understand and appreciate modern Galois Theory better and further they are able to read advanced topics in Galois Theory on their own.

In the last Chapter I discuss some of the advanced topics in Group Theory. This chapter is more geometric in nature with group action playing an important role. To bring out the importance of Cauchy's theorem, I prove Theorem 5.2.5, Corollary 5.2.9 and Lemma 5.2.8 as applications of Cauchy's Theorem. Typically these results are proved using Sylow type argument. The proofs of Sylow's theorem are geometric. I also discuss finite subgroups of $E(n)$, the euclidean motion group, usually not done in an algebra course.

Chapter 1 and sections 2.1-2.8 of Chapter 2 can be used as a one semester course at the undergraduate level. If the students are already familiar with elementary number theory Chapter 2 and 5 can be one semester course at the undergraduate level and Chapters 3-5 can be used as a one semester course at the graduate level.

There are a few prerequisites for this book. I expect the readers to be familiar with calculus, elementary linear algebra, complex numbers and their basic properties. A nodding acquaintance of topology will be an added advantage to understand some of the examples, though not essential.

Acknowledgements

1. I thank the students of IIT Kanpur, the participants of MTTS program who showed appreciation for my way of teaching algebra and urged me to write this as a text book.

2. I thank the MTTS faculty who went through my earlier notes and used them in their courses. They made suggestions to improve the text.

3. I thank my teacher Professor S Kumaresan. I can't express my gratitude to him in words. He has played a very decisive role in my mathematical development. When he saw the rough outline of my course at IIT Kanpur he liked it very much and encouraged me to write it as a textbook. In the last chapter on Field theory he clarified many points.

4. I thank my wife Mangai and my children Mullai and Sudar for their loving support while I was busy with the preparation of the book.

I welcome suggestions for improvements and they may be sent to
`santhana@iitk.ac.in`

G Santhanam

Important though the general concepts and propositions may be with which the modern industrious passion for axiomatizing and generalizing has presented us, in algbera perhaps more than anywhere else, nevertheless I am convinced that special problems in all their complexity constitute the stock and core of mathematics, and that to master their difficulties requires on the whole harder labour.

Hermann Weyl

(From the Preface to *The Classical Groups*, Princeton University Press, 1939)

Contents

Chapter 1

Elementary Number Theory

This chapter is devoted to studying fairly elementary concepts and results in number theory, namely, divisibility properties of integers, the gcd and lcm of two integers, Division algorithm, Fundamental theorem of arithmetic, linear congruences and Chinese remainder theorem.

Many of the readers may have known these concepts and results. However we study these formally in this chapter. There are very good reasons for doing so. As we will see in later chapters, the properties of integers we study here are valid in more general objects such as polynomials and this chapter prepares the readers to appreciate abstract concepts and the art of writing proofs.

1.1 Divisibility

Definition 1.1.1. Let a and b be two integers. We say that the integer a divides the integer b if there exists $m \in \mathbb{Z}$ such that $b = ma$.

If the integer a divides the integer b, we say that a is a divisor of b and write $a|b$. We also say that b is a multiple of a.

If a does not divide b then we write $a \not| b$.

We now record some immediate consequences of Definition 1.1.1.

Properties 1.1.2. Let $a, b, c \in \mathbb{Z}$. Then the following holds.

1. If $b \in \mathbb{Z}$ is nonzero, then ± 1 and $\pm b$ are divisors of b. (They are called trivial divisors of b.)
2. If $a, b \in \mathbb{N}$ and $a|b$, then $1 \leq a \leq b$.
3. If $a|b$ and $b|a$, then $a = \pm b$.
4. If $a|b$ and $b|c$, then $a|c$.
5. If $a|b$, then $a|kb$ for any $k \in \mathbb{Z}$.

6. $a|b$ if and only if $ka|kb$ for every *nonzero* integer k.

7. If $a|b$ and $a|c$, then $a|xb + yc$ for any $x, y \in \mathbb{Z}$.

Definition 1.1.3. Let a and b be two integers. If a divides b and a is not equal to $\pm 1, \pm b$, we say a is a proper divisor of b.

Exercise 1.1.4. If a and b are two nonzero integers and a is a proper divisor of b, then $|a| \leq \lfloor \frac{|b|}{2} \rfloor$ where $\lfloor \frac{|b|}{2} \rfloor$ denotes the largest integer less than or equal to $\frac{|b|}{2}$.

If a divides b, then there exists an integer q such that $b = aq$. Since a is a proper divisor of b, it follows that $|q| \geq 2$. Hence $|b| = |a||q| \geq 2|a|$ and $|a| \leq \lfloor \frac{|b|}{2} \rfloor$.

Exercise 1.1.5. 1. Show that an integer k divides $k + 2$ iff $k = 1$ or 2.

2. Find integers k such that k divides $k + 3$.

3. Let $a \geq 2$. Show that $a - 1$ divides the integer $a^n - 1$ for every $n \in \mathbb{N}$.

4. Let $a \geq 2$. Show that $a + 1$ divides $a^n + 1$, if n is odd. What happens if n is even?

We let $\langle a \rangle \equiv S_a \equiv \mathbb{Z}a := \{ka : k \in \mathbb{Z}\}$ denote the set of all multiples of a. In view of Properties 1.1.2, if $s, t \in S_a$, then $s + t \in S_a$, $-s \in S_a$ and $xs + yt \in S_a$ for all $x, y \in \mathbb{Z}$. (If you have a smattering of group theory, this means that S_a is a (sub)group of integers.) We shall see this type of sets play an important role in dealing with the divisor related investigations.

If $s \in S_a$ and $n \in \mathbb{Z}$, then $ns \in S_a$. (This along with the earlier observations says that S_a is an ideal in the ring $\mathbb{Z}$. We will see this in detail in Chapter 3)

Definition 1.1.6. Let $1 < a \in \mathbb{N}$. We say that a is *reducible* or *composite* if a has divisors other than ± 1 and $\pm a$. Note that if a is reducible, then we can write $a = mn$ where $1 < m, n < a$ etc. That is, a can *factored* into a product of *proper* divisors. The expression $a = mn$ is called a factorization of a.

A positive integer $a > 1$ is said to be *irreducible* if a is NOT reducible. That is, if d is a natural number and d divides a, then $d = 1$ or $d = a$.

Proposition 1.1.7. *Let $1 < a \in \mathbb{N}$. Then there exist irreducible numbers which divide a.*

Proof. If a itself irreducible, we are through. If not, there exists a proper positive divisor of a. Thus the set D of positive integers which are proper divisors of a is nonempty. Let p be the least element of the set. Then p is irreducible. For, if p is reducible, there exists a proper divisor, say, $d > 1$ of p. Hence $1 < d < p$ and d is also a nontrivial divisor of a, contradicting the minimality of p. □

Theorem 1.1.8. *Any positive integer $n > 1$ is either irreducible or is a product of finite number of irreducible numbers.*

Proof. We prove this result by induction on n. The induction statement $P(n)$ is *every natural number n is a product of finite number of irreducible numbers.*

Let $n = 2$. Then 2 is irreducible and we are through.

Now let us assume that all natural numbers $k < n$ are product of finite number of irreducible numbers. If n is irreducible, we are through. If not we can write n as $n = n_1 n_2$ where $n_i \leq \lfloor \frac{n}{2} \rfloor < n$ and by induction each n_i is a product of finite number of irreducible numbers. Thus n itself is a product of finite number of irreducible numbers and the proof is complete.

Another proof: If n is not irreducible, it has an irreducible divisor, say, p_1. Hence we can write $n = p_1 n_1$ where $n_1 = \frac{n}{p_1} < n$ and by induction, the natural number n_1 is a product of finite number of irreducibles. This completes the proof.

Sketch of another proof: If the result were false, there exist reducible elements in $\mathbb{N}$ which are not finite product of irreducible numbers. Let $N > 1$ be the smallest such. By Proposition 1.1.7, there exists an irreducible number $p \geq 2$ which divides N. Hence $N = kp$. Then $k < N$ and hence it is a product of finitely many irreducible numbers. As a consequence N is a product of finitely many irreducible numbers, a contradiction. □

Theorem 1.1.9 (Euclid's Theorem). *There exist infinitely many irreducible numbers in $\mathbb{N}$.*

Proof. Let $p_1, \ldots, p_k$ be a finite set of irreducible numbers of $\mathbb{N}$ and let $N := 1 + p_1 p_2 \cdots p_k$. By Proposition 1.1.7, there exists an irreducible number p dividing N. If p is one of the p_i's, then p divides $1 = (1 + p_1 \cdots p_k) - (p_1 \cdots p_k)$, a contradiction. Thus given any finite set of irreducible numbers, there is an irreducible number different from the ones we started with. This proves that there are infinitely many irreducible numbers. □

Theorem 1.1.10 (Division Algorithm). *Let a and b be two integers such that $a > 0$. Then there exist unique $q, r \in \mathbb{Z}$ such that $b = qa + r$ where $0 \leq r < a$.*

Proof. The proof is by the method of repeated subtraction.

Consider the set $S := \{b - ak : k \in \mathbb{Z}\}$. If $b > 0$, then $b = b - a.0$ in S is non-negative. If b is negative and if we take $k = b$, then we can write $b - ab = (-b)(a - 1) \geq 0$. Thus the set S contains non negative integers. Hence by well ordering principle, the set S has smallest non-negative element, say, r and we can write it as $r = b - aq$ for some $q \in \mathbb{Z}$. If $r \geq a$, then $0 \leq r - a = b - aq - a = b - a(q+1)$ is also in S and $r - a < r$, a contradiction to the minimality of r. Thus $0 \leq r < a$.

We will now prove the uniqueness of q and r. Let q_1, r_1, q_2 and r_2 be integers such that $b = aq_1 + r_1 = aq_2 + r_2$ and $0 \leq r_1, r_2 < a$.

Since $0 \leq r_1, r_2 < a$, it follows that $0 \leq |r_2 - r_1| < a$. Hence $0 \leq a|q_1 - q_2| = |r_2 - r_1| < a$ and this proves that $0 \leq |q_2 - q_1| < 1$. As a consequence $q_1 = q_2$ and $r_1 = r_2$. □

Corollary 1.1.11. *Let a and b be two integers such that $a \neq 0$. Then there exist unique integers q and r such that $b = aq + r$ where $0 \leq r < |a|$.*

Proof. We apply Division Algorithm (Theorem 1.1.10) to the integers $|a|$ and b. □

An immediate application of Theorem 1.1.10 is the following corollary.

Corollary 1.1.12. *Let* $\emptyset \neq S \subset \mathbb{Z}$ *be such that (i) if* $u \in S$*, then* $-u \in S$ *and (ii) if* $u, v \in S$*, then* $u + v \in S$*. Then* $S = \mathbb{Z}a$ *where* a *is the least non negative integer in* S*.*

Proof. If S contains only 0, there is nothing to prove. If it contains $u \neq 0$, then $\pm u \in S$ and hence S_+, the subset of positive elements in S is nonempty. If a is the least element of S_+, we apply division algorithm to conclude that any $b \in S$ is a multiple of a. (Why?) □

Properties 1.1.13. The set of positive divisors of a (nonzero) integer is finite.

> Let $n \in \mathbb{N}$ and d be a proper divisor of n. Then it follows from Exercise 1.1.4 that $d \leq n$. Let $D_n := \{d \in \mathbb{N} : d \text{ divides } n\}$. Then $D_n \subseteq \{1, 2, \ldots, n\}$. Hence D_n is a finite subset of $\mathbb{N}$.

Definition 1.1.14. Let a and b be two nonzero integers. An integer d is a common divisor of a and b if d divides both a and b.

Properties 1.1.15. The set of positive divisors of two nonzero integers a and b is finite.

> The set of divisors of a and b is $D_a \cap D_b$. Hence it is finite.

Definition 1.1.16. Let a and b be two nonzero integers. The integer $d = \max(D_a \cap D_b)$ is called the greatest common divisor of a and b. It is often denoted by (a, b) or $\gcd(a, b)$.

Question 1.1.17. Why do we restrict ourselves to integers a and b both nonzero?

> The left out cases are in some sense trivial and can be easily handled. For example, if $a \neq 0$ then $\gcd(a, 0) = |a|$.

In most of the books the greatest common divisor is defined as follows.

Definition 1.1.18. Let a and b be two integers with at least one of them nonzero. A positive integer d is called the greatest common divisor of a and b if it satisfies the following.

1. The integer d divides the integers a and b.

2. If c is a positive integer such that c divides a and b, then c divides d.

Given $a, b \in \mathbb{Z}$, we let $\langle a, b \rangle \equiv S_a + S_b := \{xa + yb : x, y \in \mathbb{Z}\}$. Then it is easy to check that $\langle a, b \rangle$ enjoys the following properties.

1. The integers a and b are in $\langle a, b \rangle$.

2. If $u, v \in \langle a, b \rangle$, then $u + v \in \langle a, b \rangle$.

3. If $u \in \langle a, b \rangle$, then $-u \in \langle a, b \rangle$.

4. If $u \in \langle a, b \rangle$, then $ku \in \langle a, b \rangle$ for all $k \in \mathbb{Z}$.

5. For all $r, s \in \mathbb{Z}$ and $u, v \in \langle a, b\rangle$, the integer $ru + sv \in \langle a, b\rangle$.

6. $\langle a, b\rangle$ contains positive integers.

In view of Corollary 1.1.12, we see that $\langle a, b\rangle = \mathbb{Z}d$ where d is the least positive integer in $\langle a, b\rangle$.

Exercise 1.1.19. Given an integer a, and an element c in S_a, show that the set $S_c \subseteq S_a$.

Theorem 1.1.20. *Given two nonzero integers a and b, let d in $\mathbb{N}$ be such that $\langle a, b\rangle = \mathbb{Z}d = S_d$. Then $d = \gcd(a, b)$ according to Definition 1.1.18.*

Proof. Let d in $\mathbb{N}$ be such that $\langle a, b\rangle = S_a + S_b = \mathbb{Z}d$. By definition of $S_a + S_b$, the integers a and b are in $S_a + S_b = \mathbb{Z}d$. Hence there exist integers k and l such that $a = kd$ and $b = ld$. This proves that d is a common divisor of both the integers a and b.

Let c be a common divisor of the integers a and b. Then there exist integers a_1 and b_1 such that $a = ca_1$ and $b = cb_1$. Since $\mathbb{Z}d = S_a + S_b$, there exist integers x and y such that $d = xa + yb = c(xa_1 + yb_1)$. Thus c divides d and the proof is complete. □

Theorem 1.1.21 (Bezout Identity). *If $d = \gcd(a, b)$, then there exist $u, v \in \mathbb{Z}$ such that $d = au + bv$.*

Proof. We have shown above that, if $\langle a, b\rangle = \mathbb{Z}d$, then d is the gcd of the two integers a and b. Hence there exist integers u and v such that $d = au + bv$. □

This theorem will turn out to be very powerful as we will see later. Later we shall also indicate an algorithm for finding the gcd as well as the integers u, v of the Bezout identity.

We will now prove the following theorem.

Theorem 1.1.22. *Definition 1.1.16 and Definition 1.1.18 for the greatest common divisor of two integers a and b are equivalent.*

Proof. Let a and b two nonzero integers. Let d and d_1 be the gcd of a and b with according to Definition 1.1.16 and Definition 1.1.18 respectively. We will show that $d = d_1$.

By Definition 1.1.18, the integer d divides d_1. Hence $d \leq d_1$. On the other hand, since $d = \max D_a \cap D_b$ the integer d must be greater than or equal to d_1. This proves that $d = d_1$. □

Let us record some facts about the gcd's.

Proposition 1.1.23. *Let a and b be integers and $b \neq 0$. Then*

1. $\gcd(ca, cb) = c\gcd(a, b)$ *for any* $c \in \mathbb{N}$.

2. $\gcd(a/d, b/d) = 1$ *if* $d = \gcd(a, b)$.

3. $\gcd(a, b + ka) = \gcd(a, b)$ *for every integer* k.

Proof. Let a and b be two nonzero integers. Let $d = \gcd(a,b)$ and $e = \gcd(ca, cb)$.

The first part is proved by repeated application of Bezout's identity.

Since d is the gcd of a and b, there exist integers u and v such that $d = au + bv$. Therefore $cd = c(au + bv) = cau + cbv$ is in $S_{ca} + S_{cb} = \mathbb{Z}e$. That is $cd = et$ for some integer t. As a consequence $e \leq cd$.

Again we use Bezout identity and write $e = cax + cby$ for some integers x and y. Therefore $e = cax + cby = c(ax + by) = cdr$ for some integer r. Hence $cd \leq e$ and the proof of first part is complete.

Let a and b be two nonzero integers.

Let $d = \gcd(a,b)$ and $d_1 = \gcd(\frac{a}{d}, \frac{b}{d})$. Then using the first part, we conclude that $dd_1 = d\gcd(\frac{a}{d}, \frac{b}{d}) = \gcd(d\frac{a}{d}, d\frac{b}{d}) = \gcd(a,b) = d$. Thus $d_1 = 1$.

We will now prove the third part. Let a and b be two nonzero integers.

Let d be a divisor of the integers a and b. Then d divides the integer $b + ka$ for every integer k. Conversely let k be an integer and e be a divisor a and $b + ka$. Then e divides $b = (b + ka) - ka$. Thus the set of common divisors of the set $\{a, b\}$ and $\{a, b + ka\}$ are same for every integer k. As a consequence $d = \gcd(a,b) = \gcd(a, b + ka)$ for any integer k. □

Exercise 1.1.24.

1. Find $(n, n+2)$ for $n \in \mathbb{N}$.
2. Let a and b be two integers such that $(a,b) = 1$. Find $(a+b, a-b)$.
3. Let a be an integer. Show that $(a+k, a)$ divides k for every $k \in \mathbb{N}$.
4. Show that $(k, k+1) = 1$ for all $k \in \mathbb{Z}$.
5. Let $k \geq 2$ be irreducible and n be an integer such that k does not divide n. Show that $(k, n) = 1$.

Definition 1.1.25. Two integers $a, b \in \mathbb{Z}$ are said to be *relatively prime* if $\gcd(a,b) = 1$.

More generally, $a_1, \ldots, a_n$ are said to be relatively prime if 1 is the greatest common divisor of all of them.

This is same as saying that there exist $x, y \in \mathbb{Z}$ such that $1 = xa + yb$. (Why?)

Exercise 1.1.26. Let $a_1, a_2, \ldots, a_k$ be nonzero integers. Show that

$$\gcd(a_1, a_2, \ldots, a_k) = \gcd(\gcd(a_1, a_2), a_3, \ldots, a_k).$$

It suffices to show this for three nonzero integers a_1, a_2 and a_3. (Why?)

Let $d = \gcd(a_1, a_2, a_3)$, $d_1 = \gcd(a_1, a_2)$ and $d_2 = \gcd(d_1, a_3)$.

The integer d divides a_i for $1 \leq i \leq 3$. In particular d divides a_1 and a_2. Therefore d divides d_1. Again by definition, it follows that d divides $d_2 = \gcd(d_1, a_3)$. Hence $d \leq d_2$.

The integer d_2 divides d_1 and a_3. Since d_1 divides a_1 and a_2, we conclude that d_2 divides a_i for $1 \leq i \leq 3$. Hence $d_2 \leq d$.

Exercise 1.1.27. Show that if $a_1, \ldots, a_n$ are relatively prime, there exist $x_1, \ldots, x_n \in \mathbb{Z}$ such that $1 = \sum_{k=1}^{n} x_k a_k$.

More generally if d is the gcd of integers $a_1, \ldots, a_n$, show that there exist integers $x_1, \ldots, x_n$ such that $d = \sum_{k=1}^{n} x_k a_k$.

Proposition 1.1.28 (Euclid). *Let a, b and c be three integers. If $c|ab$ and if $\gcd(c, a) = 1$, then $c|b$.*

Proof By Bezout's Identity(Theorem 1.1.21), there exist integers x and y such that $xa + yc = 1$. If we multiply this equation by b on both sides, we get $b = xab + ybc$. Since c divides ab, it follows that c divides b and the proof is complete. □

Remark 1.1.29. This result is no longer true if $\gcd(c, a) \neq 1$. For, example 6 divides $12 = 3 \times 4$. However 6 does not divide 4.

Definition 1.1.30. An integer $p > 1$ is said to be a *prime number* if $p|ab$, then p divides at least one of the integers a and b.

Theorem 1.1.31 (Euclid). *An integer $p \in \mathbb{N}$ is a prime iff p is irreducible.*

Proof. If p is prime and reducible, then $p = mn$ for some $m, n \in \mathbb{N}$, with $1 < m, n < p$. Since p is prime, p divides either m or n. Suppose p divides m. Then $m = pq$ for a natural number q. This means that $p > m = pq \geq p$, a contradiction.

Let p be irreducible and a, b be two integers such that $p|ab$.

Since p is irreducible, it does not have proper divisors. Suppose p does not divide a. Then $\gcd(p, a) = 1$. (Why?)

> For if d is the gcd of p and a, then d must divide p. Hence $d = 1$ or p. If $d = p$ then $d = p$ divides a, a contradiction.

By Euclid's Lemma 1.1.28, it follows that $p|b$ and the proof is complete. □

Exercise 1.1.32.

1. Let a be an integer and p be a prime number such that p divides a^2. Then p^2 divides a^2. More generally, if p divides a^k for some $k \geq 2$, then p^k divides a^k. Show that this is not true, if p is not a prime number.

2. Let a be a natural number. Show that $a^{\frac{1}{n}}$ is a rational number iff $a = b^n$ for a natural number b.

Exercise 1.1.33. If $\gcd(a, b) = 1$ and $\gcd(a, c) = 1$, then $\gcd(a, bc) = 1$.

> Let $d = \gcd(a, bc)$. Since $(a, b) = 1$, there exist integers x and y such that $xa + yb = 1$. We multiply both sides by c and get $c = acx + ybc$. Both sides of the resulting equation are divisible by d. Hence d is a common divisor of a and c which shows that $d = 1$.

Exercise 1.1.34. Let $a|c$ and $b|c$. Assume that $\gcd(a, b) = 1$. Then show that $ab|c$. Further show by an example this is not true, if $\gcd(a, b) \neq 1$.

Now we prove Euclid's Algorithm. The active principle of this is Proposition 1.1.23. In particular, if $0 < a < b$ and if $b = aq + r$ with $0 \leq r < a$, then $\gcd(b, a) = \gcd(b, r = b - qa)$.

Hence to find $\gcd(b, a)$ we can work with the pair of 'smaller' numbers (a, r) and repeat the process. Each time the remainder 'r' is non-negative and strictly decreasing. Hence this process terminates at a finite stage. The last nonzero remainder is the required $\gcd(b, a)$.

We set up the notation: Given these two nonzero integers a and b, there exist q_1 and r_1 such that $b = aq_1 + r_1$ where $0 \leq r_1 < a$. If $r_1 \neq 0$, then we divide a by r_1 and write as $a = q_2 r_1 + r_2$ with $0 \leq r_2 < r_1$. If we continue like this, we get $a > r_1 > r_2 > \ldots \geq 0$. This must eventually get us a remainder $r_k = 0$. This will be achieved in at most a steps!

We write down the last two steps:

$$\begin{aligned} r_{k-3} &= q_{k-1} r_{k-2} + r_{k-1} \quad \text{with } 0 < r_{k-1} < r_{k-2} \\ r_{k-2} &= q_k r_{k-1} + r_k \quad \text{with} \quad r_k = 0 \end{aligned}$$

The last nonzero integer r_{k-1} is $\gcd(b, a)$.

Theorem 1.1.35 (Euclid's algorithm). *Let a and b be two integers and d be the gcd of a and b. Then $d = r_{k-1}$ in the above calculations.*

Proof. By applying Proposition 1.1.23, we get that

$$d = \gcd(a, b) = \gcd(a, r_1) = \gcd(r_1, r_2) = \cdots = \gcd(r_{k-2}, r_{k-1}) = r_{k-1}.$$

□

We will now do some numerical examples to understand the algorithm and how to read the Bezout identity from Euclidean algorithm.

Example 1.1.36. 1. Let $a = 216$ and $b = 135$. As in the Euclidean algorithm, we repeatedly use division algorithm an write

$$\begin{aligned} 216 &= 135 + 81 \\ 135 &= 81 + 54 \\ 81 &= 54 + 27 \\ 54 &= 2 \times 27 + 0. \end{aligned}$$

Thus 27 is the gcd of 216 and 135. Using the algorithm we will now find integers u and v and write $27 = u \times 216 + v \times 135$.

$$\begin{aligned} 27 &= 81 - 54 \\ &= 81 - (135 - 81) \\ &= 2 \times 81 - 135 \\ &= 2 \times (216 - 135) - 135 \\ &= 2 \times 216 - 3 \times 135. \end{aligned}$$

That is $u = 2$ and $v = -3$.

2. Let $a = 2378$ and $b = 1769$. Then

$$\begin{aligned} 2378 &= 1769 + 609 \\ 1769 &= 2 \times 609 + 551 \\ 609 &= 551 + 58 \\ 551 &= 9 \times 58 + 29 \\ 58 &= 2 \times 29 + 0. \end{aligned}$$

Thus $\gcd(2378, 1769) = 29$. We will now find integers u and v to write $29 = 2378 \times u + 1769 \times v$.

$$\begin{aligned} 29 &= 551 - 9 \times 58 \\ &= 551 - 9 \times (609 - 551) \\ &= 10 \times 551 - 9 \times 609 \\ &= 10 \times (1769 - 2 \times 609) - 9 \times 609 \\ &= 10 \times 1769 - 29 \times 609 \\ &= 10 \times 1769 - 29 \times (2378 - 1769) \\ &= -29 \times 2378 + 39 \times 1769. \end{aligned}$$

That is $29 = -29 \times 2378 + 39 \times 1769$ and a choice for the pair (u, v) is $(-29, 39)$.

Given $a, b \in \mathbb{N}$, the common multiples of a, b are given by $S_a \cap S_b$.

Definition 1.1.37. Let a and b be two natural numbers. The least positive element in the intersection $S_a \cap S_b$ is called the least common multiple $\operatorname{lcm}(a, b)$ of a and b.

It is easy to see that, for all integers $x, y \in S_a \cap S_b$, their sum $x + y$ and $-x, -y$ are also in $S_a \cap S_b$. Hence $S_a \cap S_b = \langle \operatorname{lcm}(a, b) \rangle$.

Theorem 1.1.38. *Let $d := \gcd(a, b)$ and $\ell = \operatorname{lcm}(a, b)$. If $a = rd$ and $b = sd$ then $\ell = rsd$.*

Proof. If m is a common multiple of a and b, it suffices to show that $rsd|m$. (Why?)

Since d is the gcd of a and b, the integers r and s are relatively prime and there exists integers x and y such that $rx+sy = 1$. Then $m = m\cdot 1 = mrx+msy$.

Since $a = rd$ and $b = sd$ divide m, there exists integers q and t such that $m = rdq$ and $m = sdt$. This proves that

$$m = mrx + msy = sdtrx + rdqsy = rsd(tx + qy).$$

Thus the integer rsd divides m and hence rsd is the lcm of the integers a and b. $\square$

There is an identity relating gcd and lcm and we state this as the corollary.

Corollary 1.1.39. *Let a and b two positive integers. Then*

$$\gcd(a, b) \times \operatorname{lcm}(a, b) = ab.$$

Proof. It follows immediately from Theorem 1.1.38: Let a and b be two natural numbers, $d = \gcd(a, b)$ and $a = rd$, $b = sd$. Then $ab = (rd)(sd) = d(rsd)$. □

Exercise 1.1.40. Let a and b two natural numbers. Find $(a + b, \operatorname{lcm}(a, b))$.

If p is an irreducible(prime) number and divides a nonzero integer a, we let $e_p(a)$ denote the highest(positive) power of p that divides a. (Why does $e_p(a)$ exist?) In case a is fixed, we denote $e_p(a)$ by e_p.

For two nonzero integers a, b, we have $e_p(ab) = e_p(a) + e_p(b)$. Write $a = p^{e_p(a)}r$ and $b = p^{e_p(b)}s$. Then p does not divide any of r, s. Hence p does not divide rs. (Why?) This shows that $ab = p^{e_p(a)+e_p(b)}rs$.

This extends in an obvious way to product of finitely many nonzero integers.

Theorem 1.1.41 (Fundamental Theorem of Arithmetic). *Let $a > 1$ be a natural number. Then there exist distinct prime numbers p_i and natural numbers e_i for $1 \leq i \leq k$ such that $a = p_1^{e_1} \cdots p_k^{e_k}$ and this expression is unique. (Here $e_1 = e_{p_1}(a)$ etc.)*

Thus, any natural number $a > 1$ can be written in a unique way as a finite product of prime powers.

Proof. We have already proved the existence of the prime factorization.

Suppose $a = p_1^{e_1} \cdots p_k^{e_k} = q_1^{f_1} \cdots q_l^{f_l}$ where p_i and q_j's are primes and e_i and f_j are exponents of the primes p_i and q_j respectively.

Since p_1 divides $a = q_1^{f_1} \cdots q_l^{f_l}$, it must divide one of the q_j's and we assume it to be q_1. Since p_1 and q_1 are prime numbers, it follows that $p_1 = q_1$. We may now write $a = p_1^{e_1} \cdots p_k^{e_k} = p_1^{f_1} \cdots q_l^{f_l}$. If $e_1 \neq f_1$ and if $f_1 - e_1 \geq 1$, we get that $p_2^{e_2} \cdots p_k^{e_k} = p_1^{f_1 - e_1} q_2 \cdots q_l^{f_l}$. This means that p_1 divides $p_2^{e_2} \cdots p_k^{e_k}$, a contradiction.(Why?) Therefore $e_1 = f_1$ and $p_2^{e_2} \cdots p_k^{e_k} = q_2^{f_2} \cdots q_l^{f_l}$.

We may proceed as above and conclude that $k = l$, $p_i = q_{j_i}$ for some j_i and $e_i = f_{j_i}$ for $1 \leq i \leq k$. This proves the uniqueness. □

Exercise 1.1.42. Let $k \geq 2$ and $a_1, a_2, \cdots, a_k$ be natural numbers such that $\gcd(a_1, a_2, \ldots, a_k) = 1$. Assume that the product $a_1 a_2 \cdots a_k = n^m$ for some natural numbers n and m. Show that there exists natural numbers n_j such that $a_j = n_j^m$.

1.2 Diophantine Equations

Let a and b be two natural numbers. We know the equation $ax + by = c$ represents a line in $\mathbb{R}^2$. We want to find integer solutions to this equation $ax + by = c$. That is we want to find $(x, y) \in \mathbb{Z} \times \mathbb{Z}$ such that $ax + by = c$.

suppose there exist integers x and y such that $ax + by = c$. Since $ax + by$ is in $S_a + S_b = \mathbb{Z}d$, we can write $ax + by = du$ for some $u \in \mathbb{Z}$ and hence $c = du$. This means that d divides the integer c.

Conversely assume that $d = (a, b)$ divides c. Then there exists u in $\mathbb{Z}$ such that $c = du$. By Bezout identity, there exist x and y in $\mathbb{Z}$ such that $ax + by = d$ and hence $c = du = a(xu) + b(yu)$. This proves that the pair (xu, yu) is a solution of the equation $ax + by = c$.

Thus we have shown that the equation $ax + by = c$ has a solution in integers iff $d := \gcd(a, b)$ divides c.

Suppose a pair $(r, s) \in \mathbb{Z} \times \mathbb{Z}$ is a solution of the equation $ax + by = c$. Let $(x_0, y_0) \in \mathbb{Z} \times \mathbb{Z}$ be another solution. Then $a(x_0 - r) + b(y_0 - s) = 0$. Hence $a(x_0 - r) = b(s - y_0)$. If we write $a = du$ and $b = dv$, then $\gcd(u, v) = 1$. Substituting the expressions for a and b in the earlier equation, we obtain $du(r - x_0) = dv(y_0 - s)$. Hence $u(x_0 - r) = v(s - y_0)$. As a consequence u divides $s - y_0$. Therefore $s - y_0 = tu$ so that $y_0 = s - tu$ for some integer t.

We substitute this in the equation $u(x_0 - r) = v(s - y_0)$ to get $u(x_0 - r) = vtu$. This shows that $x_0 = r + tv$. Thus we have proved the following theorem.

Theorem 1.2.1. *Let a, b and c be three integers. Then the Diophantine equation $ax + by = c$ has a solution in integers iff $d = \gcd(a, b)$ divides c.*

Further, if (r, s) and (x, y) are two solutions of the equation then $x = r + tv$ and $y = s - tu$ where $a = du$, $b = dv$ and $t \in \mathbb{Z}$.

Corollary 1.2.2. *If a and b are two relatively prime integers, then the equation $ax + by = c$ has an integer solution for all integers c.*

Proof. Easy. □

Example 1.2.3. Find all the solutions of the Diophantine equation $28x + 35y = 133$.

The gcd of the two integers 28 and 35 is 7 and it divides the integer 133. Hence the equation has a solution. By clearing out the gcd, we write the equation as $4x + 5y = 19$. A solution for this equation is $x = 1$ and $y = 3$. Hence general solution (x, y) is of the form $(x, y) = (1, 3) + t(5, -4)$. Thus

$$\{(1, 3) + t(5, -4) : t \in \mathbb{Z}\}$$

is the set of solutions of the equation $28x + 35y = 133$.

Exercise 1.2.4.

1. Find all integer solutions of $15x + 35y = 23$.
2. Find all integer solutions of the equation $3x + 6y = 7$.
3. Find all solutions $(x, y) \in \mathbb{Z} \times \mathbb{Z}$ of the equation $2x + 3y = 500$ such that $\gcd(x, y) = 7$.

1.3 Congruences

We start with some simple examples.

What day of the week will it be after 100 days from today? To answer this question, first let us observe that, the days repeat with a cycle of seven. Therefore to know the day after 100 days we just divide 100 by 7 and get the remainder 2. This means that if it is Monday today, the day after after 100 days will be Wednesday.

If we have understood this example, then we can answer the question of finding the day after n days from today. For example, if it is Monday today, it will be Friday after 200 days from today.

Another example: How do we determine if a given number is a square or not. Suppose a number k is a square and let $k = l^2$. If l is even, say, $l = 2t$, then $l^2 = 4t^2$ and 4 divides k. If l is odd, say $l = 2t + 1$, then $k = l^2 = 4(t^2 + t) + 1$. This means k leaves the remainder 1 when divided by 4. Thus we have shown that if a number k is a square then it leaves the remainder 0 or 1 when divided by 4.

Let us observe that this result does not say that a number k is a square if it leaves the remainder 0 or 1 when divided by 4.

For example let $k = 225$. Then $k = (15)^2$ and $15 = 2 \times 7 + 1$. Therefore $225 = 4 \times 56 + 1$ leaves the remainder 1.

As another example let $k = 20$. Then 20 leaves the remainder 0 when divided by 4. Similarly, $k = 57$ leaves the remainder 1 when divided by 4. However neither 20 nor 57 are squares.

Let us look at the following example. Let $n = 33092575$. We write $33092575 = 330925 \times 100 + 75 = 330925 \times 100 + 18 \times 4 + 3$. This shows that the number n, when divided by 4 leaves the remainder 3. Hence n is not a square.

Exercise 1.3.1. Show that the last digit of a perfect square can not be 2, 3, 7 or 8.

With these examples with us, now we make a definition of congruence relation between integers.

Definition 1.3.2. Let $n \in \mathbb{N}$ be fixed. We say two integers a and b are congruent modulo n if the integer $a - b$ is a multiple of n.

If a and b are congruent modulo n, we write $a \equiv b \pmod{n}$.

Properties 1.3.3. Let $n \in \mathbb{N}$ be fixed.

1. Two integers a and b are congruent modulo n iff the integers a and b leave the same remainder when divided by n. That is, if $a = nq_1 + r_1$ and $b = nq_2 + r_2$, then $r_1 = r_2$.

   Let a and b be two integers. Division algorithm tells us that there exist unique integers q_1, q_2 and non-negative integers r_1, r_2 such that $a = nq_1 + r_1$ and $b = nq_2 + r_2$ where $0 \leq r_1, r_2 < n$. Then $a - b = n(q_1 - q_2) + (r_1 - r_2)$. Therefore a and b are congruent modulo n iff $n | r_1 - r_2$. Since $-n < r_1 - r_2 < n$, this is possible iff $r_1 = r_2$. This proves our claim.

2. Let$a$ be an integer and $a = nq + r$ for $q \in \mathbb{Z}$ and $0 \leq r < n$. Then a is congruent to r.

   We write $a - r = nq$ and conclude our claim.

3. Congruence relation on $\mathbb{Z}$ is an equivalence relation.

   Let a, b and c be three integers. If n divides $a - b$, then n divides $b - a = -(a - b)$ as well.

   If n divides $a - b$ and $b - c$, then n also divides their sum $a - c = (a - b) + (b - c)$. This proves that the relation is transitive too.

Let $n \in \mathbb{N}$ and a be an integer. Let q in $\mathbb{Z}$ and $r \in \mathbb{N} \cup \{0\}$ be such that $a = nq + r$. We have seen above that a is congruent to r modulo n.

Let b be an integer. Then $b \equiv a \pmod{n}$ iff $b = a + nq_1$ for some integer q_1. Since $a = nq + r$, we can write $b = n(q + q_1) + r$. This shows that the equivalence class of a, denoted by $[a]$ is $[a] = \{nk + r : k \in \mathbb{Z}\}$. Further we also infer that the elements of the equivalence class are in arithmetic progression with initial non-negative integer r and common difference n.

Thus we have shown that for every non-negative integer $r < n$, we get an equivalence class $[r] = \{nk + r : k \in \mathbb{Z}\}$ and there are n equivalence classes in all. Each of these equivalence class represents one congruence class.

Exercise 1.3.4. Let $n \in \mathbb{N}$ and a be an integer. If an integer b is in $[a]$, show that a is in $[b]$ and $[b] = [a]$.

> Suppose b is in $[a]$. Then there exists integer q such that $b = a + nq$. Hence $a = b - nq$ is in $[b]$. For any integer r, the integer $b + rn = a + n(q + r)$ is also in $[a]$. Thus $[b] \subseteq [a]$. If we now reverse the roles of a and b, we see that $[a] = [b]$.

Example 1.3.5. Let $n = 2$. Let a be an integer. If a is even then the remainder is zero when divided by 2. If a is odd then the remainder is 1 when divided by 2. Thus an integer is in the equivalence class containing zero or 1. This means that there are only two equivalence classes, namely $[0]$, the set of all even integers and $[1]$, the set of all odd integers.

Exercise 1.3.6. Let $n = 1$. Find the equivalence class of $[a]$ modulo n of an integer a.

Let us record some simple properties of congruence relation in the following theorem.

Theorem 1.3.7. *Let $n \in \mathbb{N}$ be fixed. Let a, b, c, d be integers. Then the following holds.*

1. *If $a \equiv b \pmod{n}$ and $c \equiv d \pmod{n}$, then*

   (a) $a \pm c \equiv b \pm d \pmod{n}$ and

   (b) $ac \equiv bd \pmod{n}$.

2. *If $a \equiv b \pmod{n}$, then $ac \equiv bc \pmod{n}$. Further $ac \equiv bc \pmod{nc}$ if $c \geq 1$.*

3. *If $a \equiv b \pmod{n}$, then $a^k \equiv b^k \pmod{n}$ for every $k \geq 1$.*

4. *$a \equiv 0 \pmod{n}$ iff n divides a.*

5. *If $ac \equiv bc \pmod{n}$ and $\gcd(c, n) = 1$, then $a \equiv b \pmod{n}$.*

6. *If $ac \equiv bc \pmod{n}$ and if m is a natural number such that $c = mc_1$ and $n = mn_1$, then $ac_1 \equiv bc_1 \pmod{n_1}$. In particular, if m is the gcd of c and n, then $a \equiv b \pmod{n_1}$.*

7. *If $a \equiv b \pmod{n}$ and m divides n then, $a \equiv b \pmod{m}$.*

Proof. Let a, b, c and d be integers be such that $a \equiv b \pmod{n}$ and $c \equiv d \pmod{n}$. Then there exist integers q_1 and q_2 such that $a - b = nq_1$ and $c - d = nq_2$. Therefore $a \pm c = (b + nq_1) \pm (d + nq_2) = (b \pm d) + n(q_1 \pm q_2)$ and $ac = (b + nq_1)(d + nq_2) = bd + n(q_1 + q_2 + nq_1q_2)$. This proves that $a \pm c \equiv b \pm d \pmod{n}$ and $ac \equiv bd \pmod{n}$.

Let $a \equiv b \pmod{n}$ and c be an integer. Then n divides $a-b$ and $a-b$ divides $c(a-b)$. Hence $ac \equiv bc \pmod{n}$. If $c \geq 1$, then nc is also a natural number and if $a - b = nq$ for some integer q, then $ac - bc = cnq$. Thus $ac \equiv bc \pmod{nc}$.

Let a and b be two integers and $k \geq 2$ be a natural number. Then $a^k - b^k = (a-b)(a^{k-1} + a^{k-2}b + \cdots + b^{k-1})$. Hence $a-b$ divides $a^k - b^k$. If $a \equiv b \pmod{n}$, then, since n divides $a - b$ and $a - b$ divides $a^k - b^k$, it follows that n divides $a^k - b^k$. As a consequence $a^k \equiv b^k \pmod{n}$.

Let a be an integer. Let q in $\mathbb{Z}$ and $0 \leq r < n$ be such that $a = nq + r$. This means that a is congruent to 0 modulo n iff $r = 0$.

Let a, b and c be three integers such that $ac \equiv bc \pmod{n}$ and $\gcd(c, n) = 1$. By definition the integer $n | c(a - b)$. Since $\gcd(c, n) = 1$, we use Euclid's Lemma 1.1.28 to conclude that $n | a - b$. Thus $a \equiv b \pmod{n}$.

Let a, b and c be three integers such that $ac \equiv bc \pmod{n}$ and m be a natural number such that $c = mc_1$ and $n = mn_1$. If $ac - bc = nq$ for an integer q, then $(a - b)mc_1 = mn_1q$. By cancelling m on both sides of this equation, we get that $(a - b)c_1 = n_1q$. Thus Let $ac_1 \equiv bc_1 \pmod{n_1}$.

Suppose $m = \gcd(c, n)$. Then $\gcd(c_1, n_1) = 1$ and it follows from the previous arguments that $a \equiv b \pmod{n_1}$.

The last item is easy and left as an exercise. □

Exercise 1.3.8. Let $n \geq 2$ be fixed.

1. If a and b are in $\mathbb{Z}$ such that $\gcd(a, n) = 1$ and $a \equiv b \pmod{n}$ show that $\gcd(ab, n) = 1$.

2. Let a and b be two integers such that $\gcd(a, n) = \gcd(b, n) = 1$. Then $\gcd(ab, n) = 1$.

   Since $\gcd(a, n) = 1 = \gcd(b, n)$, there exist integers u, v, x and y such that $au + nv = 1$ and $bx + ny = 1$. Therefore

   $$\begin{aligned} 1 &= (au + nv)(bx + ny) \\ &= abux + n(any + bvx + vny). \end{aligned}$$

   This proves that $\gcd(ab, n) = 1$.

3. Let a be an integer such that $\gcd(a, n) = 1$. If a_1 is in $[a]$, then $\gcd(a_1, n) = 1$.

   Let a_1 be an element of $[a]$ and let $d = \gcd(a_1, n)$. We write $a_1 = nt + a$ for an integer t in $\mathbb{Z}$. Since d is the gcd of a_1 and n, the integer d divides n and a_1. As a consequence d divides $a = nt + a - nt = a_1 - nt$. Therefore d divides a and n. This proves that $d = 1$.

Let us fix a natural number n. Then given an integer a, there exists a unique non-negative integer $r < n$ such that a is congruent to r modulo n. In practice, as we have seen in some of the examples, it is easier to work with this remainder to solve our problems in hand. This technique of assigning a smaller number congruent to a given number is also called *modular arithmetic*. We will study more about modular arithmetic in Chapter 2.

We look at some numerical examples:

Example 1.3.9. 1. Let $n = 23$. If $a = 29$, then $a = 23 + 6$. Therefore $a \equiv 6 \pmod{23}$ and $[a] = \{23q + 6 : q \in \mathbb{Z}\} = [6]$.

If $a = 291$, then we can write $a = 12 \times 23 + 15$. Hence $[a] = [15]$.

2. Let $n = 1729$. If $a = 3461$, then $a = 2 \times 1729 + 3$. Therefore $a \equiv 3 \pmod{1729}$ and $[a] = \{1729\,q + 3 : q \in \mathbb{Z}\} = [3]$.

Just as we add and multiply two integers, there is a very natural way to add and multiply two congruence classes and the arithmetic is very similar to integer arithmetic. This arithmetic is called **Modular Arithmetic**.

Let n in $\mathbb{N}$ be fixed and let $[a]$ and $[b]$ denote the congruence classes of the two integers a and b respectively. Then we define the addition and multiplication of two congruences as

$$[a] + [b] := [k + l]$$

and

$$[a] \cdot [b] := [kl]$$

where $k \in [a]$ and $l \in [b]$.

Let us now look at an example.

Example 1.3.10. Let $n = 19$. Let $a = 185$ and $b = 237$. Then 14 is in the equivalence class $[a] = [185]$ and 9 is in the equivalence class $[b] = [237]$. Therefore the sum $[a] + [b] = [14 + 9] = [23] = [4]$ and the product $[a][b] = [14 \times 9] = [126] = [12]$.

This addition and multiplication satisfies the following properties.

Proposition 1.3.11. *Let $n \in \mathbb{N}$. Then, for $a, b, c \in \mathbb{Z}$,*

1. $[a] + [b] = [a + b] = [b] + [a]$,

2. $[a] + ([b] + [c]) = ([a] + [b]) + [c]$,

3. $[a] \cdot [b] = [a \cdot b] = [b] \cdot [a]$ *and*

4. $[a] \cdot ([b] \cdot [c]) = ([a] \cdot [b]) \cdot [c]$.

Proof. Follows easily from Exercise 1.3.4. □

We denote the set of congruence classes modulo n by $\mathbb{Z}_n$. This set has exactly n elements, namely $[0], [1], \ldots, [n - 1]$.

1.3.1 Linear Congruences

We will now study the solutions of equations called linear congruences.

Definition 1.3.12. Let $n \geq 2$ and a and b be two integers such that $a \not\equiv 0 \pmod{n}$. Then the congruence $ax \equiv b \pmod{n}$ is called a linear congruence.

We want to find an integer x such that $ax \equiv b \pmod{n}$. If there exists an integer t such that $at \equiv b \pmod{n}$, then t is called a solution of the linear congruence $ax \equiv b \pmod{n}$.

In the last section, we solved Diophantine equations $ax + by = c$. We revisit them now and connect it with solving linear congruences.

Let us consider a linear congruence $ax \equiv b \pmod{n}$. Suppose an integer t is a solution of the linear congruence $ax \equiv b \pmod{n}$. That is $at \equiv b \pmod{n}$. Therefore there exists an integer k such that $at - nk = b$.

Conversely suppose there exist integers t and k such that $at - nk = b$, then t is a solution of the linear congruence $ax \equiv b \pmod{n}$.

This proves that a linear congruence $ax \equiv b \pmod{n}$ has a solution iff the Diophantine equation $ax + ny = b$ has a solution $(x, y) \in \mathbb{Z} \times \mathbb{Z}$. We have shown that a Diophantine equation $ax + by = c$ has a solution iff $d = \gcd(a, b)$ divides c. As a consequence it follows that a linear congruence $ax \equiv b \pmod{n}$ has a solution iff $d = \gcd(a, n)$ divides b.

Having answered the question on the existence of solution for a linear congruence, we shall now find out the number of solutions of a linear congruence.

Let t_0 be a solution of the linear congruence $ax \equiv b \pmod{n}$. If t is another solution, then $at_0 \equiv at \pmod{n}$. If we write $a = da_1$ and $n = dn_1$, then the congruence $at \equiv at_0 \pmod{n}$ is equivalent to the congruence $a_1(t - t_0) \equiv 0 \pmod{n_1}$. Since $\gcd(a_1, n_1) = 1$, it follows that $t - t_0 \equiv 0 \pmod{n_1}$. Thus $t = t_0 + n_1 q$ for some integer q. For each value of q between 0 and $d - 1$, we get d distinct solutions of the linear congruence $ax \equiv b \pmod{n}$ and they are $t_0 + n_1 q$ where $0 \leq q < d$.

Thus we have found all the solutions of a linear congruence and we state this as a theorem.

Theorem 1.3.13. *Let $n \geq 2$. Given two integers a and b, the linear congruence*

$$ax \equiv b \pmod{n}$$

has a solution iff $d = \gcd(a, n)$ divides b.

Further this congruence has exactly d distinct solutions modulo n.

We shall work out few examples now.

Examples 1.3.14. 1. Solve the linear congruence $3x \equiv 1 \pmod{5}$.

The integers 3 and 5 are relatively prime. Therefore the congruence has a solution and it is unique modulo 5. Since 3 and 5 are relatively prime, there exist integers u and v such that $3u + 5v = 1$. For example $u = 2$ and $v = -1$ is one such solution. Therefore, for each integer k, the integer $u = 5k + 2$ is a solution of the linear congruence $3x \equiv 1 \pmod{5}$.

2. Solve the linear congruence $7x \equiv 3 \pmod{12}$.

   The integers 7 and 12 are relatively prime. Hence there exist integers u and v such that $7u + 12v = 1$. If $u = 7$ and $v = -4$, then $7u + 12v = 1$. Hence $3u = 3 \times 7 = 21$ is a solution of the linear congruence $7x \equiv 3 \pmod{12}$. Since $21 \equiv 9 \pmod{9}$, it follows that $x = 9$ is a solution of the linear congruence $7x \equiv 3 \pmod{12}$ and the solution is unique modulo 12.

3. Solve the linear congruence $12x \equiv 15 \pmod{22}$.

   Since the gcd of 12 and 22 does not divide 15, it follows that this linear congruence does not have a solution.

4. Solve the linear congruence $8x \equiv 6 \pmod{10}$.

   Since $\gcd(8, 10) = 2$ divides 6, this congruence has two solutions. If x is a solution of $4x \equiv 3 \pmod{5}$, then x is a solution of $8x \equiv 6 \pmod{10}$. The integer $x = 2$ is a solution of the congruence $4x \equiv 3 \pmod{5}$ and it is unique modulo 5. However the congruence $8x \equiv 6 \pmod{10}$ has two solutions modulo 10. If we add 5 to the solution $x = 2$, then we get another solution, namely 7, for the congruence $8x \equiv 6 \pmod{10}$. That is 2 and 7 are the two solutions modulo 10 for this congruence.

5. Solve the linear congruence $9x \equiv 12 \pmod{15}$.

   The gcd of 9 and 15 is 3 and it divides 12. Hence the congruence $9x \equiv 12 \pmod{15}$ has a solution.
   If x is a solution of $3x \equiv 4 \pmod{5}$, then x is a solution of $9x \equiv 12 \pmod{15}$. A solution to $3x \equiv 4 \pmod{5}$ is $x = 3$. Since the integers 3 and 5 are relatively prime, $x = 3$ is the unique solution modulo 5. As in the earlier problem the congruence $9x \equiv 12 \pmod{15}$ has three solutions modulo 15 and they are 3, $3+5 = 8$ and $3+2\times5 = 13$.

Exercise 1.3.15. 1. Solve the following Diophantine equations.

(a) $4x + 61y = 9$.

(b) $5x - 43y = 19$.

(c) $15x + 24y = 333$.

2. Find all the solutions of the following linear congruences.

(a) $27x \equiv 15 \pmod{29}$.

(b) $5x \equiv 2 \pmod{26}$.

(c) $34x \equiv 60 \pmod{78}$.

(d) $144x \equiv 137 \pmod{307}$.

3. Using congruence check the divisibility of the following.

(a) $13 | 145^6 + 1$

(b) $23 | 1979^{24} - 1$

(c) $18|(1999)^{1999} - 1$

(d) $431|2^{43} - 1$

(e) $731|3^{73} - 1$

4. Find all the solutions of the linear congruence $3x - 7y \equiv 13 \pmod{17}$.

5. If x is odd and $n \geq 1$, prove that $x^{2^n} \equiv 1 \pmod{2^{n+2}}$.

1.4 Chinese Remainder Theorem

In the last section, we studied a single linear congruence equation. In this section, we will study the problem of solving simultaneous linear congruences.

The theorem is supposed to have originated in China with a problem that is solved after the proof of the theorem.

Theorem 1.4.1 (Chinese Remainder Theorem). *Let $r \geq 2$ and $n_1, n_2, \ldots, n_r$ be positive integers such that $\gcd(n_i, n_j) = 1$ for $i \neq j$. Then given any set of integers $a_1, a_2, \ldots, a_r$, there exists x in $\mathbb{Z}$ such that $x \equiv a_i \pmod{n_i}$.*

Further the integer x is unique up to the multiple of lcm of the numbers n_1, $n_2, \ldots, n_r$.

Before we start with the proof we will solve a simple simultaneous linear congruence to get the idea of the proof.

Example 1.4.2. Solve the simultaneous linear congruence

$$x \equiv 3 \pmod{5} \quad \text{and} \quad x \equiv 7 \pmod{9}.$$

Solution: Suppose x is a solution of the linear congruence

$$x \equiv 3 \pmod{5} \quad \text{and} \quad x \equiv 7 \pmod{9}.$$

Then $x = 5t + 3$ and $x = 9s + 7$ for some integers t and s. This means that $5t - 9s = 4$. This is a linear Diophantine equation. Since 5 and 9 are relatively prime there exist integers u and v such that $5u + 9v = 1$. Now we can multiply this equation by 4 and get the solution for the linear congruence. However we will go about solving it slightly differently.

Since $5u + 9v = 1$, we see that the integers u and v satisfy the linear congruences $5u \equiv 1 \pmod{9}$ and $9v \equiv 1 \pmod{5}$. Therefore, if we multiply $5u$ by 7 and $9v$ by 3 and let $x = 7 \times 5u + 3 \times 9v$, then

$$\begin{aligned} x &= 7 \times 5u + 3 \times 9v \\ &\equiv 3 \times 9v \pmod{5} \\ &\equiv 3 \pmod{5} \end{aligned}$$

and

$$\begin{aligned} x &= 7 \times 5u + 3 \times 9v \\ &\equiv 7 \times 5u \pmod{9} \\ &\equiv 7 \pmod{9}. \end{aligned}$$

The equation $5u + 9v = 1$ has $(u, v) = (2, -1)$ as a solution. Therefore $x = 7 \times 10 - 3 \times 9 = 43$ is a solution.

If x_0 is a solution, then $x_k = x_0 + k \times \text{lcm}(5, 9) = x_0 + 45k$ is also a solution for $k \in \mathbb{Z}$. Therefore $x = 43$ is the unique solution modulo the lcm. □

We will see that this method of solving two simultaneous linear congruences tells us how to solve more than two simultaneous linear congruence equations by solving one more example.

Example 1.4.3. Solve the simultaneous linear congruence

$$x \equiv 3 \pmod 5 \qquad x \equiv 7 \pmod 9 \qquad \text{and} \qquad x \equiv 11 \pmod{13}.$$

Solution: The three integers 5, 9 and 13 are mutually relatively prime and hence there exist integers u_1, u_2 and u_3 such that $5u_1+9u_2+13u_3 = 1$. Following the earlier example, we may let $x = 3(9u_2 + 13u_3) + 7(5u_1 + 13u_3) + 11(5u_1 + 9u_2)$. Then $x \equiv 3 \pmod 5$ only if $7(5u_1 + 13u_3) + 11(5u_1 + 9u_2) \equiv 0 \pmod 5$. A priori we can't prove this.

Let $N_1 = 9\times 13$ and $N_2 = 5\times 13$ and $N_3 = 5\times 9$. Then $\gcd(N_1, N_2, N_3) = 1$ and hence there exist integers u_1, u_2 and u_3 such that $N_1u_1 + N_2u_2 + N_3u_3 = 1$. If we now let $x = 3 \times N_1u_1 + 7 \times N_2u_2 + 11 \times N_3u_3$, then

$$x \equiv 3 \pmod 5 \qquad x \equiv 7 \pmod 9 \qquad \text{and} \qquad x \equiv 11 \pmod{13}.$$

For the sake of completeness, we will find out integers u_1, u_2 and u_3 such that $N_1u_1 + N_2u_2 + N_3u_3 = 1$ and x.

$$\begin{aligned} 1 &= \gcd(117, 65, 45) \\ &= \gcd(\gcd(117, 65), 45) \\ &= \gcd(13 \gcd(9, 5), 45) \\ &= (13 \gcd(5, 9)) \times 7 + 45 \times (-2) \\ &= (13(5 \times 2 + 9 \times (-1))) \times 7 + 45 \times (-2) \\ &= 117 \times (-7) + 65 \times 14 + 45 \times (-2). \end{aligned}$$

Therefore $x = 3 \times (117 \times (-7)) + 7 \times (65 \times 14) + 11 \times (45 \times (-2)) = 2923$ is a solution of the given congruences. The lcm of the integers 5, 9 and 13 is 585 and the integer $2923 \equiv 583 \pmod{585}$. Hence $x = 583$ is the smallest positive integer satisfying the congruences. □

Proof of Theorem 1.4.1

Proof. Let $N := \prod_{j=1}^{r} n_j$, $N_i := \frac{N}{n_i}$ and $d_i = \gcd(N_i, n_i)$ for $1 \leq i \leq r$. If $d_i \geq 2$ and p is a prime divisor of d_i, then p divides N_i and n_i, a contradiction. This proves that the integers and N_i and n_i are co-prime for each $i \leq r$. Hence there exists integers x_i such that

$$N_i x_i \equiv 1 \pmod{n_i}$$

for $1 \leq i \leq r$.

Let us observe that $N_i x_i \equiv 0 \pmod{n_j}$ for $1 \leq i \neq j \leq r$. Now let $x := \sum_{i=1}^{r} a_i N_i x_i$. Then $x \equiv a_i \pmod{n_i}$ for $1 \leq i \leq r$. □

We shall now use this theorem to solve the following Exercises. Let us first start with a problem that is supposed to have originated in China.

Exercise 1.4.4. A sack of gold coins was stolen by nine thieves. When they shared the coins equally among them 2 coins remained. One thief was killed in the ensuing melee in deciding as to who can get the last two coins. When the remaining thieves decided to share the coins equally among them one coin remained and again a brawl started in deciding who will get the last coin. One of the thieves was killed in the brawl. The remaining seven thieves could share the coins equally among them. Find the number of coins in the sack.

Let x denote the number of gold coins in the sack.
When divided equally among

- nine thieves they were left with 2 gold coins,
- eight thieves they were left with 1 gold coin and
- seven thieves no coin was left.

Thus we need to solve the following simultaneous linear congruence

$$\begin{aligned} x &\equiv 2 \pmod 9 \\ x &\equiv 1 \pmod 8 \\ x &\equiv 0 \pmod 7. \end{aligned}$$

We will now use Chinese Remainder Theorem and solve the simultaneous linear congruences

$$x \equiv \begin{cases} 2 & \pmod 9 \\ 1 & \pmod 8 \\ 0 & \pmod 7 \end{cases}.$$

Let $N_1 = 8 \times 7 = 56$, $N_2 = 9 \times 7$ and $N_3 = 9 \times 8$. Then there exist integers $u_1 = 32$, $u_2 = -1$ and $u_3 = -24$ such that $N_1u_1 + N_2u_2 + N_3u_3 = 1$. If we now let $x = a_1N_1u_1 + a_2N_2u_2 + a_3N_3u_3 = 2 \times 56 \times 32 + 1 \times 63 \times (-1) + 0 \times 72 \times (-24) = 3521$ then x is a solution of the congruences. Since $x \equiv 497 \pmod{504}$, the least number of gold coins in the sack is 497. □

There is another problem of Indian origin. This problem is attributed to Brahmagupta.

Exercise 1.4.5. When eggs in a basket are removed 2, 3, 4, 5, 6 at a time, there remains respectively, 1, 2, 3, 4, 5 eggs. When seven eggs are removed at a time none are left over. Find the smallest number of eggs in the basket.

In this problem, if we let x denote the number of eggs in the basket, then we need to solve the linear congruences

$$\begin{aligned} x &\equiv 1 \pmod 2 \\ x &\equiv 2 \pmod 3 \\ x &\equiv 3 \pmod 4 \\ x &\equiv 4 \pmod 5 \\ x &\equiv 5 \pmod 6 \\ x &\equiv 0 \pmod 7) \end{aligned}$$

In this case the integers $2, 3, 4, 5, 6, 7$ are not mutually co-prime. So we can not apply Chinese Remainder Theorem 1.4.1.

Let us observe that, if x is an integer such that $x \equiv 3 \pmod 4$, then $x \equiv 1 \pmod 2$. A solution $x \geq 3$ to the congruence $x \equiv 1 \pmod 2$ is also a solution to the congruence $x \equiv 3 \pmod 4$.

Similarly if x is an integer such that $x \equiv 5 \pmod 6$, then $x \equiv 2 \pmod 3$ and a solution to the congruence $x \equiv 2 \pmod 3$ is also a solution to $x \equiv 5 \pmod 6$. Hence it is enough to solve the simultaneous linear congruence equations

$$x \equiv \begin{cases} 2 \pmod 3 \\ 3 \pmod 4 \\ 4 \pmod 5 \\ 0 \pmod 7. \end{cases}$$

The integers $n_1 = 3, n_2 = 4, n_3 = 5$ and $n_4 = 7$ are mutually co-prime and we can apply Theorem 1.4.1. Let $N = 3 \times 4 \times 5 \times 7 = 420$ and $N_i = \frac{N}{n_i}$. Then, following the proof of Theorem 1.4.1, we solve the linear congruence $N_i u_i \equiv 1 \pmod{n_i}$ and find out that $u_1 = 2$, $u_2 = 1$, $u_3 = 4$ and $u_4 = 2$. Hence $x = 2 \times 140 \times +3 \times 105 + 4 \times 84 \times 4 + 0 \times 60 \times 2 = 2219$ is a solution to the linear congruence

$$x \equiv \begin{cases} 2 \pmod 3 \\ 3 \pmod 4 \\ 4 \pmod 5 \\ 0 \pmod 7. \end{cases}$$

Since the lcm of $3, 4, 5$ and 7 is 420 and $2219 \equiv 119 \pmod{420}$, $x = 119$ is the solution we are looking for the original congruence

$$x \equiv \begin{cases} 1 \pmod 2 \\ 2 \pmod 3 \\ 3 \pmod 4 \\ 4 \pmod 5 \\ 5 \pmod 6 \\ 0 \pmod 7. \end{cases}$$

□

Exercise 1.4.6. 1. Solve the congruence $17x \equiv 3 \pmod{2 \cdot 3 \cdot 5 \cdot 7}$ by solving the system of linear congruence

$$17x \equiv \begin{cases} 3 \pmod 2 \\ 3 \pmod 3 \\ 3 \pmod 5 \\ 3 \pmod 7. \end{cases}$$

2. Find the smallest integer a such that

$$2|a, \quad 3|a+1, \quad 4|a+2, \quad 5|a+3, \quad 6|a+4.$$

3. (a) Find three consecutive integers a, $a+1$ and $a+2$ such that each of them have a square factor.
   (b) Find three consecutive integers a, $a+1$ and $a+1$ such that a is divisible by a square, $a+1$ is divisible by a cube and $a+2$ by a fourth power.
4. (a) Find an integer x with remainders $1, 2, 5, 5$ when divided by $2, 3, 6, 12$ respectively.
   (b) Find an integer x which leave the remainders $2, 3, 4, 5$ when divided by $3, 4, 5$ and 6 respectively.
   (c) Find an integer x having remainders 3, 11 and 15 when divided by 10, 13 and 17 respectively.
5. Find two incongruent solution of the system

$$\begin{aligned} 2x &\equiv 3 \pmod 5 \\ 4x &\equiv 2 \pmod 6 \\ 3x &\equiv 2 \pmod 7). \end{aligned}$$

1.5 Miscellaneous Exercises

Exercise 1.5.1. 1. Let $n \geq 2$ and $2 \leq k \leq n$. Show that $n! + k$ is composite and find a divisor of $n! + k$. Why do we need $k \geq 2$?

2. Let a and b two relatively prime integers such that $ab = c^2$. Show that $a = d^2$ and $b = e^2$ for two integers d and e.

3. Find five consecutive composite numbers.

4. Let $n \geq 2$. Can you find n consecutive composite numbers.

5. Let a and b be two integers with $a \neq 0$. Then there exist unique pair of integers q and r such that $b = aq + r$ and $-\frac{|a|}{2} \leq r \leq \frac{a}{2}$. This is called the least remainder algorithm.

6. If $p = 2^r + 1$ is a prime number, show that $r = 2^k$ for some $k \geq 0$. A prime number of this form is called *Fermat Prime.*

7. Let r be a rational number such that r^2 is an integer. Show that r is an integer.

Definition 1.5.2. An integer n is called square free if $n = ab$ for two integers a and b, then $\gcd(a, b) = 1$.

Exercise 1.5.3. Let n be a square free integer. Show that $a^2 - nb^2 \neq 0$ for any two integers a and b in $\mathbb{Z} \setminus \{0\}$.

Exercise 1.5.4. Let a, b be two positive integers and $n \geq 2$ be a natural number. Show that

$$\gcd(n^a - 1, n^b - 1) = n^d - 1$$

where $d = \gcd(a, b)$.

Let $a = bq_1 + r_1$ where $0 \leq r_1 < b$. Then

$$\begin{aligned}\gcd(n^a - 1, n^b - 1) &= \gcd(n^a - 1 - n^{a-b}(n^b - 1), n^b - 1) \\ &= \gcd(n^a - 1 - n^a + n^{a-b}, n^b - 1) \\ &= \gcd(n^{a-b} - 1, n^b - 1) \\ &= \gcd(n^{a-b} - 1 - n^{a-2b}(n^b - 1), n^b - 1) \\ &= \gcd(n^{a-2b} - 1, n^b - 1) \\ \vdots &= \vdots \qquad \vdots \qquad \vdots \\ &= \gcd(n^{r_1} - 1, n^b - 1).\end{aligned}$$

If we now write $b = r_1 q_2 + r_2$ for $0 \leq r_2 < r_1$, then proceeding as above, we get

$$\gcd(n^{r_1} - 1, n^b - 1) = \gcd(n^{r_1} - 1, n^{r_2} - 1).$$

Inductively, we arrive at

$$\begin{aligned}\gcd(n^{r_1} - 1, n^b - 1) &= \gcd(n^{r_1} - 1, n^{r_2} - 1) \\ &= \gcd(n^{r_3} - 1, n^{r_2} - 1) \\ \vdots &= \vdots \qquad \vdots \\ &= \gcd(n^d - 1, 0) \\ &= n^d - 1.\end{aligned}$$

Another solution: Let $d = \gcd(a, b)$ and we write $a = dr$, $b = ds$ with $\gcd(r, s) = 1$. It is easy to see that $n^d - 1$ divides $n^a - 1$ and $n^b - 1$. Let x and y in $\mathbb{Z}$ be such that $d = ax + by$. Then

$$\begin{aligned}n^d - 1 &= n^{ax+by} - 1 \\ &= n^{ax+by} - n^{by} + n^{by} - 1 \\ &= n^{by}(n^{ax} - 1) + \left(n^{by} - 1\right) \\ &= (n^a - 1)\left[n^{a(u-1)} + n^{a(u-2)} + \cdots + 1\right] n^{by} \\ &\quad + \left(n^b - 1\right)\left[n^{b(v-1)} + n^{b(v-2)} + \cdots + 1\right].\end{aligned}$$

Let $c = \gcd(n^a - 1, n^b - 1)$. Then c divides $n^a - 1$ and $n^b - 1$. Hence c divides $(n^a - 1)\left[n^{a(u-1)} + n^{a(u-2)} + \cdots + 1\right] n^{by}$ $+ (n^b - 1)\left[n^{b(v-1)} + n^{b(v-2)} + \cdots + 1\right] = n^d - 1$. This proves that $c = n^d - 1$ is the gcd of $n^a - 1$ and $n^b - 1$.

For further reading of elementary number theory, we refer the readers to [4] and [9].

Chapter 2

Group Theory I

In this chapter we are going to acquaint ourselves with one of the most important algebraic concepts, namely Group.

As mentioned in the introduction, in the first two sections, we study simple properties of groups and subgroups. In Section 2.3, we prove one of the most important results in finite groups, namely Lagrange's Theorem (Theorem 2.3.1). The significance of this result is brought out by proving Theorem 2.3.9 and Theorem 2.3.11 which are proved using more advanced concepts in groups.

In the subsequent section, we study homomorphism between groups and related results. In Section 2.7, we study quotient groups and its applications. In Section 2.8, we study important classes of groups, namely symmetric groups and Dihedral groups. Dihedral groups are studied as group of symmetries of a regular n-sided polygon.

2.1 Introduction to Groups

2.1.1 Binary Operations

Let S be a set. A *binary operation* on S is a function $f : S \times S \to S$ and it is called a law of *composition or multiplication.* If a, b and c are three elements in S, then we can form the composition $f(a, f(b, c))$, $f(f(a, b), c)$ etc., It is cumbersome and also looks ugly to write like this when we have to compose more than two elements. Therefore we write $f(a, b)$ as $a \cdot b$ and $f(a, f(b, c))$ is written as $a \cdot (b \cdot c)$ etc.,

Some of the examples of binary operation are the following.

Example 2.1.1. 1. Let $S = \mathbb{N}$, the set of natural numbers. Then, since sum of two natural numbers is a natural number, usual addition $+$ is a binary operation on $\mathbb{N}$. Multiplication of natural numbers is another binary operation on $\mathbb{N}$.

2. Addition is a binar ' operation on integers $\mathbb{Z}$ too. On integers, apart from addition and multiplication, we have another binary operation, namely subtraction: If m and n are two integers, then $m - n$ is also an integer.

If m and n are two natural numbers, then their difference $m - n$ may not be an integer. Hence subtraction restricted to natural numbers $\mathbb{N}$ is not a binary operation on $\mathbb{N}$.

Thus a binary operation need not be a binary operation on its subsets.

3. If $S = \mathbb{Q}^*$ is the set of nonzero rational numbers, then division is a binary operation on $\mathbb{Q}^*$. That is $(a, b) \mapsto \frac{a}{b}$ is a binary operation on S.

   Observe that division is a binary operation of $\mathbb{R}^*$ and $\mathbb{C}^*$ as well. However it is not a binary operation on $\mathbb{N}$ and $\mathbb{Z}^*$, the set of nonzero integers.

4. Let $M(n, \mathbb{R})$ denote the set of all $n \times n$ matrices over $\mathbb{R}$. Then matrix multiplication $(A, B) \mapsto AB$ is a binary operation on $M(n, \mathbb{R})$.

   Since the product of two invertible matrices is invertible, the matrix multiplication is a binary operation on the set of all invertible matrices $\mathrm{GL}(n, \mathbb{R}) := \{A \in M(n, \mathbb{R}) : A \text{ is invertible } \}$.

   Observe that addition of matrices is a binary operation on $M(n, \mathbb{R})$ and it is not a binary operation on $\mathrm{GL}(n, \mathbb{R})$.

5. If A and B are two subsets of a non-empty set X, then $A \cup B$ and $A \cap B$ are also subsets of X. Thus

   (a) $(A, B) \mapsto A \cup B$

   (b) $(A, B) \mapsto A \cap B$

   are two binary operations on $S = \mathcal{P}(X)$, the set of all subsets of X. We can show that

   (a) $(A, B) \mapsto A \setminus B$,

   (b) $(A, B) \mapsto B \setminus A$ and

   (c) $(A, B) \mapsto (A \setminus B) \cup (B \setminus A)$ (This is denoted by $A \Delta B$ and it is called the *symmetric difference* of A and B.)

   are also binary operations on S.

6. The cross product of two vectors is a binary operation on $\mathbb{R}^3$. That is $(a, b) \mapsto a \times b$ is a binary operation on $\mathbb{R}^3$. It is an easy exercise to verify that $a \times b = -(b \times a)$ for all vectors a and b in $\mathbb{R}^3$. Further $a \times (b \times c)$ need not be equal to $(a \times b) \times c$ for vectors a, b and c in $\mathbb{R}^3$.

A binary operation on a set S is called *associative* if $a \cdot (b \cdot c) = (a \cdot b) \cdot c$ for all elements a, b and c in S.

Addition is an associative binary operation on $\mathbb{N}$, $\mathbb{Z}$, $\mathbb{Q}$, $\mathbb{R}$ and $\mathbb{C}$.

Let a, b and c be three integers. Then $a + (b + c) = (a + b) + c$. However $a - (b - c) = a - b + c \neq (a - b) - c$. Thus subtraction is not an associative binary operation on $\mathbb{Z}$. Neither is it an associative binary operation on $\mathbb{Q}$, $\mathbb{R}$ or $\mathbb{C}$.

The cross product of vectors in $\mathbb{R}^3$ is not associative. (Why?)

We leave it as an exercise to show that the binary operations $(A, B) \mapsto A \setminus B$ and $(A, B) \mapsto B \setminus A$ are not associative on $\mathcal{P}(X)$. However the symmetric difference, $(A, B) \mapsto A\Delta B$, is associative and further $A\Delta B = B\Delta A$ for all A and B in S.

The matrix multiplication on $M(n, \mathbb{R})$ is associative and hence it is associative on $\mathrm{GL}(n, \mathbb{R})$ too. (Why?)

Let S be a set with a binary operation "$\cdot$". We say that an element e is an *identity* for the binary operation, if $ea = a$ and $ae = a$ for all a in S.

In our examples, the set of integers, rationals, real numbers and complex numbers all have an identity for the addition.

The set of natural numbers does not have an identity for the addition. However for the multiplication of natural numbers 1 is an identity.

If X is a set and S is the set of all functions from X to itself, then composition of functions is an associative binary operation on S and the identity function is an identity for the composition of functions.

Matrix multiplication has an identity, the identity matrix I_n in $M(n, \mathbb{R})$.

Properties 2.1.2. A binary operation on a set S can have at most one identity.

Proof. If e and e' are two identities, then, since e is an identity, $e' = e \cdot e'$. Similarly, since e' is an identity, $e = e \cdot e'$. Thus $e = e'$. □

Suppose a binary operation on a set S has an identity e. We say an element a in S is *invertible*, if there exists an element b in S is such that $a \cdot b = e = b \cdot a$. We also call the element b an *inverse* of a.

Properties 2.1.3. Let S be a set with a binary operation . and identity e. Then every element a in S has at most one inverse.

Proof. Let a be an element in S and let b, b' in S be such that $a \cdot b = a \cdot b' = e$. Then $b = b \cdot (a \cdot b') = (b \cdot a) \cdot b' = e \cdot b' = b'$. This proves that if an element a in S has an inverse then it is unique. □

For the binary operation $+$ on integers, 0 is the identity and every integer a is invertible: $a + (-a) = 0$.

The multiplication on $\mathbb{N}$, $\mathbb{Z}$, $\mathbb{Q}$, $\mathbb{R}$ and $\mathbb{C}$ has 1 as identity. 1 is the only invertible element in $\mathbb{N}$, ± 1 are the only two invertible elements in $\mathbb{Z}$. In $\mathbb{Q}$, $\mathbb{R}$ and $\mathbb{C}$ any nonzero number is invertible.

If A is a $n \times n$ matrix, then it has an inverse iff A is an invertible matrix. Thus every element A in $\mathrm{GL}(n, \mathbb{R})$ is invertible.

2.1.2 Groups

Definition 2.1.4 (Group). A set G with a binary operation $\cdot : G \times G \to G$ is called a group, if

1. $a \cdot (b \cdot c) = (a \cdot b) \cdot c$ for all elements a, b and c in G,
2. there exists an element $e \in G$ such that $a \cdot e = e \cdot a = a$ for all $a \in G$ and
3. for every element $a \in G$ there exists $b \in G$ such that $b \cdot a = e$.

Remark 2.1.5. 1. The axiom 2 ensures that the set G is not empty.

2. An element e in G satisfying axiom 2 is called the identity element in G.

3. An element b in G satisfying axiom 3 is called the inverse of the element a in G.

We derive some simple consequences of the definition and look at some examples.

Proposition 2.1.6. *Let G be a group. Then*

1. *identity element is unique and*

2. *inverse of an element is unique.*

Proof. Let e and e' be two elements in G such that $a \cdot e = e \cdot a = a$ and $a \cdot e' = e' \cdot a = a$ for all $a \in G$. Then, since e is an identity, the pair $(e, e') \mapsto e$ under the binary operation. Similarly, since e' is also an identity, the pair $(e, e') \mapsto e'$. But the image of elements under a function (in our case the binary operation) is unique. Hence $e = e'$.

We will first show that one sided inverse is a two sided inverse: Let a be an element in G. Then there exists an element $b \in G$ such that $b \cdot a = e$. Let $c \in G$ be such that $c \cdot b = e$. Then $c = c \cdot e = c \cdot (b \cdot a) = (c \cdot b) \cdot a = e \cdot a = a$. As a consequence $a \cdot b = c \cdot b = e$.

Given $a \in G$, let b, b' in G be such that $a \cdot b = a \cdot b' = e$. Then $b = b \cdot (a \cdot b') = (b \cdot a) \cdot b' = e \cdot b' = b'$. This proves that the inverse of an element is unique. □

Remark 2.1.7. Given an element a in G, the unique element b in G such that $ab = e = ba$ is denoted by a^{-1}.

Examples 2.1.8.

1. Let $G = \mathbb{Z}$, the set of integers. Then G is a group with usual addition of integers.

   Similarly the sets $\mathbb{Q}$, $\mathbb{R}$ and $\mathbb{C}$ are all groups with usual addition.

2. Let $G = \mathbb{N}$. Then $(\mathbb{N}, +)$ is **not** a group. This set does not have identity for this binary operation.

3. Let $G = \mathbb{Z}^+$, the set of non-negative integers. Then G is not a group under addition. The integer 0 is the identity. However no nonzero element has inverse for this binary operation.

4. Let $n \geq 2$ and $G = \mathbb{Z}_n$ where $\mathbb{Z}_n = \{[k] : k \in \mathbb{Z}\}$ is as defined in Chapter 1. Then G is a group under the addition defined on $\mathbb{Z}_n$.

   Let $[k]$, $[l]$ and $[m]$ be three elements in $\mathbb{Z}_n$. Then, from the associativity of usual addition of integers, it follows that $([k] + [l]) + [m] = [k + l] + [m] = [(k + l) + m] = [k + (l + m)] = [k] + [l + m] = [k] + ([l] + [m])$.
   $[0]$ is the identity element in $\mathbb{Z}_n$: Let $[k]$ be an element in G. Then $[k] + [0] = [k + 0] = [k]$.
   If $[k]$ is a nonzero element in $\mathbb{Z}_n$ and if $k = qn + r$ for q in $\mathbb{Z}$ and $0 < r < n$, then $k - r = qn$. Hence $[k] + [-r] = [k - r] = [qn] = [0]$. Thus G is a group under this addition.

Let $[k]$ and $[l]$ be two elements of $\mathbb{Z}_n$. Then, it follows from the definition of addition that $[k]+[l]=[k+l]=[l+k]=[l]+[k]$.

5. Let $G := \{A \in M_n(\mathbb{R}) : \det(A) \neq 0\}$. Then G is a group under matrix multiplication.

   Let A and B be two elements in G. We know that $\det(AB) = \det(A)\det(B)$. Hence $\det(AB) \neq 0$ and $(A, B) \mapsto AB$ is a binary operation on G.

   The matrix multiplication is associative. Hence $A\cdot(B\cdot C) = (A\cdot B)\cdot C$ for any three matrices A, B and C in G.

   The identity matrix I_n is the identity element of this binary operation.

   If $\det(A) \neq 0$ for a matrix A, then A is invertible and if we denote the inverse by A^{-1}, then A^{-1} is also in G.

   Hence G is a group under matrix multiplication. This group G is denoted by $\mathrm{GL}(n, \mathbb{R})$ or $\mathrm{GL}_n(\mathbb{R})$ and called the *general linear group* of order n.

6. Let X be set and let $G := S(X) := \{f : X \to X : f \text{ is a bijection }\}$. For two functions f and g in $S(X)$, we define $f \circ g : X \to X$ by $(f \circ g)(x) := f(g(x))$. Then it is easy to see that $f \circ g$ is a bijection of X and $S(X)$ becomes a group under this binary operation. This group is called the *symmetric group or the permutation group* on the set X.

   If X is a finite set with n-elements, we also write $S(X)$ as S_n and call this the *permutation group or symmetric group* on n symbols. If X is a finite set with n-elements, we take X to be the set $I_n := \{1, 2, 3, \ldots, n\}$.

   Let us take $X = \{1, 2, \ldots, n\}$. We denote the elements of S_n as σ, τ, θ etc.

   If σ is a permutation of X and $\sigma(i) = a_i$, we write σ as

$$\sigma = \begin{pmatrix} 1 & 2 & 3 & \cdots & n \\ a_1 & a_2 & a_3 & \cdots & a_n \end{pmatrix}$$

   Let τ be an element of S_n. If $\tau(a_i) = b_i$, where $\sigma(i) = a_i$, then $\tau \circ \sigma$ is written as

$$\tau \circ \sigma = \begin{pmatrix} 1 & 2 & 3 & \cdots & n \\ b_1 & b_2 & b_3 & \cdots & b_n \end{pmatrix}$$

   We work out some examples of composition of two permutations and find inverse of a permutation for specific values of n:

   - Let $n = 3$.

     If $\sigma = \begin{pmatrix} 1 & 2 & 3 \\ 2 & 3 & 1 \end{pmatrix}$ and $\tau = \begin{pmatrix} 1 & 2 & 3 \\ 2 & 1 & 3 \end{pmatrix}$, then $\sigma \circ \tau = \begin{pmatrix} 1 & 2 & 3 \\ 3 & 2 & 1 \end{pmatrix}$ and $\tau \circ \sigma = \begin{pmatrix} 1 & 2 & 3 \\ 1 & 3 & 2 \end{pmatrix}$.

     The inverse $\sigma^{-1} = \sigma^2 = \begin{pmatrix} 1 & 2 & 3 \\ 3 & 1 & 2 \end{pmatrix}$ and the inverse of τ is τ itself.

- Let $n = 4$.
  If $\sigma = \begin{pmatrix} 1 & 2 & 3 & 4 \\ 3 & 4 & 2 & 1 \end{pmatrix}$ and $\tau = \begin{pmatrix} 1 & 2 & 3 & 4 \\ 3 & 4 & 1 & 2 \end{pmatrix}$, then the product $\sigma \circ \tau = \begin{pmatrix} 1 & 2 & 3 & 4 \\ 2 & 1 & 3 & 4 \end{pmatrix}$ and the product $\tau \circ \sigma = \begin{pmatrix} 1 & 2 & 3 & 4 \\ 1 & 2 & 4 & 3 \end{pmatrix}$.
  The inverse σ^{-1} of σ is $\begin{pmatrix} 1 & 2 & 3 & 4 \\ 4 & 3 & 1 & 2 \end{pmatrix}$.
- Let $n = 6$.
  Let $\sigma = \begin{pmatrix} 1 & 2 & 3 & 4 & 5 & 6 \\ 2 & 4 & 3 & 6 & 5 & 2 \end{pmatrix}$ and $\tau = \begin{pmatrix} 1 & 2 & 3 & 4 & 5 & 6 \\ 1 & 2 & 4 & 5 & 3 & 6 \end{pmatrix}$ be two elements in S_6. Then $\tau \circ \sigma = \begin{pmatrix} 1 & 2 & 3 & 4 & 5 & 6 \\ 2 & 5 & 4 & 6 & 3 & 1 \end{pmatrix}$ and $\sigma \circ \tau = \begin{pmatrix} 1 & 2 & 3 & 4 & 5 & 6 \\ 2 & 4 & 6 & 5 & 3 & 1 \end{pmatrix}$. Let us now compute the inverse of σ, τ and $\sigma \circ \tau$. Since $\sigma^{-1} \circ \sigma(i) = i = \tau^{-1} \circ \tau$, we get that $\sigma^{-1} = \begin{pmatrix} 1 & 2 & 3 & 4 & 5 & 6 \\ 6 & 1 & 3 & 2 & 5 & 4 \end{pmatrix}$, $\tau^{-1} = \begin{pmatrix} 1 & 2 & 3 & 4 & 5 & 6 \\ 1 & 2 & 5 & 3 & 4 & 6 \end{pmatrix}$ and $(\sigma \circ \tau)^{-1} = \begin{pmatrix} 1 & 2 & 3 & 4 & 5 & 6 \\ 6 & 1 & 5 & 2 & 4 & 3 \end{pmatrix}$. We leave it as an exercise to find out the inverse of $(\tau \circ \sigma)^{-1}$.
  We can also show that $\sigma^4 = Id$ and $\tau^3 = Id$. Therefore $\sigma^{-1} = \sigma^3$ and $\tau^{-1} = \tau^2$.
  It will be a nice exercise to find the smallest k such that $(\sigma \circ \tau)^k = Id$ and $(\tau \circ \sigma)^k = Id$.
  In general let σ be an element of S_n and for each i let $b_i \in \{1, 2, \ldots, n\}$ be such that $\sigma(b_i) = i$. Then $\sigma^{-1}(i) = b_i$ and σ^{-1} is written as $\sigma^{-1} = \begin{pmatrix} 1 & 2 & 3 & \cdots & n \\ b_1 & b_2 & b_3 & \cdots & b_n \end{pmatrix}$.

7. Let a be a nonzero real number and b a real number. We define a map

$$\varphi_{a,b} : \mathbb{R} \to \mathbb{R} \quad \text{by} \quad \varphi_{a,b}(x) := ax + b \quad \text{for} \quad x \in \mathbb{R}.$$

This map is one-one and on-to: Suppose x and y are in $\mathbb{R}$ such that $ax + b = ay + b$. Then $ax = ay$. Since $a \neq 0$, we see that $x = y$.

If z is a real number, then $\varphi_{a,b}\left(\frac{z-b}{a}\right) = z$. This proves that the map $\varphi_{a,b} : \mathbb{R} \to \mathbb{R}$ is one-one and on-to for every element $(a, b) \in \mathbb{R}^* \times \mathbb{R}$.

Let (a, b) and (c, d) be two elements in $\mathbb{R}^* \times \mathbb{R}$. Then $\varphi_{a,b} \circ \varphi_{c,d}(x) = \varphi_{a,b}(cx + d) = a(cx + d) + b = \varphi_{ac,ad+b}(x)$ for x in $\mathbb{R}$.

We let

$$G := \{\varphi_{a,b} : (a, b) \in \mathbb{R}^* \times \mathbb{R}\}.$$

Then G is closed under the binary operation of composition of maps: Let $\varphi_{a,b}$ and $\varphi_{c,d}$ be two elements of G. Then we have shown that $\varphi_{a,b} \circ \varphi_{c,d} = \varphi_{ac,ad+b}$. Since $a \neq 0$ and $c \neq 0$, we see that $ac \neq 0$ and hence the map $\varphi_{ac,ad+b}$ is also in G.

We will now show that G is a group under this binary operation.

(a) Let (a,b), (c,d) and (e,f) be elements in $\mathbb{R}^* \times \mathbb{R}$. Then

$$\begin{aligned}
\varphi_{a,b} \circ (\varphi_{c,d} \circ \varphi_{e,f})(x) &= \varphi_{a,b} \circ (cex + cf + d) \\
&= a(cex + cf + d) + b \\
&= acex + acf + ad + b \\
&= ac(ex + f) + ad + b \\
&= \varphi_{ac,ad+b}(ex + f) \\
&= (\varphi_{a,b} \circ \varphi_{c,d})(ex + f) \\
&= (\varphi_{a,b} \circ \varphi_{c,d}) \circ \varphi_{e,f}(x)
\end{aligned}$$

for all x in $\mathbb{R}$. This proves associative law.

(b) If $\varphi_{a,b}$ is an element in G, then $\varphi_{a,b} \circ \varphi_{1,0}(x) = \varphi_{a,b}(x) = \varphi_{1,0} \circ \varphi_{a,b}(x)$ for all points x in $\mathbb{R}$. Thus $\varphi_{1,0}$ is the identity element in G.

(c) If $\varphi_{a,b}$ is an element in G, then $\varphi_{\frac{1}{a},-\frac{b}{a}}$ is the inverse of $\varphi_{a,b}$ can be shown as follows:
For $x \in \mathbb{R}$, $\varphi_{\frac{1}{a},-\frac{b}{a}} \circ \varphi_{a,b}(x) = \varphi_{\frac{1}{a},-\frac{b}{a}}(ax + b) = x = \varphi_{1,0}(x)$.

Thus G is a group. This group is called the one-dimensional *affine group* and it also denoted as $\mathbb{R}^* \ltimes \mathbb{R}$.

There is another *abstract* description of this group described below.

By abuse of notation we denote $G := \mathbb{R}^* \times \mathbb{R}$. For (a,x) and (b,y) in G, we define

$$(a,x) \cdot (b,y) := (ab, x + ay).$$

Observe that $(ab, x + ay)$, the product of (a,x) and (b,y) in G is also in G.

We will now show that G is a group under this binary operation and we denote this group also by $\mathbb{R}^* \ltimes \mathbb{R}$.

(a) Let (a,x), (b,y) and (c,z) be elements of $\mathbb{R}^* \ltimes \mathbb{R}$. Then

$$\begin{aligned}
(a,x) \cdot ((b,y) \cdot (c,z)) &= (a,x) \cdot (bc, y + bz) \\
&= (a(bc), a(bz + y) + x) \\
&= ((ab)c, abz + ay + x) \\
&= ((ab, ay + x) \cdot (c,z)) \\
&= ((a,x) \cdot (b,y)) \cdot (c,z).
\end{aligned}$$

This proves the associative law.

(b) Let (a,x) be in $\mathbb{R}^* \ltimes \mathbb{R}$. Then

$$(a,x) \cdot (1,0) = (a, a0 + x) = (a,x) = (1,0) \circ (a,x).$$

Hence $(a,0)$ is the identity in $\mathbb{R}^* \ltimes \mathbb{R}$.

(c) Let (a,x) and (b,y) be in $\mathbb{R}^* \ltimes \mathbb{R}$ such that $(a,x) \cdot (b,y) = (1,0)$. But

$$(a,x) \cdot (b,y) = (ab, x + ay).$$

This means that $b = 1/a$ and $y = -x/a$. Therefore $(1/a, -x/a)$ is the inverse of the element (a,x) in $\mathbb{R}^* \ltimes \mathbb{R}$.

Hence $\mathbb{R}^* \ltimes \mathbb{R}$ is a group.

Exercise 2.1.9. 1. Let $G = \mathbb{R}^* := \mathbb{R} \setminus \{0\}$, the set of nonzero real numbers. Show that G is a group under the multiplication. Is the set $\mathbb{R}^+$, the set of positive real numbers, a group under multiplication?

2. Show that the sets $\mathbb{Q}^*$, the nonzero rationals, $\mathbb{C}^*$, the nonzero complex numbers are groups under usual multiplication. Is the set of nonzero integers, a group under multiplication?

3. Let $n \geq 1$ and $G := \mathbb{R}^n$. For $x = (x_1, x_2, \ldots, x_n)$ and $y = (y_1, y_2, \ldots, y_n)$ in G, we define $x + y := (x_1 + y_1, x_2 + y_2, \ldots, x_n + y_n)$. Show that G is a group under this addition.

4. Let m and $n \in \mathbb{N}$ and $M_{m,n}(\mathbb{R}) := \{A := (a_{ij}) : a_{ij} \in \mathbb{R}, 1 \leq i \leq m, 1 \leq j \leq n\}$, the set of all matrices of order $m\times, n$ over real numbers. Prove that $M_{m,n}(\mathbb{R})$ is a group under the usual matrix addition.

5. We know that matrix multiplication is a binary operation on $M(n, \mathbb{R})$ with the identity matrix I_n as the identity of this binary operation. Show that $M(n, \mathbb{R})$ is **not** a group under this binary operation.

   If A is a $n \times n$ matrix with $\det(A) = 0$, then A does not have an inverse.

6. Let $G := \{A \in M_n(\mathbb{R}) : \det(A) = 1\}$. Show that G is group under matrix multiplication.

   This group is denoted by $\mathrm{SL}(n, \mathbb{R})$ or $\mathrm{SL}_n(\mathbb{R})$ and it is called the *special linear group*.

7. Let $G := \{A \in M(n, \mathbb{R}) : A^t A = I_n\}$. Show that G is a group under the matrix multiplication. This group is called the *orthogonal group* and it is denoted by $\mathrm{O}(n, \mathbb{R})$ or simply $\mathrm{O}(n)$.

   If $n = 2$, then this is the set of all rotations about the origin and reflections about the lines passing through origin in $\mathbb{R}^2$.

8. Let $G := \{A \in \mathrm{O}(n, \mathbb{R}) : \det(A) = 1\}$. Show that G is a group. It is called the *special orthogonal group* and denoted by $\mathrm{SO}(n, \mathbb{R})$.

   If $n = 2$, then $\mathrm{SO}(2, \mathbb{R})$ is the set of all rotations of $\mathbb{R}^2$.

9. This exercise is an extension of Example 2.1.8(7).

   Let $n \geq 2$ and $G := \mathrm{GL}(n, \mathbb{R}) \times \mathbb{R}^n$. Let (A, v) and (B, w) be two elements in G. Define
   $$(A, v) \cdot (B, w) := (AB, Aw + v)$$
   Prove that G is a group under this binary operation.

   This group denoted by $\mathrm{GL}(n, \mathbb{R}) \ltimes \mathbb{R}^n$ is called the n-dimensional *affine group*.

(a) Let (A, v), (B, w) and (C, u) be in $\mathrm{GL}(n, \mathbb{R}) \ltimes \mathbb{R}^n$. Then

$$\begin{aligned}(A,v)\cdot((B,w)\cdot(C,u)) &= (A,v)\cdot(BC, Bu+w)\\ &= (A(BC), A(Bu+w)+v)\\ &= ((AB)C, ABu+Aw+v)\\ &= ((AB, Aw+v)\cdot(C,u))\\ &= ((A,v)\cdot(B,w))\cdot(C,u).\end{aligned}$$

This proves the associative law.

(b) Let (A, v) be in $\mathrm{GL}(n, \mathbb{R}) \ltimes \mathbb{R}^n$. Then

$$(A,v)\cdot(I,0) = (A, A0+v) = (A,v) = (I,0)\circ(A,v).$$

Hence $(I, 0)$ is the identity in $\mathrm{GL}(n, \mathbb{R}) \ltimes \mathbb{R}^n$.

(c) Let (A, v) and (B, w) be in $\mathrm{GL}(n, \mathbb{R}) \ltimes \mathbb{R}^n$ such that the product $(A, v) \cdot (B, w) = (I, 0)$. But

$$(A,v)\cdot(B,w) = (AB, Aw+v).$$

Equating $(AB, Aw + v) = (I, 0)$ we get $B = A^{-1}$ and $w = -A^{-1}v$. Therefore $(A^{-1}, -A^{-1}v)$ is the inverse of the element (A, v) in $\mathrm{GL}(n, \mathbb{R}) \ltimes \mathbb{R}^n$.

Hence $\mathrm{GL}(n, \mathbb{R}) \ltimes \mathbb{R}^n$ is a group.

Observe that we could have let

$$G := \{\varphi_{A,v} : A \in \mathrm{GL}(n, \mathbb{R}) \text{ and } v \in \mathbb{R}^n\}$$

and defined the group operation as composition of functions.

10. Let $G = \{(A, v) \in \mathrm{GL}(n, \mathbb{R}) \ltimes \mathbb{R}^n : A^tA = Id\}$. Prove that G is also a group under the binary operation defined in Exercise 2.1.8(9). This group is called the group of *rigid motions* of $\mathbb{R}^n$ or the *Euclidean group of motions* of $\mathbb{R}^n$.

11. Let

$$G := \{A = (a_{ij}) \in M(n, \mathbb{R}) : a_{ii} = 1 \text{ and } a_{ij} = 0 \text{ for } i < j\}$$

Show that G is a group under the multiplication of matrices, called the group of *upper triangular matrices.*

As a special case let us take $n = 2$. Then

$$G := \left\{A = \begin{pmatrix}1 & x\\ 0 & 1\end{pmatrix} \text{ for } x \in \mathbb{R}\right\}.$$

If $A = \begin{pmatrix}1 & x\\ 0 & 1\end{pmatrix}$ and $B = \begin{pmatrix}1 & y\\ 0 & 1\end{pmatrix}$, then $AB = \begin{pmatrix}1 & x+y\\ 0 & 1\end{pmatrix}$ is also in G. Thus matrix multiplication is a binary operation on G. It is now very easy to check that G is a group.

12. (a) Let X be a set and $G := \{f : f : X \to \mathbb{R} \text{ is a function }\}$. On this set define an binary operation as follows: If f and g are in G, we define a new function $f + g : X \to \mathbb{R}$ by $(f + g)(x) := f(x) + g(x)$ for all $x \in X$.

Show that with respect to this binary operation the set G becomes a group.

i. Let f, g and h be in G. Then by definition $f+(g+h)$ is the function defined by

$$(f+(g+h))(x) := f(x)+(g+h)(x) = f(x)+(g(x)+h(x))$$

and

$$((f+g)+h)(x) := (f(x)+g(x))+h(x)$$

for all $x \in X$. Since $(\mathbb{R},+)$ is a group, it follows that $f(x)+(g(x)+h(x)) = (f(x)+g(x))+h(x)$ for all $x \in X$. Hence $f+(g+h)=(f+g)+h$.

ii. Let $\theta : X \to \mathbb{R}$ be the function defined by $\theta(x)=0$ for all $x \in \mathbb{R}$. Then for any function $f : X \to \mathbb{R}$, we have $(f+\theta)(x) = f(x)+\theta(x) = f(x)+0 = f(x) = (\theta+f)(x)$ for all x in X. (How did we guess the function θ?).

iii. Let $f \in G$. Define a function $g : X \to \mathbb{R}$ by $g(x) = -(f(x))$, the additive inverse of the real number $f(x)$. Then it is easy to check that as functions $f+g=g+f=\theta$. (How did we guess g?). This proves that G is a group.

Let us observe that to show that G is a group, we needed the group structure $(\mathbb{R},+)$. We also denote this group by $\mathcal{F}(X,\mathbb{R})$.

We change the target and get different group.

(b) Let X be a set and $G := \mathcal{F}(X,\mathbb{R}^*) := \{f \in \mathcal{F}(X,\mathbb{R}) : f(x) \neq 0 \text{ for all } x \in X\}$. Observe that if f and g are two functions taking values in $\mathbb{R}^*$, their sum $f+g$ may be zero at some point x in X. On the other hand, if $f,g : X \to \mathbb{R}^*$ are two functions, then for every point x in X, the product $f(x)\cdot g(x) \neq 0$. Hence the "product" defined by

$$(f \cdot g)(x) := f(x)\cdot g(x)$$

for all x in X is a binary operation on G. Show that G is a group under this binary operation.

(c) We generalize the construction in Exercise 2.1.9(12a). Let G be a group and X be a set. Let

$$\mathcal{F}(X,G) := \{f : X \to G\}.$$

Show that $\mathcal{F}(X,G)$ is a group under the (obvious) binary operation $(f,g) \mapsto f.g$ where $f.g : X \to G$ is the function defined by

$$(f.g)(x) := f(x)\cdot g(x).$$

- Two special cases are $G=(\mathbb{R},+)$ and $G=(\mathbb{R}^+,\cdot)$.
  When $G=(\mathbb{R},+)$ what we get is the Exercise 2.1.9(12a).
  If $G=(\mathbb{R}^+,\cdot)$, the set $\mathcal{F}(X,G)$ is the set of all real valued functions taking only positive values.

13. A special case of the previous Exercise 2.1.9(12a) is the following. Let $X=\mathbb{N}$. Then $\mathcal{F}(\mathbb{N},\mathbb{R})$ is the set of all real sequences. Prove that this is a group under addition.

If (x_n) and (y_n) are two sequences then $(x_n) + (y_n) := (x_n + y_n)$ is in $\mathcal{F}(\mathbb{N}, \mathbb{R})$.

- The zero sequence (0) is the identity.
- If (x_n) is in $\mathcal{F}(\mathbb{N}, \mathbb{R})$, then $-(x_n)$, the inverse of (x_n) is the sequence $(-x_n)$.

14. Let $X = \mathbb{R}$ and $C(X, \mathbb{R}) := \{f \in \mathcal{F}(X, \mathbb{R}) : f \text{ is continuous}\}$. Show that the set $C(X, \mathbb{R})$ becomes a group, under the addition of functions defined as in Exercise 2.1.9(12a) above.

    Use the algebra of continuous functions to conclude that the set $C(X, \mathbb{R})$ is a group.

15. Let $D(X, \mathbb{R}) := \{f \in \mathcal{F}(X, \mathbb{R}) : f \text{ is differentiable in } X\}$. Then using the algebra of differentiable functions, show that $D(\mathbb{R}, \mathbb{R})$ is a group under the addition of functions defined in Exercise 2.1.9(12a).

16. Let $C(\mathbb{N}, \mathbb{R}) := \{(a_n) \in \mathcal{F}(\mathbb{N}, \mathbb{R}) : \lim_n a_n \text{exists}\}$. That is $C(\mathbb{N}, \mathbb{R})$ is the set of all convergent sequences of real numbers. Then using the algebra of convergent sequences, show that $C(\mathbb{N}, \mathbb{R})$ is a group under the addition defined as in Exercise 2.1.9(14).

17. Let $\mathbb{T} := \{z \in \mathbb{C} : |z| = 1\}$, the set of complex numbers of unit modulus. Show that $\mathbb{T}$ is a group under the multiplication of complex numbers.

18. Let $n \geq 1$ be fixed and $G_n := \{z \in \mathbb{C} : z^n = 1\}$. Prove that G_n is group under the multiplication of complex numbers.

    Let z and w be in G_n. Then using the generalized associative law of multiplication of complex numbers, we get $(zw)^n = z^n w^n = 1$. Hence zw is in G_n. We leave the other details to show that G_n is a group.

    The group G_n is known as the group of n-th roots of unity. They are also vertices of a regular n-sided polygon inscribed on the unit circle with 1 as the one of the vertices.

19. Let G_n be as in Exercise 2.1.9(18) above and $G := \cup_{n=1}^{\infty} G_n$. Show that G is a group under the multiplication of complex numbers.

    The group G is the union of all the roots of unity. Geometrically it is the union of all regular polygons inscribed on the unit circle.

    If z is a n-th root of unity and w is a m-th root of unity, then $(zw)^{mn} = 1$. Thus zw is a mn-th root of unity.

20. Let $G = \mathbb{Z}$ and $\oplus : G \times G \to G$ be defined by $a \oplus b := a + b + 2$. Show that $(G, \oplus)$ is a group.

    For any two integers a and b, the sum $a + b + 2$ is also an integer. Hence $\oplus$ is a binary operation on $\mathbb{Z}$.

    If a, b and c are integers, then $(a \oplus b) \oplus c = (a+b+2) \oplus c = (a+b+2) + c + 2$. Now we use the associativity of usual addition in integers to write $(a+b+2)+c+2 = a+2+(b+c+2) = a \oplus (b+c+2) = a \oplus (b \oplus c)$ and conclude that $\oplus$ is associative on $\mathbb{Z}$.

Let x in $\mathbb{Z}$ be such that $a \oplus x = a + x + 2 = a$ for all integers a. This shows that $x = -2$ is the identity for this law of composition.
Let a be an integer and b be an integer such that $a \oplus b = -2$. That is $a + b + 2 = -2$. This shows that $b = -(a+4)$ is the inverse of a.
Thus G is a group.

21. Let $G = \mathbb{R}$. For x and $y \in \mathbb{R}$, define $x \oplus y := x + y - 1$ and show that $(G, \oplus)$ is a group.

22. Let $G = \mathbb{R}^*$. For x and $y \in \mathbb{R}^*$, define $x \odot y := 2xy$. Prove that $(G, \odot)$ is a group.
More generally, if $\alpha \neq 0$ is a real number, show that the set $G = \mathbb{R}^*$ becomes a group under the binary operation $(x, y) \mapsto \alpha xy$.

23. Show that $G := \mathbb{R} \setminus \{-1\}$ is a group under the binary operation

$$a * b = a + b + ab \quad \text{for } a, b \in G.$$

24. Let G be a group and fix an element $c \in G$. Define a binary operation

$$* : G \times G \to G \qquad \text{by } a * b = acb.$$

Show that $(G, *)$ is a group.

25. Let X be a non-empty set and $\mathcal{P}(X)$ be the set of all subsets of X. If A and B are two subsets of X, the map

$$(A, B) \mapsto (A \setminus B) \cup (B \setminus A)$$

is a binary operation on $\mathcal{P}(X)$ and this binary operation makes $\mathcal{P}(X)$ in to a group. This binary operation is called the symmetric difference of A and B and it is denoted by $A \Delta B$.

26. Let G and K be two groups. There is a natural binary operation on the cartesian product

$$G \times K := \{(g, k) : g \in \text{ and } k \in K\}$$

defined by

$$(g_1, k_1).(g_2, k_2) := (g_1 g_2, k_1 k_2)$$

for (g_i, k_i) in $G \times K$. Show that $G \times K$ becomes a group under this binary operation. This group is called the product group of the groups G and K.

The element (e, e) is the identity element of the group $G \times K$. If (g, k) is in $G \times K$, then (g^{-1}, k^{-1}) is the inverse of (g, k).

Note that for each $n \geq 2$, the groups $\mathbb{R}^n$ are product groups.

There is another important class of groups which we introduce now.
Let $n \geq 2$ and $U_n := \{[k] \in \mathbb{Z}_n : \gcd(k, n) = 1\}$. We have shown that the multiplication defined on $\mathbb{Z}_n$ in Chapter 1 is a binary operation on U_n. We will now show that U_n is a group.

Proposition 2.1.10. *The set U_n is a group for $n \geq 2$.*

Proof. Let a, b and c be three nonzero integers such that $\gcd(a, n) = \gcd(b, n) = \gcd(c, n) = 1$. We know from Exercise 1.1.33 of Chapter 1 that $\gcd(ab, n) = 1$, if $\gcd(a, n) = \gcd(b, n) = 1$. Hence $\gcd(abc, n) = 1$ and $[a]([b][c]) = [a][bc] = [a(bc)] = [(ab)c] = ([a][b])[c]$ for all elements $[a]$, $[b]$ and $[c]$ in U_n.

It is easy to see that the identity element is $[1]$.

Let $[a]$ be an element in U_n. Then there exist integers x and y in $\mathbb{Z}$ such that $ax + ny = 1$. Therefore $[1] = [ax + ny] = [ax] = [a][x]$. This proves that U_n is a group.

Further $[a][b] = [ab] = [ba] = [b][a]$ for all elements $[a]$ and $[b]$ in U_n. □

Definition 2.1.11. The group U_n is called the group of units modulo n.

Example 2.1.12.

1. Let $n = 5$. Then $U_5 := \{[1], [2], [3], [4]\}$.
2. Let $n = p$ a prime number. Then $U_p = \{[k] : 1 \leq k < p\}$.
3. Let $n = 6$. Then $U_6 = \{[1], [5]\}$.
4. If $n = 8$ then $U_8 = \{[1], [3], [5], [7]\}$.
5. If $n = 10$ then $U_{10} = \{[1], [3], [7], [9]\}$.

We will now use the group structure of U_n to prove an important result in elementary number theory called Wilson's theorem.

Theorem 2.1.13 (Wilson's theorem). *Let $p \geq 2$ be a prime number. Then*

$$(p-1)! \equiv -1 \pmod p$$

Proof. For a prime number p, the group $U_p = \{[k] : 1 \leq k \leq (p-1)\}$. First let us observe that $(p-1) \equiv -1 \pmod p$ and hence $[(p-1)]$ is its own inverse in U_p. Therefore it suffices to prove that $(p-2)! \equiv 1 \pmod p$.

Let $2 \leq k \leq (p-2)$ be a natural number. Then there exists $[l] \in U_p$ such that $[kl] = [1]$. That is $kl \equiv 1 \pmod p$. Therefore $l \leq (p-2)$.

Let $S := \{k \in \mathbb{N} : k \neq 1, (p-1)\}$. Then S has $(p-3)$ elements. For every element $k \in S$, there exists unique $[l] \in U_p$ such that $kl \equiv 1 \pmod p$. Therefore we can write S as

$$S = \{k_1, k_2, \cdots, k_{\frac{p-3}{2}}, l_1, l_2, \cdots, l_{\frac{p-3}{2}}\}$$

where $k_i l_i \equiv 1 \pmod p$. Hence

$$2 \cdot 3 \cdots (p-2) = \prod_{i=1}^{\frac{p-3}{2}} (k_i\, l_i) \equiv 1 \pmod p$$

and

$$(p-1)! \equiv -1 \cdot (p-2)! \equiv -1 \pmod p.$$

This completes the proof. □

We will now prove some simple but important properties of groups.

If a_i, for $1 \leq i \leq n$, are elements in a group G, we write the product

$$(a_1(a_2(a_3 \cdots) \cdots) a_{n-1}) a_n) \text{ as } a_1 \cdot a_2 \cdots a_n.$$

Proposition 2.1.14. *Let G be a group. Then the following holds.*

1. *If a is an element in G, then $(a^{-1})^{-1} = a$.*
2. *If a and b are in G, then $(ab)^{-1} = b^{-1}a^{-1}$.*
3. *If $n \in \mathbb{N}$ and $a_1, a_2, \ldots, a_n$ are in G, then the product*

$$a_1 \cdot a_2 \cdots a_n = (a_1 \cdot a_2 \cdots a_k)(a_{k+1} \cdot a_{k+2} \cdots a_n)$$

for any $k < n$. This property is called the generalized associative law.

Proof. Let a be an element of G. Then $a \cdot a^{-1} = e$. Now it follows from the definition that $(a^{-1})^{-1} = a$.

Let a and b be two elements in G and c be the inverse of the product ab. Then, using associative law, we get $e = c(ab) = (ca)b$. This means that $ca = b^{-1}$. Hence $c = c(a \cdot a^{-1}) = (ca)(a^{-1}) = b^{-1}a^{-1}$. This proves that $(ab)^{-1} = b^{-1}a^{-1}$ for all a and b in G.

We will now prove the generalized associative law by induction. If $n = 3$, then the generalized associative law is the associative law of the binary operation.

Let $n \geq 4$ and assume the result to be true for all $m < n$. That is if $a_1, a_2, \cdots a_m$ are elements in G, then for any $k < m$, $(a_1 (a_2(a_3 \cdots a_m)) = (a_1 \cdots a_k)(a_{k+1} \cdots a_m)$.

Let $a_1, a_2, \cdots a_n$ be elements in G. Then by induction hypothesis, we can write $(a_1 (a_2(a_3 \cdots a_n)) = (a_1 ((a_2 \cdots a_k)(a_{k+1} \cdots a_n)))$ for any $k < n$. Now using associative law, we write the product $a_1 ((a_2 \cdots a_k)(a_{k+1} \cdots a_n)) = (a_1(a_2 \cdots a_k)) (a_{k+1} \cdots a_n)$. Since $k < n$, using induction hypothesis again we conclude $(a_1(a_2 \cdots a_k)) (a_{k+1} \cdots a_n) = (a_1 \cdot a_2 \cdots a_k)(a_{k+1} \cdots a_n)$. This completes the proof of generalized associative law of an associative binary operation. □

Let G be a group and x be an element of G. If $k \in \mathbb{N}$, then we denote by x^k, the product $\overbrace{x \cdot x \cdots x}^{k-\text{times}}$ and by x^{-k} by the product $\overbrace{x^{-1} \cdot x^{-1} \cdots x^{-1}}^{k-\text{times}}$.

If the group operation on G had additive notation, then x^k is denoted by kx and x^{-1} by $-x$.

If $k = 0$, x^0 is defined to be the identity element e.

If $k < 0$ then $-k$ is positive and we define $x^k = \overbrace{x^{-1} \cdot x^{-1} \cdots x^{-1}}^{(-k)-\text{times}}$.

If x is an element in G and k, l are are two integers, then it follows from the generalized associative law that $x^{k+l} = x^k x^l$. This is called the *law of indices*.

Exercise 2.1.15. Let G be a group. If $a_1, \ldots, a_n$ are elements of G, show that

$$(a_1 a_2 \cdots a_n)^{-1} = a_n^{-1} a_{n-1}^{-1} \cdots a_1^{-1}.$$

Deduce that the inverse of x^k is x^{-k} for every x in G and $k \in \mathbb{Z}$.

Another important property in a group G is the *laws of cancellation.*

Proposition 2.1.16. *Let G be a group. Then the following hold.*

1. *If a, b and c are in G such that $ab = ac$, then $b = c$. This is called the left cancellation law.*
2. *If a, x and y are in G such that $xa = ya$, then $x = y$. This is called the right cancellation law.*
3. *If a and b are two elements in G, then the equations $ax = b$ and $ya = b$ have unique solutions in G.*
4. *If x in G is such that $x^2 = x$, then $x = e$.*

Proof. Let a, b and c in G be such that $ab = ac$. Then $b = (a^{-1}a)b = a^{-1}(ab) = a^{-1}(ac) = c$. This proves the left cancellation law.

For the proof of right cancellation law, we multiply x by the identity e on the right to prove $x = x(aa^{-1}) = (xa)a^{-1} = (ya)a^{-1} = y(aa^{-1}) = y$.

Let a and b be two elements of G. If x in G is such that $ax = b$, then $a^{-1} \cdot (a \cdot x) = a^{-1}b$. This proves that $x = a^{-1}b$.

If x and z are two solutions of the equation $ax = b$, then $ax = b = az$. If we now use left cancellation law, we get $x = z$, proving the uniqueness.

The solution for the equation $ya = b$ is ba^{-1}. This is proved exactly as above.

Let x in G such that $x^2 = x$. We rewrite this as $x^2 = x.x = x = e.x$. Then it follows from cancellation law that $x = e$. □

In the Example 2.1.8 and Exercise 2.1.9, we notice that some of the groups have the following property: For any two elements a and b in G, the product $ab = ba$. These groups are called *abelian groups.*

Definition 2.1.17. A group G is called abelian if $xy = yx$ for all elements x and y in G.

Examples 2.1.18.

1. The groups $G = \mathbb{Z}$, $\mathbb{Q}$, $\mathbb{R}$ and $\mathbb{C}$ are all abelian under addition.
2. The groups $G = \mathbb{Q}^*$, $\mathbb{R}^*$ and $\mathbb{C}^*$ are all abelian under multiplication.
3. The group $G = (\mathbb{Z}_n, +)$ is abelian.
4. The group $(U_n, \cdot)$ is abelian.
5. The group $G = \mathrm{GL}(n, \mathbb{R})$ is non-abelian.

   We will first show that $\mathrm{GL}(2, \mathbb{R})$ is non-abelian.

   Let $A = \begin{pmatrix} 1 & 3 \\ 2 & 4 \end{pmatrix}$ and $B = \begin{pmatrix} 0 & 1 \\ 1 & 0 \end{pmatrix}$. Then $AB = \begin{pmatrix} 3 & 1 \\ 4 & 2 \end{pmatrix}$ and $BA = \begin{pmatrix} 2 & 4 \\ 1 & 3 \end{pmatrix}$ which are not equal.

   Let $n \geq 3$. Then we let $A_1 = \begin{pmatrix} A & O_{2\times(n-2)} \\ O_{(n-2)\times 2} & I_{n-2} \end{pmatrix}$ and $B_1 = \begin{pmatrix} B & 0 \\ 0 & I_{n-2} \end{pmatrix}$; here $O_{2\times(n-2)}$, $O_{(n-2)\times 2}$ are zero matrices and I_{n-2} is the identity matrix of order $n - 2$. Then $A_1B_1 \neq B_1A_1$.

6. A group G with four elements is abelian.

   Let G be a group with four elements. Let $a(\neq e)$ be an element in G. We choose an element $b \neq e, a, a^{-1}$ in G. Then $ab \neq e$ and $G = \{e, a, b, ab\}$. The element ba is also in G and further it is not equal to e, a or b. Therefore $ab = ba$. This proves that G is abelian.

   We can say more about the structure of the group G. The element a^2 is in G and it is not equal to either a or ab. Therefore $a^2 = b$ or $a^2 = e$, the identity element of G.

   Suppose $a^2 = b$. Then $G := \{e, a, a^2, a^3\}$. The element a^4 can't be equal to a, a^2 or a^3. Therefore $a^4 = e$ and we can describe G as $G := \{e, a, a^2, a^3 : a^4 = e\}$.

   Suppose $a^2 = e$. The element b^2 is also in G. As in the case of the element, it follows that $b^2 = a$ or $b^2 = e$. If $b^2 = a$, then we can write G as $G := \{e, b, b^2, b^3 : b^4 = e\}$.

   Suppose $b^2 = e$. Since $a^2 = e$, it follows that $(ab)^2 = (ab)(ab) = a(ba)b = a^2b^2 = e$. Hence $G := \{e, a, b, ab : a^2 = e = b^2 = ab\}$.

   Thus we have shown that any group G with four elements is either of the form
   $$G := \{e, a, a^2, a^3 : a^4 = e\}$$
   or
   $$G := \{e, a, b, ab : a^2 = b^2 = (ab)^2 = e\}.$$

Exercise 2.1.19. 1. Let G be a group such that $x^2 = e$ for every x in G. Then G is abelian.

   Let us first observe that $x^2 = e$ implies that $x = x^{-1}$ for all x in G. Let x and y be two elements in G. Then $x = x^{-1}$ and $y = y^{-1}$. Therefore $xy = x^{-1}y^{-1} = (yx)^{-1} = yx$. Hence G is abelian.

2. Let X be a non-empty set and $G = \mathcal{P}(X)$ as in Example 2.1.8(25). Show that G is abelian.

   For every element A in $\mathcal{P}(X)$ we have $A \Delta A = \emptyset$, the identity element of $\mathcal{P}(X)$.

3. Let G be a group with $n(\geq 2)$ elements such that $a^2 = e$ for all elements a in G. Then there exists $k \geq 1$ such that $n = 2^k$.

   Let G be a group with $n \geq 2$ elements such that $a^2 = e$ for all elements a in G. Observe that for every element $a \neq e$ in G the subset $\{e, a\}$ is a group.

   Let $a_1 \neq e$ be an element in G and $G_1 := \{e, a_1\}$. If $G = G_1$, then we are through. If not choose an element $a_2 \in G \setminus G_1$ and let $G_2 := \{e, a_1, a_2, a_1a_2\}$. Then G_2 is a group with 2^2 elements. If $G = G_2$, we are through. Otherwise we proceed by induction.

   Suppose we have constructed a group G_{r-1} with 2^{r-1} elements of G. If G_{r-1} is not equal to G, we choose an element a_r in $G \setminus G_{r-1}$ and let $G_r := \{a, aa_r : a \in G_{r-1}\}$. Then G_r is a group with 2^r elements. Since G is a finite set, this process terminates after k-steps for some $k \geq 1$. Hence G is a group with $n = 2^k$ elements.

4. Let G be a group with five elements. Is G an abelian group?

Since the group G has five elements, there exists an element a in G such that $a \neq a^{-1}$. Then the elements e, a and a^2 are in G. If $a^3 = e$, choose an element $b \neq e, a, a^2$. Then $e, a, a^2 = a^{-1}, b, ab, a^2b$ are six distinct elements in G, a contradiction. Therefore $a^3 \neq e$. Similarly, we can show that $a^4 \neq e$. Thus $G := \{e, a, a^2, a^3, a^4 : a^5 = e\}$ and hence G is abelian.

Exercise 2.1.20. Show that the following statements are equivalent in a group G:

1. G is abelian.

2. $(xy)^n = x^n y^n$ for all x, $y \in G$ and all $n \in \mathbb{Z}$.

3. $(xy)^n = x^n y^n$ for all x, $y \in G$ and for three consecutive integers n.

4. $(xy)^2 = x^2 y^2$ for all x, $y \in G$.

5. $(xy)^{-1} = x^{-1} y^{-1}$ for all x, $y \in G$.

Exercise 2.1.21. Let a and b two elements in a group G such that $ab \neq ba$. Show that the elements of the set $\{1, a, b, ab, ba\}$ are all distinct. Hence conclude that any non-abelian group G has at least six elements.

Exercise 2.1.22. Which of the following binary operations turn G in to a group? Which are abelian?

1. Let $G = \mathbb{Q}$ and $\star : G \times G \to G$ be defined by $a \star b = \frac{a+b}{2}$.

2. Let $G = \mathbb{Q}^*$ and $\star : G \times G \to G$ defined by $a \star b = a/b$.

An important example of a non-abelian group is given below.

Example 2.1.23 (Quaternion Group). Let $G := \{1, -1, i, -i, j, -j, k, -k\}$ with the binary operation defined as follows: The element 1 is the identity. The products are defined by

$$i^2 = j^2 = k^2 = -1 \text{ and } i \cdot j = k.$$

The rule $i \cdot j = k$ together with the relations $i^2 = j^2 = k^2 = -1$ implies the following.

$$k \cdot j = (i \cdot j) \cdot j = i \cdot j^2 = -i$$

and

$$(j \cdot k)(k \cdot j) = 1$$

This implies that $j \cdot k = i$. Using this we can compute all other products and prove that this is a group. This group is not abelian.

This group can also be described as follows:

Let

$$G := \{\pm I_4, \pm A, \pm B, \pm C\}$$

where $A = \begin{pmatrix} 0 & -1 & 0 & 0 \\ 1 & 0 & 0 & 0 \\ 0 & 0 & 0 & -1 \\ 0 & 0 & 1 & 0 \end{pmatrix}, B = \begin{pmatrix} 0 & 0 & -1 & 0 \\ 0 & 0 & 0 & 1 \\ 1 & 0 & 0 & 0 \\ 0 & -1 & 0 & 0 \end{pmatrix}$, and

$C = \begin{pmatrix} 0 & 0 & 0 & -1 \\ 0 & 0 & -1 & 0 \\ 0 & 1 & 0 & 0 \\ 1 & 0 & 0 & 0 \end{pmatrix}$.

It is easy to check that

1. $AB = C$, $BC = A$ and $CA = B$,
2. $A^2 = B^2 = C^2 = -I_4$ and $A^4 = B^4 = C^4 = I_4$.

Using these relations, we can show that G is a group under matrix multiplication. Observe that G is not abelian.

2.2 Subgroups

Let G be a group and H be a non-empty subset of G. Then for any two elements h_1 and h_2 in H, the product h_1h_2 is in G. i.e., $\cdot : H \times H \to G$. If $\cdot(H \times H) \subseteq H$, then we get a binary operation $\cdot : H \times H \to H$. Further if H becomes a group under this binary operation, we say that H is a *subgroup* of G.

Let H be a subgroup of G and e_H be the identity of H. If x is an element of H, then $x = x.e_H$. On the other hand $x = x.e_G$ as well; here e_G is the identity of G. Hence $x.e_H = x = x.e_G$ and using cancellation law in G, we conclude that $e_H = e_G$. Thus identity element of H is same as the identity element of G.

Let x be an element of H with y in H as its inverse. If z in G is the inverse of x, then $x.y = e_H = e_G = x.z$. Again we use cancellation law in G to conclude that $y = z$. This proves that, for every element h in H, the inverse of h in H is same as the inverse of h in G.

Before looking at examples of subgroups, we prove a characterization for a subset H of a group G to be a subgroup of G.

Proposition 2.2.1. *Let G be a group and H be a non-empty subset of G. Then the following are equivalent.*

1. *H is a subgroup of G.*
2. *For all elements a and b in H, the elements ab and a^{-1} are in H.*
3. *For all elements a and b in H, the element ab^{-1} is in H.*

Proof. We will prove that (1) $\Rightarrow$ (2) $\Rightarrow$ (3) $\Rightarrow$ (1).

Let H be a subgroup of G. By definition H is a group with respect to the binary operation "$\cdot$". Therefore, if a and b are two elements in H, then the product ab and a^{-1} the inverse of a are in H. This proves (1) $\Rightarrow$ (2).

We will now prove (2) $\Rightarrow$ (3). Let H be a subset of G such that for any two elements a and $b \in H$ the product ab and a^{-1} are in H.

Let a and b be two elements in H. Then by hypothesis b^{-1} is in H. Therefore ab^{-1} is also in H.

We will show that (3) $\Rightarrow$ (1). Let H be a subset of G such that ab^{-1} is in H for all elements a and b in H. We will now show that H is a subgroup of G.

Since H is non-empty, choose any $a \in H$. Then $e = aa^{-1}$ is in H. If a is in H then $a^{-1} = ea^{-1}$ is also in H. Hence the identity element and inverse of an element is in H. If we now show that H is closed under multiplication we are through. (why?) Let a and $b \in H$. Then, since b^{-1} is also in H, it follows that $ab = a(b^{-1})^{-1}$ is also in H. It is easy to verify the associativity: Let a, b and c be in H. Then we know that $(ab)c$ and $a(bc)$ are in H. Since $a(bc) = (ab)c$ in G, associativity follows. Hence H is a subgroup of G. □

Examples 2.2.2. 1. Let $G = (\mathbb{Z}, +)$. Then for every integer k, the set $H_k = k\mathbb{Z} := \{kl : l \in \mathbb{Z}\}$ is a subgroup of G.

2. Let m and n be two natural numbers. Then $H_m \cap H_n$ is a subgroup of $\mathbb{Z}$.

3. Let m and n be two natural numbers. The set

$$H_m + H_n := \{mx + ny : x, y \in \mathbb{Z}\}$$

is a subgroup of $\mathbb{Z}$.

4. Consider the inclusions $\mathbb{Z} \subseteq \mathbb{Q} \subseteq \mathbb{R} \subseteq \mathbb{C}$. Each of these groups under addition are subgroups in the bigger group they are contained in.

5. Consider the inclusion $\mathbb{Q}^* \subseteq \mathbb{R}^* \subseteq \mathbb{C}^*$. Each of these groups under multiplication are subgroups in the bigger group they are contained in.

6. Consider the inclusion of the groups $\mathrm{SO}(n, \mathbb{R}) \subseteq \mathrm{O}(n, \mathbb{R}) \subseteq \mathrm{GL}(n, \mathbb{R})$ and the inclusion $\mathrm{SL}(n, \mathbb{R}) \subseteq \mathrm{GL}(n, \mathbb{R})$. These are all subgroups of the bigger groups in which they are contained in.

Exercise 2.2.3. 1. Let H and K be two subgroups of G. Show that $H \cap K$ is a subgroup of G.

2. Let $G = \mathrm{GL}(n, \mathbb{C})$ and $H := \{A \in \mathrm{GL}(n, \mathbb{C}) : (\overline{A})^t A = Id\}$. Show that H is a subgroup of G. This group is called the *unitary group* and its elements are called *unitary matrices.* This group is denoted as $U(n)$.

3. Let $G := \{a + b\sqrt{2} : a, b \in \mathbb{Z}\}$. Show that G is a group under addition and hence it is a subgroup of $\mathbb{R}$. How about the set $\{a + b\sqrt{2} : a, b \in \mathbb{Q}\} \subseteq \mathbb{R}$?

4. Let $n \geq 2$ and $G = S_n$, the permutation group on n-symbols. Show that the following subsets of G are subgroups of G.

   (a) $H := \{\sigma \in G : \sigma(n) = n\}$.

   (b) $H_A := \{\sigma \in G : \sigma(a) = a \text{ for all } a \in A\}$ where A is a subset of $\{1, 2, \cdots, n\}$.

   (c) $K := \{\sigma \in G : \sigma(a) \in A \text{ for all } a \in A\}$.

   Further show that H_A is a subgroup of K.

5. More generally, let $G = S(X)$ be the permutation group on a set X. If A is a subset of X, we let $H = \{\sigma \in G : \sigma(a) = a \text{ for all } a \in A\}$ and $K = \{\sigma \in G : \sigma(a) \in A \text{ for all } a \in A\}$. Prove that H and K are subgroups of G and H is a subgroup of K.

6. Let $G = \mathbb{C}^*$. Show that for every natural number n, the group G_n, the set of n-th roots of unity is a subgroup of G.

7. Let $G := \cup_{n\in\mathbb{N}} G_n$. Show that each G_n is a subgroup of G and G is a subgroup of $\mathbb{T}$

8. Let $\mathcal{C}_{00}(N, \mathbb{R}) := \{(a_n) \in \mathcal{F}(\mathbb{N}, \mathbb{R}) : a_n = 0 \text{ for all but finitely many}\}$. Then $\mathcal{C}_{00}(N, \mathbb{R})$ is a subgroup of $C(\mathbb{N}, \mathbb{R})$ in Example 2.1.8(16).

9. Show that $\mathcal{C}_0(N, \mathbb{R}) := \{(a_n) \in C(\mathbb{N}, \mathbb{R}) : \lim_n a_n = 0\}$ is a subgroup of $C(\mathbb{N}, \mathbb{R})$ and $\mathcal{C}_{00}(N, \mathbb{R})$ is a subgroup of $\mathcal{C}_0(N, \mathbb{R})$.

10. Let X be a set and H be a subgroup of G. Let $\mathcal{F}(X, G)$ be as in Exercise 2.1.9(12c). Show that $\{f \in \mathcal{F}(X, G) : f(x) \in H\}$ is a subgroup of $\mathcal{F}(X, G)$.

11. Let $G := \{(a_n) \in \mathcal{F}(\mathbb{N}, \mathbb{R}) : \sum_n a_n < \infty\}$ and $H := \{(a_n) \in G : \sum_n |a_n| < \infty\}$. Show that G is a subgroup of $\mathcal{F}(\mathbb{N}, \mathbb{R})$ and H is a subgroup of G.

Exercise 2.2.4. 1. Let $H := \{(x, y) \in \mathbb{R}^2 : ax + by = 0\}$ where $(a, b) \neq (0, 0)$. Show that H is a subgroup of $\mathbb{R}^2$.

The set H is non-empty, since $(0, 0)$ is always a solution of the homogeneous linear equation $ax + by = 0$.

We know that if (x_1, y_1) and (x_2, y_2) are two solutions of a linear homogeneous equation, then $(x_1 + x_2, y_1 + y_2)$ is again a solution of the same linear homogeneous equation. Therefore the addition $(x_1, y_1) + (x_2, y_2) = (x_1 + x_2, y_1 + y_2)$ in $\mathbb{R}^2$ is a binary operation on H.

Associativity and the existence of inverse follow very easily.

Observe that H is a line passing through origin in $\mathbb{R}^2$. We have shown above all lines passing through the origin in $\mathbb{R}^2$ are subgroups of $\mathbb{R}^2$.

2. Let $H := \{(x, y, z) \in \mathbb{R}^3 : ax + by + cz = 0\}$ where $(a, b, c) \in \mathbb{R}^3$ is nonzero. Show that H is a subgroup of $\mathbb{R}^3$.

Proof is similar to above exercise. This example shows that all planes passing through origin in $\mathbb{R}^3$ are subgroups of $\mathbb{R}^3$.

3. More generally let $n \geq 2$ and $(a_1, a_2, a_3, \ldots, a_n)$ is a nonzero vector in $\mathbb{R}^n$. Show that $H := \{(x_1, x_2, x_3, \ldots, x_n) \in \mathbb{R}^n : a_1x_1 + a_2x_2 + \cdots + a_nx_n = 0\}$ is a subgroup of $\mathbb{R}^n$. Is the set $\{(x_1, x_2, x_3, \ldots, x_n) \in \mathbb{R}^n : a_1x_1 + a_2x_2 + \cdots + a_nx_n = 1\}$, a subgroup of $\mathbb{R}^n$?

Have you seen this elsewhere? Relate it with the solutions of homogeneous and non-homogeneous linear equations.

4. Let H be a finite subgroup of $\mathbb{C}^*$. Show that H is a subgroup of $\mathbb{T}$.

   Let H be a finite subgroup of $\mathbb{C}^*$. Then for every element $z \in H$, there exists a natural number k such that $z^k = 1$. Hence $|z| = 1$. Thus z is in $\mathbb{T}$ and $H \subseteq \mathbb{T}$.

5. Let $G = (\mathbb{R}, +)$. Show that G does not have non-trivial finite subgroups. Do the groups $(\mathbb{Q}, +)$ and $(\mathbb{C}, +)$ have non-trivial finite subgroups?

6. Let $G = (\mathbb{R}^+, \cdot)$. Find non-trivial subgroups of G. Does there exist a non-trivial finite subgroup of G?

7. Let $G = (\mathbb{R}^*, \cdot)$. Find non-trivial finite subgroups of G.

   Let H be a finite subgroup of $\mathbb{R}^*$. Let $x \neq 1$ be an element in H. If H is finite then there exists a natural number k such that $x^k = 1$. Hence $x = \pm 1$. This proves that the only finite subgroup of $\mathbb{R}^*$ are $H = \{1\}$ and $H = \{\pm 1\}$.

Exercise 2.2.5. 1. Let H be a subgroup of G. Show that the product $HH := \{ab : a, b \in H\}$ is a subgroup of G and it is equal to H.

2. Let H be a finite subset of a group G such that $xy \in H$ if x and y are in H. Show that H is a subgroup of G. Further show that this result need not be true in case of infinite subsets of a group G.

   Using Proposition 2.2.1, it is enough to show that for every element $h \in H$, the inverse h^{-1} is also in H.

   Let $h \in H$. By hypothesis the set $\{h^k : k \in \mathbb{N}\}$ is contained in H. Since H is finite, there exist two distinct natural numbers k and l such that $h^k = h^l$. If we assume that $k > l$, then, $h^{k-l} = e$ is in H. If $k - l = 1$, then $h = e$ and its inverse is itself. Otherwise, $k - l - 1 \geq 1$ and by hypothesis, h^{k-l-1} is in H. Since $h^{k-l-1}h = e$, it follows that $h^{-1} = h^{k-l-1}$ and hence h^{-1} is in H. This proves that H is a subgroup of G.

   Let $G = (\mathbb{Z}, +)$ and $H = \mathbb{N}$. Then sum of two natural numbers is a natural number. But $\mathbb{N}$ is not a subgroup of G.

3. Let G be a finite group and $x(\neq e)$ be an element in G. Show that the set $\{x^k : k \in \mathbb{N}\}$ is a subgroup of G. Show by an example that this result need not be true if the group is infinite.

Exercise 2.2.6. 1. Let $G = \mathbb{R}^3$. Show that a line passing through origin is a subgroup of $\mathbb{R}^3$.

2. Let $G = \mathbb{R}^n$. Let W be a vector subspace of G. Show that W is a subgroup of $\mathbb{R}^n$.

3. Let H_1 and H_2 be two subgroups of G. Show that $H_1 \cap H_2$ is a subgroup of G.

4. Let $\{H_i : i \in I\}$ be a family of subgroups of G. Show that $\cap_{i \in I} H_i$ is a subgroup of G.

5. Let $G = \mathbb{Z}$ and $H_k = k\mathbb{Z}$ for every $k \in \mathbb{N}$. What is $\cap_{k=1}^{\infty} H_k$?

6. Let (k_n) be a sequence of natural numbers and $H_{k_j} := k_j\mathbb{Z}$. Find $\cap_{j=1}^{\infty} H_{k_j}$.

While the previous exercises show that the intersection of subgroups is a subgroup, we may ask if the union of subgroups is again a subgroup. This is answered in the following exercises.

Exercise 2.2.7. 1. Show by an example that union of two subgroups of a group G need not be a subgroup of G.

Let $H_1 := \{(x, 0) : x \in \mathbb{R}\}$ and $H_2 := \{(0, y) \in \mathbb{R}^2 : y \in \mathbb{R}\}$, the x-axis and the y-axis respectively. Then both H_1 and H_2 are subgroups of $\mathbb{R}^2$. But $H_1 \cup H_2$ is not a subgroup of $\mathbb{R}^2$. For example, the elements $(1, 0)$ and $(0, 1)$ are in $H_1 \cup H_2$. But $(1, 1) = (1, 0) + (0, 1)$ is not in $H_1 \cup H_2$.

2. Let H_1 and H_2 be two subgroups of G. Show that $H_1 \cup H_2$ is a subgroup of G iff $H_1 \subseteq H_2$ or $H_2 \subseteq H_1$.

Suppose $H := H_1 \cup H_2$ is a subgroup of G and neither of them contained in the other. Let $x \in H_1 \setminus H_2$ and $y \in H_2 \setminus H_1$. Since H is a subgroup of G, the product xy is in H. Assume xy is in the subgroup H_1. Then $y = x^{-1}(xy)$ is also in H_1, a contradiction. If xy is in H_2, then $x = (xy)y^{-1}$ is in H_2, again a contradiction. This shows that either $H_2 \subseteq H_1$ or $H_1 \subseteq H_2$.

3. Let H_i for $1 \leq i \leq 3$ be three subgroups of G. Is $\cup_{i=1}^{3} H_i$ a subgroup of G?

Sometimes yes! Let $G = \mathbb{Z}_2 \times \mathbb{Z}_2$. Let $H_1 := \{(0, 0), (1, 0)\}$, $H_2 = \{(0, 0), (0, 1)\}$ and $H_3 := \{(0, 0), (1, 1)\}$. Then each of these H_i's are subgroups of G and $\cup_{i=1}^{3} H_i = G$ is a subgroup of G!

4. Let G be a group of infinite order. Can we write G as a union of finitely many subgroups of G?

5. Let G be a group and $\{H_i : i \in I\}$ be a family of subgroups of G. Is

$$\cup_{i \in I} H_i := \{x \in G : \text{ there exists } i \text{ in } I \text{ such that } x \text{ is in } H_i\}$$

a subgroup of G?

Think of writing $\mathbb{R}^2$ as a union of subgroups.

6. Let G be a finite group. Can we write G as a union of its subgroups?

Exercise 2.2.8. Let G be an abelian group. For every natural number k, we let $H_k := \{x \in G : x^k = e\}$. Show that H_k is a subgroup of G.

Let $k \geq 2$. The identity element e is in H_k. If $H_k = \{e\}$, there is nothing to prove.

Assume that H_k has at least two elements. Let x and y be two elements of H_k. Since G is abelian, $xy = yx$. Now it follows from generalized associative law that $(xy)^k = x^k y^k = e$.

If $x^k = e$, then we know that $x^{-k} = e$ too. Thus H_k is a subgroup of G.

If G is non-abelian, then the set H_k need not be a subgroup can be seen in the following example.

Example 2.2.9. Let $G = S_3$ and $k = 2$. Then

$$H_2 = \left\{ Id, \sigma = \begin{pmatrix} 1 & 2 & 3 \\ 2 & 1 & 3 \end{pmatrix}, \tau = \begin{pmatrix} 1 & 2 & 3 \\ 1 & 3 & 2 \end{pmatrix} \right\}.$$

Let us now observe that $\sigma \circ \tau = \begin{pmatrix} 1 & 2 & 3 \\ 2 & 3 & 1 \end{pmatrix}$. Observe that that $(\sigma \circ \tau)^2 = \begin{pmatrix} 1 & 2 & 3 \\ 3 & 1 & 2 \end{pmatrix} \neq Id$ and $(\sigma \circ \tau)^3 = Id$. Hence H_2 is not a subgroup of G.

We have seen that, for every integer k, the set $H_k := \{kl : l \in \mathbb{Z}\}$ is a subgroup of $(\mathbb{Z}, +)$. We will now prove that every subgroup of $\mathbb{Z}$ is of this form.

Theorem 2.2.10. *Let H be a subgroup of $(\mathbb{Z}, +)$. Then there exists a non-negative integer k such that $H = k\mathbb{Z}$.*

Proof. If the subgroup H is trivial subgroup, then $H = (0) = 0\mathbb{Z}$ and we are through. Therefore we will assume that $H \neq (0)$.

Let H be a proper subgroup of G. Let a be a nonzero element in H. Since H is a group, $-a$ is also in H. If $a < 0$ then $-a > 0$. Therefore a proper subgroup H of $\mathbb{Z}$ contains positive integers. Let $S := \{m \in H : m > 0\}$. Then S is a non-empty subset of natural numbers. By well ordering principle, there exists a smallest natural number $k \in S$. Now, we will prove that every element in H is an integer multiple of k.

Let m be a nonzero integer in H. By division algorithm, there exist unique integers q and r such that $m = qk + r$ and $0 \leq r < k$. Since H is a subgroup of $\mathbb{Z}$, the integer $r = m - qk$ is in H. Since k is the smallest positive integer in H and $r < k$, it follows that the non-negative integer $r = 0$. This shows that $m = qk$. □

Corollary 2.2.11. *Let m and d be two natural numbers. Then $m\mathbb{Z} \subseteq d\mathbb{Z}$ iff d divides m.*

Proof. Easy. □

We have seen in Example 2.2.2 that for any two integers $m \neq n$ the sets $H_m \cap H_n$ and $H_m + H_n$ are subgroups of $\mathbb{Z}$. But Theorem 2.2.10 tells us that there exist integers l and d such that

$$H_m \cap H_n = l\mathbb{Z} \quad \text{and} \quad H_m + H_n := d\mathbb{Z}.$$

We will now find the integers l and d in the following corollary.

Corollary 2.2.12. *Let m and n be two integers. If $l = lcm(m, n)$ and $d =$ gcd(m, n), then*

$$H_m \cap H_n = l\mathbb{Z} \quad \textit{and} \quad H_m + H_n := d\mathbb{Z}.$$

Proof. It is an easy exercise to show that $H_m \cap H_n$ and $H_m + H_n$ are subgroups of $\mathbb{Z}$.

Let $l = \text{lcm}(m, n)$ and $d = \gcd(m, n)$. We will now show that $H_m \cap H_n = l\mathbb{Z}$ and $H_m + H_n = d\mathbb{Z}$.

By definition of l, the integer l is in $H_m \cap H_n$. If t is in $H_m \cap H_n$, then there exist integers k_1 and k_2 such that $l = k_1 m = k_2 n$. This shows that t is a multiple of the integers m and n. Since l is the least common multiple of m and n, it follows that $\leq t$. Hence $H_m \cap H_n = l\mathbb{Z}$.

Let t be an element of $H_m + H_n$. Then there exist integers k and l such that $t = mk + nl$. However we know that any integral linear combination of the integers m and n is an integral multiple of $d = \gcd(m, n)$. Therefore $H_m + H_n \subseteq d\mathbb{Z}$. On the other hand, since d is in $H_m + H_n = d\mathbb{Z}$, it follows that $H_d \subseteq H_m + H_n$. Hence $H_m + H_n = d\mathbb{Z}$. □

Just as in the case of integers, we can find all subgroups of $\mathbb{Z}_n$ too.

Theorem 2.2.13. *Let $G = (\mathbb{Z}_n, +)$ and H be a proper subgroup of G. Then there exists unique non-negative integer $d < n$ such that the subgroup $H = d\mathbb{Z}_n := \{[d][l] : [l] \in \mathbb{Z}_n\} = \{[dl] : l \in \mathbb{Z}\}$.*

Proof. If the subgroup H is trivial subgroup, then $H = (0)$ and we are through. Therefore we will assume that $H \neq (0)$.

Let H be a proper subgroup of G. Let $[a]$ be a nonzero element in H. Since H is a group, $[-a]$ is also in H. Let $S := \{m \in \mathbb{N} : [m] \in H\}$. Then S is a finite subset of natural numbers and let $d = \min S$. Now, we claim that $H = \{[dl] : l \in \mathbb{Z}\}$.

Let $[x]$ be a nonzero element in H and let us choose an integer m in the equivalence class $[x]$. By division algorithm, there exist unique integers q and r such that $m = qd + r$, where $0 \leq r < d$. Since H is a subgroup of $\mathbb{Z}_n$, it follows that $[r] = [m - qd]$ is in H. Since d is the smallest positive integer such that $[d]$ is in H and $r < d$, it follows that the non-negative integer $r = 0$. This shows that $m = qd$ and hence $[x] = [m] = [qd]$. This completes the proof. □

Let H and K be two subgroups of G. We have seen that $H \cap K$ is a subgroup of G and $H \cup K$ is not a subgroup of G unless one of them is contained in the other. Now we ask if the product of the sets H and K defined as $HK := \{hk : h \in H \text{ and } k \in K\}$ a subgroup of G or not.

In general the subset HK need not be a subgroup of G. If G is abelian, then HK is a subgroup of G.

If G is abelian, then $hk = kh$ for all elements h in H and k in K. Therefore if $a = h_1k_1$ and $b = h_2k_2$ are elements in HK, then $ab = h_1k_1h_2k_2 = h_1h_2k_1k_2$ and $b^{-1} = k^{-1}h^{-1} = h^{-1}k^{-1}$ is also in HK. Hence HK is a subgroup of G.

If $H = l\mathbb{Z}$ and $K = k\mathbb{Z}$ are subgroups of $\mathbb{Z}$ and if $d = \gcd(k, l)$, then $H + K = l\mathbb{Z} + k\mathbb{Z} = d\mathbb{Z}$ is a subgroup of $\mathbb{Z}$.

If G is not abelian, then HK need not be a subgroup of G.

For example let

$$H := \left\{ \begin{pmatrix} 1 & x \\ 0 & 1 \end{pmatrix} : x \in \mathbb{R} \right\}$$

and

$$K := \left\{ \begin{pmatrix} 1 & 0 \\ y & 1 \end{pmatrix} : x \in \mathbb{R} \right\}.$$

We have shown earlier that H and K are subgroups of $\mathrm{GL}(2, \mathbb{R})$. The product HK is the set

$$\left\{ \begin{pmatrix} 1+xy & x \\ y & 1 \end{pmatrix} : x, y \in \mathbb{R} \right\}.$$

This set is not a subgroup of G: Let $A = \begin{pmatrix} a & c \\ b & d \end{pmatrix} = \begin{pmatrix} 1+xy & x \\ y & 1 \end{pmatrix}$. Then $a - d = xy = bc$. The inverse of A is $\begin{pmatrix} 1 & -x \\ -y & 1+xy \end{pmatrix}$ and it is not in HK unless x or y is zero. □

However if the two sets HK and KH are equal, then HK is a subgroup of G as shown below.

Let $a = h_1k_1$ and $b = h_2k_2$ be two elements in HK. We will show that ab^{-1} is in HK: Since $HK = KH$, we can write $b^{-1} = k_2^{-1}h_2^{-1} = h_3k_3$ for some h_3 in H and k_3 in K. Hence

$$\begin{aligned} ab^{-1} &= (h_1k_1)(h_3k_3) \\ &= h_1(k_1h_3)k_3 \\ &= h_1(h_4k_4)k_3 \\ &= (h_1h_4)(k_1k_4) \end{aligned}$$

is also in $HK = KH$. Thus HK is a subgroup of G.

In the next section we will discuss conditions under which the sets HK and KH are equal, if H and K are subgroups of G.

Order

Let x be an element of a group G. Then x^k is in G for all $k \in \mathbb{Z}$. If G is finite then not all the elements x^k are distinct. Hence there exist integers k and l such that $k \neq l$ and $x^k = x^l$. Consequently $x^{k-l} = x^{l-k} = e$. This proves that there exists a natural number t such that $x^t = e$. Let $S := \{t \in \mathbb{N} : x^t = e\}$. By well ordering principle the set S has a smallest element say n. Thus in a finite group G, given an element x in G, there exists a smallest natural number $n = n(x)$ such that $x^n = e$.

If G is an infinite, then this need not be true.

Example 2.2.14.

1. Let $G = \mathbb{Z}$ and $x = 1$. Then, for every nonzero integer k, the product $x^k = \overbrace{1 + 1 + \cdots + 1}^{k-\text{times}} = k$ is never equal to zero.

2. If $G = \mathrm{GL}(2, \mathbb{R})$ and $A = \begin{pmatrix} 1 & 1 \\ 0 & 1 \end{pmatrix}$ then, for every natural number k the matrix

$$A^k = \begin{pmatrix} 1 & k \\ 0 & 1 \end{pmatrix} \text{ and } A^{-k} = \begin{pmatrix} 1 & -k \\ 0 & 1 \end{pmatrix}$$

is never identity.

If $G = \mathbb{C}^*$, the multiplicative group of complex numbers, we see that there are elements $z \in G$ such that $z^k = 1$ for some $k \in \mathbb{N}$ and $z^k \neq 1$ for any $k \in \mathbb{N}$. Let $z = e^{2\pi i\theta} \in G$ be such that $z^k = 1$ for some natural number k. Since $z^k = e^{2\pi ik\theta}$ it follows that $z^k = 1$ iff $k\theta$ is an integer. Let $k\theta = m$. Therefore $z^k = 1$ for some k iff θ is a rational number $\frac{m}{k}$. If we choose θ to be an irrational number then $z^k \neq 1$ for any $k \in \mathbb{N}$.

Exercise 2.2.15. Let $G = \mathbb{C}^*$ and $z \in G$ be such that $|z| \neq 1$. Then $z^k \neq 1$ for any $k \in \mathbb{N}$.

Definition 2.2.16 (Order of an element).

1. Let G be a group. If an element x in G is such that $x^k = e$ for some $k \in \mathbb{N}$, we say that the element x has finite order in G.
2. If an element x in G has finite order, then $n := \min\{k \in \mathbb{N} : x^k = e\}$ is defined as the order of the element x and it is also denoted as $o(x)$ or $|x|$.
3. If $x^k \neq e$ for every $k \in \mathbb{N}$, we say that the order of x is infinite.

Remark 2.2.17. Let us observe that the only element of order 1 is the identity.

Examples 2.2.18.

1. Let G be a finite group. Then every element in G is of finite order.
2. Let $G = \cup_n G_n$ be as in Example 2.1.8(19). Then every element of G is of finite order. Notice that the group G itself is not finite.
3. Let $G = \mathbb{Z}$, the integers under addition. Then every nonzero element is of infinite order.
4. Let $G = \mathbb{C}^*$. As we have seen above, the group G has elements of finite order as well as infinite order.

   We have seen in Exercise 2.2.15 that if an element z in $\mathbb{C}^*$ is such that $|z| \neq 1$, then $z^k \neq 1$ for every $k \in \mathbb{N}$. Hence such elements are of infinite order. For every natural number n, the element $e^{\frac{2\pi i}{n}}$ is of order n.

5. Let $G = \mathbb{R}^*$. Then ± 1 are the only elements of finite order.

   Let x in G be such that $x^k = 1$ for some $k \in \mathbb{N}$. Then $|x|^k = |x^k| = 1$. Therefore $|x| = 1$. Hence $x = \pm 1$ are the only elements of finite order in $\mathbb{R}^*$.

6. Let $G = \mathrm{GL}(2, \mathbb{R})$ and $A = \begin{pmatrix} 1 & 1 \\ 0 & 1 \end{pmatrix}$. Then $A^k = \begin{pmatrix} 1 & k \\ 0 & 1 \end{pmatrix}$ is not identity for any $k \geq 2$. Thus A has infinite order.
7. Let $G = \mathrm{GL}(2, \mathbb{R})$ and $A = \begin{pmatrix} 0 & 1 \\ 1 & 0 \end{pmatrix}$. Then A is of order 2. Geometrically A is the reflection about the line $y = x$. Hence $A^2 = I_2$, the identity matrix.

   More generally, for $\alpha \neq 0$, let $A_\alpha = \begin{pmatrix} 0 & \alpha \\ \frac{1}{\alpha} & 0 \end{pmatrix}$. Then A is of order 2. Is this matrix a reflection about a line in $\mathbb{R}^2$?

8. Let X be a non-empty set and $\mathcal{P}(X)$ be the group as defined in Example 2.1.8(25). In this group every element is of order 2. If X is an infinite set, then the group $\mathcal{P}(X)$ is infinite and every element in $\mathcal{P}(X)$ is of order 2.

Exercise 2.2.19. 1. Find the orders of

(a) [2], [35], [8], [17] and [34] in $\mathbb{Z}_{36}$.

(b) [2], [25], [8], [7] and [33] in $\mathbb{Z}_{13}$.

2. Let G be a group and $g \in G$. Show that $H := \{k \in \mathbb{Z} : g^k = e\}$ is a subgroup of $\mathbb{Z}$.

3. Let G be a group and x be an element of G. If $\cap_{x \in H} H$ is infinite, then show that the order of x is infinite.

4. Let $G = S^1 := \{e^{2\pi i\theta} : \theta \in \mathbb{R}\}$. Show that

(a) an element $e^{2\pi i\theta}$ is of finite order, if θ is a rational number and

(b) an element $e^{2\pi i\theta}$ is of infinite order, if θ is an irrational number.

5. Let $G := \{x \in \mathbb{R} : 0 \leq x < 1\}$ and $\star$ be the binary operation defined by

$$x \star y := x + y - [x + y]$$

for x and y in G; here $[x + y]$ denotes the greatest integer less than or equal to $x + y$. Show that

(a) $(G, \star)$ is an abelian group,

(b) all rational numbers in G are of finite order, and

(c) all irrational numbers in G are of infinite order.

6. Let $m \geq 2$ and $G = \mathbb{Z}_m$. Let $[x]$ be an element of G. Let $0 < k < m$ be the natural number such that $[k] = [x]$. Since G is finite, the element $[x]$ is of finite order, say l. Therefore $\overbrace{[k] + [k] + \cdots + [k]}^{l-\text{times}} = [kl] = [0]$ and kl is a multiple of m. Thus m must divide the integer kl and we conclude the following:

(a) If k and m are relatively prime, then $l = m$. In particular every nonzero element is of order m, if m is prime.

(b) If m is a composite number, then every proper divisor d of m is of order $\frac{m}{d}$.

For example, let $n = 8$. If $[x] = [2]$, then the order of $[x]$ is 4. If $[x] = [5]$ and its order is l, then 8 divides $5l$. Since 8 and 5 are relatively prime, 8 must divide l and the smallest such l is 8. Thus the order of 5 is 8, the order of the group itself.

We will now see some simple properties about order.

Proposition 2.2.20. *Let G be a group and $x(\neq e)$ be an element in G of finite order, say k. If $x^t = e$ for some integer t, then $t = kr$ for some integer r.*

Proof. If $x = e$, then $k = 1$ and the result is true.

Assume that $x \neq e$ and let the order of x be k. Then $x^r \neq e$ for $0 < r < k$. Let $t \in \mathbb{Z}$ be such that $x^t = e$. By division algorithm, there exist integers, q and r such that $t = kq + r$, where $0 \leq r < k$. Therefore $e = x^t = x^{kq+r} = (x^k)^q x^r = x^r$. This proves that $r = 0$ and the proof is complete. □

Proposition 2.2.21. *Let G be a group. Then the following holds.*

1. *If an element $x \in G$ is of infinite order, then $x^k \neq x^l$ for $k \neq l \in \mathbb{Z}$.*

2. *If $o(x) = kl$ for an element x in G, then $o(x^k) = l$.*

3. *$o(x) = o(x^{-1})$ for every element $x \in G$.*

4. *If x is an element of G, then $o(x) = o(gxg^{-1})$ for all g in G.*

Proof. Let x be an element of G of infinite order. If $x^k = x^l$ for two integers $k \neq l$, then $x^{k-l} = e$, a contradiction.

Let x be an element of order kl. By the generalized associative law, $x^{kr} = (x^k)^r$ for every $r \in \mathbb{N}$. Therefore $(x^k)^l = x^{kl} = e$. This shows that the order of x^k is less than or equal to l. Let t be the order of x^k. If $t < l$, then $kt < kl$ and $e = (x^k)^t = x^{kt}$, a contradiction. Hence $|x^k| = l$.

Let x be an element of G. Then $(x^k)^{-1} = x^{-k} = (x^{-1})^k$. This shows that, if the order of x is k then order of x^{-1} is finite and less than or equal to k. Suppose order of x^{-1} is l. Then the equation $e = (x^{-1})^l = (x^l)^{-1}$, proves that the order of x is less than or equal to l. Thus the orders of x and x^{-1} are same.

If order of x is infinite, then from the argument in the preceding paragraph it follows that order of x^{-1} is also infinite.

Let x and g be two elements of G. We claim that $(gxg^{-1})^k = gx^kg^{-1}$ for every natural number $k \geq 2$.

If $k = 2$, then using generalized associative law, we get

$$\begin{aligned}(gxg^{-1})^2 &= (gxg^{-1})(gxg^{-1}) \\ &= (gx)(gg^{-1})(xg^{-1}) \\ &= (gx)(xg^{-1}) \\ &= gx^2g^{-1}.\end{aligned}$$

Let us now assume that $(gxg^{-1})^k = gx^kg^{-1}$ for some $k \geq 2$. Then we use law of indices and generalized associative law to get

$$\begin{aligned}(gxg^{-1})^{k+1} &= (gxg^{-1})(gxg^{-1})^k \\ &= (gxg^{-1})(gx^kg^{-1}) \\ &= (gx)(g^{-1}g)(x^kg^{-1}) \\ &= (gx)(x^kg^{-1}) \\ &= gx^{k+1}g^{-1}.\end{aligned}$$

Thus $(gxg^{-1})^k = gx^kg^{-1}$ for all natural numbers k.

Let us now assume that the element x is of finite order say k. Then $(gxg^{-1})^k = gx^kg^{-1} = geg^{-1}$. Hence $|gxg^{-1}|$ is finite and $|gxg^{-1}| \leq |x|$.

Now suppose the order of gxg^{-1} is finite, say l. Then we use the relation $x = g^{-1}(gxg^{-1})g$ to conclude that $x^l = g^{-1}(gxg^{-1})^lg = e$. This shows that $|x| \leq |gxg^{-1}| = l$. Hence $|x| \leq |gxg^{-1}| = l$, if $|x|$ is finite.

We have shown in the preceding paragraphs that $gx^kg^{-1} = e$ iff $x^k = e$. Therefore, if the order of x is not finite, $(gxg^{-1})^k = gx^kg^{-1} \neq e$ for any natural number k. Hence the order of gxg^{-1} is not finite either.

This completes the proof of the proposition. □

Exercise 2.2.22. Let G be a group.

1. If an element x is of odd order, say, $|x| = 2n + 1$, then $x^j \neq x^{-j}$ for $1 \leq j \leq 2n$.

   Suppose $x^j = x^{-j}$ for some $j \leq 2n$. Then $x^{2j} = e$. In view of Proposition 2.2.20, it follows that $2n + 1$ divides $2j$. Therefore there exists $k \in \mathbb{N}$ such that $2j = k(2n + 1)$. Since $2j$ is even, it follows that $k \geq 2$. On the other hand the inequality $2j \leq 4n$, shows that $2(2n + 1) \leq k(2n + 1) = 2j \leq 4n$. This is a contradiction.

2. If an element x is of even order, say, $|x| = n = 2k$ and $x^i = x^{-i}$ then $i \equiv k \pmod{n}$.

   Let i be an integer such that $x^i = x^{-i}$. Hence $x^{2i} = e$.

   By division algorithm, there exist unique integers q and j such that $i = nq + j$ where $0 \leq j < n$. If we now show that $j = k$, we are through.

   Let us observe that $e = x^{2i} = x^{2j}$. This shows that n divides $2j$ and if $2j = nt$, then $nt = 2j < 2n$. Thus $2j = n$ and $j = k$.

Definition 2.2.23. Let G be a group. If G is a finite group, then the number of elements of G is called the order of the group G and it is denoted by $o(G)$ or $|G|$.

If G is infinite, then G is called a group of infinite order.

Examples 2.2.24. 1. Let $G = G_n$ be the group of n-th roots of unity as in Example 2.1.8(18). Then $|G| = n$.

2. The order of $G = \mathbb{Z}$ is infinite and order of every nonzero element in G is also infinite.

3. Let $G = \cup_n G_n$ as in Example 2.1.8(19). Then G is a group of infinite order in which every element is of finite order.

4. Let $X = \mathbb{N}$ and
$$G = \prod_{n=1}^{\infty} \mathbb{Z}_2 := \{f : \mathbb{N} \to \mathbb{Z}_2\}.$$
   Then G is a group of infinite order and every element in G is of order 2.

5. Let G be a finite group. Then every element in $\mathcal{F}(\mathbb{N}, G)$ is of finite order.

Exercise 2.2.25. Let x and y be two distinct elements of finite order in a group G such that $xy = yx$. Show that the order of xy is finite.

> Let x and y be two distinct elements in G of order $|x| = m$ and $|y| = n$ respectively. Let $z = xy$. Since $xy = yx$, it follows that $z^{mn} = (xy)^{mn} = x^{mn}y^{mn} = (x^m)^n(y^n)m = e$. Thus the order of xy is finite.

If two elements x and y in a group G don't commute, then the order of xy could be infinite. This can be seen in the following exercises.

Exercise 2.2.26. Let $G = \mathrm{GL}(2, \mathbb{R})$.

1. Let $A = \begin{pmatrix} 0 & -1 \\ 1 & 0 \end{pmatrix}$ and $B = \begin{pmatrix} 1 & -1 \\ 1 & 0 \end{pmatrix}$. Show that the order of the elements A and B is finite, but the order of AB is infinite.

   Observe that $A^2 = -I_2$ and $B^3 = -I_2$. Therefore order of A is 4 and order of B is 6. But $AB = \begin{pmatrix} -1 & 0 \\ 1 & -1 \end{pmatrix}$ and for every $k \in \mathbb{N}$, $(AB)^k = \begin{pmatrix} -1 & 0 \\ (-1)^{k-1}k & (-1)^k \end{pmatrix}$. Therefore the order of AB is not finite.

2. Let $A = \begin{pmatrix} 0 & 1 \\ 1 & 0 \end{pmatrix}$ and $B = \begin{pmatrix} 0 & \alpha \\ 1/\alpha & 0 \end{pmatrix}$ for $0 \neq \alpha \neq 1$. Show that the orders of A and B is 2 and the order of AB is not finite.

   It is very easy to see that A and B are of order 2. The product $AB = \begin{pmatrix} 1/\alpha & 0 \\ 0 & \alpha \end{pmatrix}$ is not of finite order.

3. Let $A = \begin{pmatrix} 0 & -1 \\ 1 & 0 \end{pmatrix}$ and $B = \begin{pmatrix} 0 & -\alpha \\ 1/\alpha & 0 \end{pmatrix}$ for $0 \neq \alpha \neq 1$. Show that the order of AB is not finite.

Proposition 2.2.27. *Let x and y be two distinct elements in a group G such that $xy = yx$. Assume $|x| = m$ and $|y| = n$. Then there exists an element z in G such that $|z| = \mathrm{lcm}(m, n)$.*

Proof. Let x and y be two distinct elements in G such that $|x| = m$ and $|y| = n$.

We will first prove the result when $\gcd(m, n) = 1$. In this case $l = \mathrm{lcm}(m, n) = mn$.

Let $z = xy$. Since $xy = yx$, it follows that $z^{mn} = (xy)^{mn} = x^{mn}y^{mn} = (x^m)^n(y^n)^m = e$. Thus the order of xy is finite and it is less than or equal to l. Let k be the order of the element $z = xy$. Then $z^k = (xy)^k = e$ and we have $x^k = y^{-k}$. Since x^k is in $\langle x \rangle$, y^{-k} is in $\langle y \rangle$ and $x^k = y^{-k}$, it follows that $x^k = y^{-k} \in \langle x \rangle \cap \langle y \rangle$. Hence $|x^k|$ divides $|x| = m$ and $|y| = n$. We have assumed the integers m and n to be relatively prime. This proves that $|x^k| = 1$. That is $y^{-k} = x^k = e$. The equation $x^k = e$ shows that k is a multiple of m and the equation $y^k = e$ shows that k is a multiple of n as well. Further k is the smallest such integer. Therefore $k = mn$ and the proof is complete in this case.

Let us now assume that $d = \gcd(m, n) \neq 1$. Then $\gcd(\frac{m}{d}, n) = 1$ and the order of the element x^d is $\frac{m}{d}$. Therefore, if we now let $z = x^d y$, then $|z| = \frac{m}{d}n = \mathrm{lcm}(m, n)$. This completes the proof. □

Corollary 2.2.28. *Let G be a finite abelian group and x be an element of maximal order. Then $|y|$ divides $|x|$ for every element y in G.*

Proof. Let $|x| = m$ and let y in G be an element of order, say $|y| = n$.

If n does not divide m, then $l := \operatorname{lcm}(m, n) > m$ and we know from the Proposition above that there is an element z in G of order l, a contradiction. □

Exercise 2.2.29.

1. Let $G := U_8$. Find the order of U_8 and also the orders of each of its elements.
2. Let $G := U_7$. What is the order of G? Find an element of order 2 and 3? Is the group cyclic?
3. Let $G := U_{10}$. What are the order of G and the orders of each of its elements? Compare this with the first exercise. What do you observe?
4. Let $p \geq 2$ be a prime and $G := U_p$. What is the order of $[p-1]$?
5. Let $n \geq 3$ be odd and $G := U_n$. What is the order of the element $[(\frac{n-1}{2})]$? What can be said if n is even?
6. Let $p \geq 2$ be a prime and $G := U_p$. what is the order of G? Let q be a prime number dividing $p-1$. Find an element of order q in U_p.
7. Let $n \geq 2$. When is $\mathbb{Z}_n^*$ a group?

Exercise 2.2.30. Let G be an abelian group and $F(G) := \cup_{k \geq 1} H_k$ where $H_k : \{x \in G : |x| = k\}$. Show that $F(G)$ is a subgroup of G.

By an example show that $F(G)$ need not be equal to G.

If an element a is of finite order in a group G, then a^{-1} is also of finite order. Therefore, if a is in $F(G)$, then a^{-1} is also in $F(G)$.

Let a and b two elements in $F(G)$ such that $|a| = n$ and $|b| = m$ for some n and $m \in \mathbb{N}$. Since G is abelian, it follows that, $(ab)^{nm} = a^{nm}b^{nm} = e$. Hence $F(G)$ is a subgroup of G.

Let $G = \{z \in \mathbb{C} : |z| = 1\}$. Then $F(G) = \cup_{n \geq 2} G_n$, where each $G_n := \{z \in G : z^n = 1\}$ is the group of n-th roots of unity for every $n \geq 2$. The subgroup $F(G) \neq G$ is seen as follows: Let z be an element of G. Then $z = e^{i\theta}$ for some θ in $\mathbb{R}$. If z is of finite order, say k, then $z^k = e^{ik\theta} = 1$. This is true iff $k\theta = 2\pi n$ for some integer. Thus an element $z = e^{i\theta}$ is of finite order iff $\theta = 2\pi \frac{n}{k}$ for some rational number $\frac{n}{k}$. If we now choose $\theta \neq 2\pi r$ for any rational number r, then $\zeta = e^{i\theta}$ is in G, but not in $F(G)$.

We will show by an example that the set $F(G)$ is not a subgroup of G, if G is non-abelian.

Example 2.2.31. Let $G = \mathrm{GL}(2, \mathbb{R})$. Then $F(G)$ is not a subgroup of G.

Let $A = \begin{pmatrix} 1 & 0 \\ 1 & -1 \end{pmatrix}$ and $B = \begin{pmatrix} -1 & 0 \\ 1 & 1 \end{pmatrix}$ be two elements of $\mathrm{GL}(2, \mathbb{R})$.

Then A and B are of order 2 and $AB = \begin{pmatrix} -1 & 0 \\ -2 & -1 \end{pmatrix}$ is of infinite order.

We give another example. Let $A = \begin{pmatrix} 0 & 1 \\ 1 & 0 \end{pmatrix}$ and $B = \begin{pmatrix} 0 & \frac{1}{2} \\ 2 & 0 \end{pmatrix}$. Then A and B are of order 2. But the product $AB = \begin{pmatrix} 2 & 0 \\ 0 & \frac{1}{2} \end{pmatrix}$ is of infinite order.

Exercise 2.2.32. 1. Does the group $G = (\mathbb{R}^*, \cdot)$ have an infinite proper subgroup? If yes can we find all of them?

2. Let $G = (\mathbb{R}^*, \cdot)$ and H be a subgroup of G. Show that H is either contained in the subgroup $(0, \infty)$ or $H = K \cup (-K)$ where $K = H \cap (0, \infty)$ and $-K := \{-x : x \in K\}$. Observe that K is a subgroup of G.

3. Let G be an infinite group. Show that G has infinitely many proper subgroups.

   Let $x_1 \neq e$ be an element of G. If the order of x_1 is infinite, then, for each $k \geq 2$, the subgroups $H_k := \langle x_1^k \rangle$ are proper subgroups of G.

   If the order of x_1 is finite, we choose $x_2 \notin G \setminus \langle x_1 \rangle$. If order of x_2 is finite, we choose $x_3 \notin G \setminus (\langle x_1 \rangle \cup \langle x_2 \rangle)$. We repeat this process again. If at any stage, we get an element x_n with infinite order, then we get infinitely many subgroups $\langle x_n^k \rangle$ for $k = 2, 3, \ldots$. Otherwise we have infinitely many finite subgroups $\langle x_k \rangle$ for $k = 1, 2, \ldots$.

2.2.1 Cyclic Groups and Cyclic Subgroups

Let G be a group and $x \neq e$ be an element of G. We would now like to know if there exists a subgroup H of G such that x is in H. An obvious choice is G itself. However we would like smallest subgroup containing the element x.

Suppose there is a subgroup H of G containing the element x. Then, since H is a subgroup, it follows that x^k is in H for all integers k. That is, if there exists a subgroup H of G containing the element x, then $\{x^k : k \in \mathbb{Z}\}$ is a subset of H. However we know that the set $\{x^k : k \in \mathbb{Z}\}$ itself is a group: This set is non-empty and the element x is in $\{x^k : k \in \mathbb{Z}\}$. The identity element $x^0 = e$ is also there. By law of indices $x^k x^l x^m = x^{k+l+m}$ is also in that set for any three integers k, l and m. Thus $\{x^k : k \in \mathbb{Z}\}$ is a group. This subgroup is denoted by $\langle x \rangle$ and it is called the *cyclic* subgroup generated by the element x.

It is possible that the subgroup $\langle x \rangle = G$ for some element x in G. In this case, we call G a cyclic group and x is called a generator of G.

Definition 2.2.33. A group G is called a cyclic group, if there exists an element x in G such that $G = \langle x \rangle$.

Exercise 2.2.34. Let G be a group and x be an element of G. Then show that $\langle x \rangle = \cap_H H$, where the intersection runs over all subgroups H of G containing the element x.

Example 2.2.35. 1. Let $G = (\mathbb{Z}, +)$, the group of integers under addition. Let $x = 1$. Then every integer $k \in \mathbb{Z}$ can be written as the "product" $x^k = \overbrace{1 + 1 + \cdots + 1}^{k-\text{times}} = k$. Thus the group $(\mathbb{Z}, +)$ is a cyclic group. We can also take $x = -1$ and write every integer k as sum of -1's.

2. Let $n \geq 2$ and $G = (\mathbb{Z}_n, +)$. Then G is a cyclic group and the element $\bar{1}$ is a generator. This is easy to verify.

   Let $k < n$ be a natural number such that $\gcd(k, n) = 1$. Then there exists integers u and v such that $ku + nv = 1$.

   If l is an integer, then $l = kul + nvl$ and therefore $[l] = [kul]$. We can write this as $[l] = \overbrace{[k] + [k] + \cdots + [k]}^{ul-\text{times}}$. Thus every natural number $k < n$ and relatively prime to n is a generator of $(\mathbb{Z}_n, +)$.

3. Let $G = (\mathbb{Z}, +)$. We know from Theorem 2.2.10 that every subgroup of G is cyclic.

4. Let $G = (\mathbb{Z}_n, +)$. We know from Theorem 2.2.13 tells us that every subgroup of G is cyclic.

We will see more examples of cyclic groups and study their properties as we go along.

Exercise 2.2.36. 1. Let G be a cyclic group. Let x be a generator of G. Show that x^{-1} is also a generator of G.

2. Let $G := \langle x \rangle$ be an infinite cyclic group. Show that x and x^{-1} are the only generators of G.

3. If G is infinite cyclic, then G has infinitely many cyclic subgroups each of which is infinite.

   Let $G := \langle x \rangle$ and $k \geq 1$. Then, for every $n \geq 2$, the cyclic subgroup $H_{nk} := \langle x^{nk} \rangle$ generated by x^{nk} is strictly contained in the cyclic subgroup $H_{(n-1)k} := \left\langle x^{(n-1)k} \right\rangle$. (Why?)

Let us recall that if H is a subgroup of $\mathbb{Z}$ then $H = k\mathbb{Z}$ for some $k \in \mathbb{Z}$. If we read the proof carefully, we notice that the arguments actually show that a subgroup of a cyclic group is cyclic. We prove this now.

Theorem 2.2.37. *Let G be a cyclic group. Then every subgroup of G is also cyclic.*

Proof. Let $G = \langle x \rangle$ be a cyclic group and H be a proper subgroup of G. If y is in H and $y = x^l$ for $l \in \mathbb{Z}$, then $y^{-1} = x^{-l}$ is also in H. Therefore there exists $l \in \mathbb{N}$ such that x^l is in H. Now we let $S = \{l \in \mathbb{N} : x^l \in H\}$. Then S is a non-empty subset of $\mathbb{N}$. By well ordering principle it has a minimum and let $k = \min\{l : l \in S\}$.

We claim that the element x^k generates H. That is $H = \langle x^k \rangle$.

Let y be an element in H. Then there exists an integer t such that $y = x^t$. By division algorithm, there exists integers q and r such that $t = kq + r$ where $0 \leq r < k$ and $y = x^t \because x^{kq+r} = x^{kq}x^r$. Since H is a subgroup, we see that $x^r = yx^{-kq}$ is also in H. Therefore $r = 0$ and $y = x^t = x^{kl}$. Thus every element y in H is a power of x^k and $H = \langle x^k \rangle$. □

Lemma 2.2.38. *If $g(\neq e)$ is an element in a group G, then $g^k = g^r$ for two integers k and r iff g is of finite order and in this case the order of g divides $k - r$.*

Proof. Let g be an element of finite order, say n. Then $g^{k+n} = g^k$ for any integer k.

Let $g \neq e$ be an element in the group G and k and r be two integers such that $g^k = g^r$. Then $g^{k-r} = e$. This proves that g is of finite order, say, n and n divides $k - r$. □

Theorem 2.2.39. *Let G be an infinite cyclic group. Then the group G has exactly two generators which are inverses of each other.*

Proof. Let $G := \langle x \rangle$ be an infinite cyclic group generated by x. Let y be another generator of G. Then there exist integers k and l such that $y = x^k$ and $x = y^l$. We combine these two equations to get $x = y^l = x^{kl}$. Hence $x^{kl-1} = e$. Since the order of x is infinite, we see that $kl = 1$. This equation has exactly two solutions $k = l = 1$ and $k = l = -1$. Accordingly we get $x = y$ or $x = y^{-1}$. □

Proposition 2.2.40. *Let $G := \langle g \rangle$ be an infinite cyclic group. For any two natural numbers k and l, we let $H = \langle g^k \rangle$ and $K = \langle g^l \rangle$ denote the subgroups generated by g^k and g^l respectively. Then $H \subseteq K$ iff l divides k.*

Proof. Since the subgroup $H \subseteq K$, there exists an integer t such that $g^k = g^{tl}$. Since G is infinite cyclic, we see that $k = lt$ and l divides k.

Conversely if l divides k, then $k = lt$ for some $t \in \mathbb{Z}$. Therefore $g^k = g^{lt} = (g^l)^t$. Hence $H \subseteq K$. □

Theorem 2.2.41. *Let $G := \langle x \rangle$ be a cyclic group. Then an element $x^k \in G$ is also a generator of G iff either of the following holds.*

1. *The group G is infinite and $k = \pm 1$.*

2. *The group G is finite and the integers k and $|G|$ are relatively prime.*

Proof. Suppose G is infinite. We claim that if k is an integer and $|k| \geq 2$, the subgroup $\langle x^k \rangle$ is a proper subgroup of G: Assume the contrary. Then the element x is in $\langle x^k \rangle$. Hence there exist an integer l such that $x = x^{kl}$. This proves that x is of finite order, a contradiction. Therefore if G is infinite cyclic, then it has only two generators.

Let $G := \langle x \rangle$ be a finite cyclic group. Assume that the order of x is n. Let $H := \langle x^k \rangle$. Then $H = G$ iff the element x is in H. Assume that x is in H. Hence there exists an integer l such that $x = x^{kl}$. Therefore there exists an integer m such that $kl - 1 = mn$ and hence $kl - mn = 1$. This proves that k and n are relatively prime.

Conversely, assume that k and n are relatively prime. Then there exists integers t and m such that $1 = kt + mn$. Hence

$$x = x^{kt+mn} = x^{kt}x^{mn} = (x^k)^t(x^n)^m = (x^k)^t.$$

Thus x is in the subgroup H and hence G is contained in H. This proves that x^k also generates the group G. □

Corollary 2.2.42. *Let G be a finite cyclic group and let the order of G be prime. Then G does not have any proper subgroups.*

Proof. Let $G := \langle x \rangle$ be a cyclic subgroup and H a subgroup of G. Since a subgroup of a cyclic subgroup is cyclic, it follows that $H = \langle x^k \rangle$ for some integer k if $H \neq \{e\}$. Since the order of the group G is prime, the integer k is relatively prime to the order of G. Hence $H = G$. □

Theorem 2.2.43. *Let G be a finite cyclic group of order n and k be a divisor of n. Then there is a unique subgroup H of G of order k.*

Proof. Let $G = \langle x \rangle$ be a finite cyclic group and let $|x| = |G| = n$.

If n is prime then G does not have any proper subgroup. Let us now assume that d is composite and let d be a proper divisor of n. Then the subgroup $H_d := \langle x^d \rangle$ generated by x^d has order $\frac{n}{d}$. If we now show that any element of order $\frac{n}{d}$ is in H_d, then we are through.

Let x^k for $k < n$ be an element of order $\frac{n}{d}$ in G. Then $x^{k\frac{n}{d}} = e$. Hence n divides $k\frac{n}{d}$ and we can write $k\frac{n}{d} = nq$ for some integer q. This shows that $k = dq$ and $x^k = (x^d)^q$ is in H_d. □

Corollary 2.2.44. *Let G be a finite cyclic group and x, y be two elements of G. Then $\langle x \rangle = \langle y \rangle$ iff $o(x) = o(y)$.*

Proof. Easy. □

As an easy corollary of the theorem, we prove the following proposition.

Proposition 2.2.45. *Let G be a group such that G does not have proper subgroups. Then G is a cyclic group of prime order.*

Proof. Let $x (\neq e)$ be an element in G and $H = \langle x \rangle$ be the cyclic subgroup generated by x. Since G does not have proper subgroups, it follows that $G = H$ is cyclic.

Fix a $k \geq 2$. By hypothesis $\langle x \rangle = G = \langle x^k \rangle$. Therefore there exists $l \in \mathbb{Z}$ such that $x = x^{kl}$. This shows $x^{kl-1} = e$ and hence x is of finite order. Furthermore the order of x must divide $kl - 1$. If we denote by d the order of x, there exists integer r such that $kl - 1 = dr$. We rewrite this as $kl - dr = 1$ to conclude that k and d are relatively prime.

Since this is true for all $k < d$, it follows that d is prime and the proof is complete. □

Theorem 2.2.46. *Let $G = \mathbb{C}^*$. Then the following holds.*

1. *Every finite subgroup H of G is cyclic.*

2. *If $k \geq 2$, there is a unique cyclic subgroup H of order k in $\mathbb{C}^*$.*

Proof. Assume $H \neq \{1\}$. Let $z \neq 1$ be an element in H. Since H is finite, the order of z is finite, say n. Further observe that $|z| = 1$. (Why?) If we denote $z = \cos 2\pi t + i \sin 2\pi t$, then using De moivres theorem we get $\cos 2\pi nt + i \sin 2\pi nt = z^n = 1$. Therefore nt must be an integer, say, m. Hence $t = \frac{m}{n}$ and we have shown that an element z is in H iff $= \cos 2\pi r + i \sin 2\pi r$ for some rational number r.

If $w = e^{2\pi it}$ is in H, then $w^{-1} = e^{-2\pi it}$ is also in H. Hence there exists positive rational number t such that $e^{2\pi it}$ is in H. Since n is the order of H, there exist only finitely many rational numbers $t = \frac{m}{n}$ such that $e^{2\pi i t} \in H$. Therefore $r = \min\{t > 0 : e^{2\pi it}(\neq 1) \in H\}$ exists.

Let $r = \frac{m}{n}$ and $z = \cos 2\pi r + i \sin 2\pi r$. We will now show that every element w in H is of the form $w = z^k$ for an integer k such that $0 \leq |k| < n$. Let $w = \cos 2\pi t + i \sin 2\pi t$ be an element in H and let $t = \frac{p}{q} > 0$. Then $t \geq r$. Therefore there exists a natural number k such that $k \leq \frac{t}{r} < k + 1$. In view of this $0 \leq t - kr < r$. Let $\theta = t - kr$. Then $e^{i\theta} = e^{i(t-kr)} = wz^{-k}$ is also in H. Since $r = \min\{t > 0 : e^{2\pi it} \neq 1 \in H\}$ we get $\theta = 0$. Hence $t = kr$ and $w = z^k$. This proves that H is a cyclic subgroup of $\mathbb{C}^*$.

The second part is easy and left as an exercise for the readers. □

Theorem 2.2.47. *Any finite subgroup of $SO(2, \mathbb{R})$ is cyclic.*

Proof. The proof follows by adapting the proof of Theorem 2.2.46.

However we give another proof using homomorphism which we discuss in Section 2.5.

Let G be a finite subgroup of G and A be an element of G. Then there exists a unique $\theta \in [0, 2\pi)$ such that $A = \begin{pmatrix} \cos\theta & -\sin\theta \\ \sin\theta & \cos\theta \end{pmatrix}$. We denote this by A_θ.

Let $f : \mathrm{SO}(2, \mathbb{R}) \to \mathbb{C}^*$ be defined by $f(A_\theta) = e^{2\pi i\theta}$. Then f is a one-one group homomorphism. Therefore $f(H)$ is a finite subgroup of $\mathbb{C}^*$ and hence cyclic. Since $f : H \to f(H)$ is an isomorphism, the subgroup H is also cyclic. □

Exercise 2.2.48. Show that the groups $\mathbb{Q}$, $\mathbb{R}$ and $\mathbb{C}$ are not cyclic.

Exercise 2.2.49.

1. Let G be a group. Show that the product group $G \times G$ is not cyclic.

2. Let G and H be cyclic. Then $G \times H$ is cyclic iff G and H are finite groups and orders of G and H are relatively prime.

3. Let m and n be two relatively prime numbers. Show that the group $\mathbb{Z}_m \times \mathbb{Z}_n$ is cyclic.

Exercise 2.2.50.

1. Does the group $G = (\mathbb{R}^*, \cdot)$ have proper cyclic subgroups other than $H = \{\pm 1\}$? If yes can we find all of them?

2. Find some proper cyclic subgroups of $\mathbb{Q}^*$ and $\mathbb{C}^*$.

3. Give an example of a group G which is not cyclic but each of its subgroups is cyclic.

4. Let G be an infinite group. Show that G has infinitely many proper subgroups.

   Let $x_1 \neq e$ be an element of G. If the order of x_1 is infinite, then, for each $k \geq 2$, the subgroups $H_k := \langle x_1^k \rangle$ are proper subgroups of G.

If the order of x_1 is finite, we choose $x_2 \notin G \setminus \langle x_1 \rangle$. If order of x_2 is finite, we choose $x_3 \notin G \setminus (\langle x_1 \rangle \cup \langle x_2 \rangle)$. We repeat this process again. If at any stage, we get an element x_n with infinite order, then we get infinitely many subgroups $\langle x_n^k \rangle$ for $k = 2, 3, \ldots$. Otherwise we have infinitely many finite subgroups $\langle x_k \rangle$ for $k = 1, 2, \ldots$.

Now we ask the following question: suppose x and y are two distinct elements of a group G. Does there exist a smallest subgroup containing the elements x and y and if so can we describe them as we did in the case of one element?

We observe that such a subgroup H must contain $\{x^r y^s : r, s \in \mathbb{Z}\}$. If $xy = yx$, then we can very easily conclude that the set $\{x^r y^s : r, s \in \mathbb{Z}\}$ is a subgroup of G. Therefore the smallest subgroup containing two commuting elements x and y is the subgroup $\{x^r y^s : r, s \in \mathbb{Z}\}$ and we denote this subgroup by $\langle x, y \rangle$.

In general such simple description is not possible !

Let

$$G = \mathrm{S}L(2, \mathbb{Z}) := \left\{ A = \begin{pmatrix} a & b \\ c & d \end{pmatrix} \in \mathrm{S}L(2, \mathbb{R}) : a, b, c, d \in \mathbb{Z} \right\}.$$

Then $\mathrm{S}L(2, \mathbb{Z})$ is a subgroup of $\mathrm{S}L(2, \mathbb{R})$.

Let $A = \begin{pmatrix} 0 & -1 \\ 1 & 0 \end{pmatrix}$ and $B = \begin{pmatrix} 1 & 1 \\ 0 & 1 \end{pmatrix}$. Let us observe that $AB \neq BA$. We can show that the smallest subgroup containing the matrices A and B is $\mathrm{S}L(2, \mathbb{Z})$. However we can't write $\mathrm{S}L(2, \mathbb{Z}) = \{A^r B^S : r, s \in \mathbb{Z}\}$.

Similarly the matrices $A := \begin{pmatrix} 0 & -1 \\ 1 & 0 \end{pmatrix}$ and $B = \begin{pmatrix} 0 & -1 \\ 1 & 1 \end{pmatrix}$ also generate $\mathrm{S}L(2, \mathbb{Z})$. We will not prove this and refer to [12] for a proof.

2.3 Lagrange's Theorem

This section is devoted to proving one of the important result in finite groups, namely Lagrange's Theorem and some applications of this important result.

Theorem 2.3.1 (Lagrange). *Let G be a finite group and H a subgroup of G. Then the order of H divides the order of G.*

Let G be a group and H a subgroup of G. We define a relation R on G by xRy iff $x^{-1}y \in H$. It is easy to check that R is an equivalence relation on G.

Let $x, y \in G$ be such that $x^{-1}y \in H$. Then there exists an element $h \in H$ such that $x^{-1}y = h$. That is $y = xh \in xH$. Let y be an element in xH, say $y = xh$. Then $x^{-1}y = h \in H$. Hence x and y are equivalent. Therefore, if x is in G, then the set of all elements equivalent to x is the set xH. The equivalence classes xH are called the *left cosets* of the subgroup H.

Example 2.3.2. 1. Let $G = \mathbb{Z}$ and $H = d\mathbb{Z}$ be a subgroup of G. Then the equivalence class of an integer k is the set of all integers $\{k+dm : m \in \mathbb{Z}\}$. This is the arithmetic progression $k + dm$ with common difference d and starting with k. Hence we get d distinct left cosets of H.

2. Let $n \geq 2$ be a composite number and let $G := \{z \in \mathbb{C}^* : z^n = 1\}$. If k is a divisor of n, let $H := \{z \in G : z^k = 1\}$. Let $w = e^{\frac{2\pi i}{n}}$. Then the elements of the group G are w^r for $0 \leq r \leq n-1$ and $\zeta^t := e^{\frac{2\pi it}{k}}$ for $0 \leq t \leq k-1$ are the elements of the subgroup H. Note that if $n = kl$, then $\zeta = z^l$.

   Observe that the elements of G are the vertices of a regular n-sided polygon with 1 as one of its vertices and the subgroup consists of vertices of a regular k sided polygon with 1 as a vertex. If z^r is an element of the group G, then the coset

$$z^r H := \{e^{\frac{2\pi ir}{n}} \zeta^t : 0 \leq t \leq k-1\} = \left\{e^{\frac{2\pi i(r+lt)}{n}} : 0 \leq t \leq k-1\right\}$$

   is the set of vertices of a regular k-sided polygon, got by rotating the polygon H by an angle $e^{\frac{2\pi ir}{n}}$. Thus the cosets are the set of all regular k-sided polygons inscribed inside the n-sided polygon.

3. Let $G = \mathbb{R}^2$ and $H = \{(x,y) \in \mathbb{R}^2 : x = y\}$. Let (a,b) be in $\mathbb{R}^2$. Then the equivalence class containing the element (a,b) is the set $(a,b) + H = \{(a+x, b+x) : x \in \mathbb{R}\}$. This set is a line passing through the point (a,b) and parallel to the line H. Thus the equivalence classes or the left the cosets of the subgroup H are lines parallel to H.

4. Let $G = \mathbb{R}^3$ and $H = \{(x,y,z) \in \mathbb{R}^3 : z = 0\}$. The equivalence classes (left cosets of H) are planes parallel to the XY-plane.

Exercise 2.3.3. Let x be an element in G. Show that the map $h \mapsto xh$ is a bijection from H to xH.

Let $\varphi : H \to xH$ be the map defined by $\varphi(h) := xh$. Then φ is one-one: Let h_1 and h_2 in H be such that $\varphi(h_1) = \varphi(h_2)$. This means that $xh_1 = xh_2$ and by cancellation law it follows that $h_1 = h_2$.

Clearly this map is onto.

Let G be a group and H a subgroup of G. Let x and y be two elements in G. Since the left cosets are equivalence classes, it follows that either $xH = yH$ or they are disjoint. Further, since an equivalence relation on a set partition the set in to distinct equivalence classes, we can write the group G as a disjoint union of the left cosets of G. That is $G = \cup_{x \in G}\, xH$.

Remark 2.3.4. We could have started with the equivalence relation x *is equivalent to* y *iff* yx^{-1} *is in* H. Then the equivalence classes are Hx called the *right cosets* of H. We can show $G = \cup_{x \in G}\, Hx$.

We are now ready for a proof of Lagrange's Theorem.

Proof. Let G be a finite group and H a subgroup of G. By what we have seen above we write $G = \cup_{x \in F}\, xH$ where F is a finite set. Since this union is disjoint, it follows that

$$\begin{aligned} |G| &= \sum_{x\in F} |xH| \\ &= \sum_{x\in F} |H| \\ &= |F||H| \end{aligned}$$

This proves that $|H|$ divides $|G|$. □

2.3.1 Applications of Lagrange's Theorem

Proposition 2.3.5. *Let G be a finite group and x be an element of G. Then the order of x divides the order of G.*

Proof. Let x be an element in G and $H := \langle x \rangle$ be the cyclic subgroup generated by x. Then the order of x is equal to $|H|$ and by Lagrange's theorem the order of H divides the order of G. This proves our claim. □

An application of this result is the following theorem due to Euler.

Theorem 2.3.6 (Euler's theorem). *Let $n \geq 2$ and $a \in \mathbb{N}$ be such that* $\gcd(a, n) = 1$. *Then $a^{\varphi(n)} \equiv 1 \pmod{n}$ where $\varphi(n)$ is the order of the group U_n.*

Proof. If a is less than n, then a is in U_n. Now by Proposition 2.3.5, the order of a divides $\varphi(n)$, the order of U_n. Hence $a^{\varphi(n)} \equiv 1 \pmod{n}$.

If $a > n$, we use division algorithm to write $a = nq + r$ for $q \in \mathbb{Z}$ and $0 \leq r < n$. Therefore $a \equiv r \pmod{n}$ and $a^t \equiv r^t \pmod{n}$ for every $t \in \mathbb{N}$. In particular for $t = \varphi(n)$, we get

$$a^{\varphi(n)} \equiv r^{\varphi(n)} \equiv 1 \pmod{n}.$$

This completes the proof. □

As a corollary to Euler's theorem(Theorem 2.3.6), we prove Fermat's Little Theorem.

Corollary 2.3.7 (Fermat's Little Theorem). *Let $p \geq 2$ be a prime number. Then*

$$a^p \equiv a \pmod{n}$$

for any integer a.

Proof. First notice that $\varphi(p) = p - 1$ for a prime number p.

If $a = 0$, then $a^k = 0$ for all k and the result is obvious.

Let a be a nonzero integer. By division algorithm there exist unique integers k and $0 < r < p$ such that $a = pk+r$. Now it follows from Euler's Theorem 2.3.6, that

$$a^{p-1} \equiv r^{p-1} \equiv 1 \pmod{p}.$$

From this we conclude that

$$a^p = a \cdot a^{p-1} \equiv a \pmod{p}.$$

□

We will now work out some exercises using Lagrange's theorem.

Exercise 2.3.8. 1. Let G be a finite group of prime order. Show that G is cyclic.

> Let $x(\neq e)$ be an element in G and let H be the cyclic subgroup generated by x. Since the order of G is prime and the order of H divides the order of G, it follows that $H = G$. Hence G is cyclic.

2. Let G be a group of odd order. Show that G can't have a subgroup of even order. In particular, it can't have elements of even order.

3. Let H and K be two finite subgroups of a group G. Suppose $\gcd(|H|, |K|) = 1$. Show that $H \cap K = \{e\}$.

We will now prove some more important applications of Lagrange's theorem, which are usually proved as applications of Sylow Theorems.

Theorem 2.3.9. *Let G be a finite group such that G has at most one subgroup of any size. Then G is cyclic.*

Proof. Let G be a finite group satisfying the hypothesis of the theorem and let $|G| = N$, say. If G has a subgroup H, then the order of H divides the order of G.

For a divisor d of N, we let $S_d := \{x \in G : |x| = d\}$. If S_d is non-empty, then there is cyclic group H_d of order d. By hypothesis, there exists at most one subgroup of order d. Therefore H_d is unique, if S_d is non-empty. Further S_d is a subset of H_d. We know that the number of generators of a cyclic group of order d is $\varphi(d)$. Therefore $|S_d| = \begin{cases} \varphi(d) & \text{if } S_d \text{ is non-empty} \\ 0 & \text{if } S_d \text{ is empty} \end{cases}$. In any case $|S_d| \leq \varphi(d)$ for every divisor d of N.

Let us now observe that the sets S_d are disjoint and every element in G is in one of S_d's. In view of this $G = \cup_{d|N} S_d$, a disjoint union. Therefore $|G| = \sum_{d|N} |S_d| \leq \sum_{d|N} \varphi(d) = N$ and hence $|S_d| = \varphi(d)$ for every divisor d of N. In particular $|S_N| = \varphi(N)$. Since 1 is relatively prime to N, it follows that $\varphi(N) \geq 1$. Hence S_N is non-empty. This means that G is a cyclic group.

> We will now give a group theoretic proof of $\sum_{d|N} \varphi(d) = N$. Let $C \equiv C_N$ denote the cyclic group of order N. If d is a divisor of N, then we know that there is exactly one subgroup of order d and the number of generators is $\varphi(d)$. Now, any $g \in C$, will be a generator of a unique cyclic subgroup, namely the one generated by g. Hence $\sum_{d|N} \varphi(d) = N$.

□

Corollary 2.3.10. *Let G be a group and H be a finite subgroup satisfying the hypothesis of Theorem 2.3.9. Then H is cyclic.*

Proof. The proof is same as in the proof of Theorem 2.3.9. □

There is another important corollary of Theorem 2.3.9, which we state as a theorem.

Theorem 2.3.11. *Let G be a finite group such that, for every natural number d dividing the order of G, the equation $x^d = e$ has at most d solutions in G. Then G is cyclic.*

Proof. Let d be a divisor of $|G|$ and $S_d := \{x \in G : |x| = d\}$. We claim that $|S_d| = \varphi(d)$ if it is non-empty.

Suppose $x \in G$ is an element of order d. If r is relatively prime to d, then x^r is also of order d: Suppose $x^{qr} = e$ for some q. Then d divides qr. By our assumption r and d are relatively prime. Hence d divides q. Since $|x^r|$ divides $|x|$, the integer q divides d. Therefore $q = d$.

Conversely if y is an element in S_d, then $y = x^r$ for some r relatively prime to d. For, otherwise, the equation will have at least $d+1$ solutions, namely y and x^i for $1 \leq i \leq d$.

Now the proof follows from Theorem 2.3.9. □

Corollary 2.3.12. *Let p be a prime number and $G := \mathbb{Z}_p^*$ be the multiplicative group of nonzero elements of $\mathbb{Z}_p$. Then G is cyclic.*

Proof. Let $d \geq 1$. Then any polynomial equation $f(X) = a_d X^d + a_{d-1} X^{d-1} + \cdots + a_1 X + a_0 \equiv 0 \pmod p$ with coefficients $a_i \in \mathbb{Z}_p^*$ has at most d-solutions. (See [9] or [4]).

In particular for every integer d such that $d | p-1$ the polynomial $f(X) = X^d - 1$ has at most d solutions in $\mathbb{Z}_p^*$. Hence it follows from Theorem 2.3.11 that the group $\mathbb{Z}_p^*$ is cyclic. □

We come back to the cosets again. Let G be a group and H be a subgroup of G. Then the set of all left cosets of H in G is denoted by $G/H = \{aH : a \in G\}$.

Let G be a finite group and H a subgroup of G. Then $|G| = |G/H||H|$ and $|G/H|$ is called the index of H in G.

Examples 2.3.13.

1. Let $k \geq 2$. If $G = \mathbb{Z}$ and $H = k\mathbb{Z}$, then $G/H = \{[0], [1], \ldots, [k-1]\}$.
2. Let $G = \mathbb{R}^*$ and $H = (0, \infty)$. Then $G/H = \{H, -H\}$.
3. Let $G = \mathbb{R}^*$ and $H = \{\pm 1\}$. Then $G/H = \{\{\pm x\} : x \in \mathbb{R}\}$.
4. Let $G = \mathbb{C}^*$ and $H = \mathbb{T}$. Then $G/H := \{rH : r > 0\}$. Observe that each left coset rH is the circle of radius $r > 0$ and center 0.
5. Let $G = \mathbb{C}^*$ and $H = \mathbb{R}_+^*$. Then $G/H = \{\{re^{it} : r > 0\} : e^{it} \in \mathbb{T}\}$. In this case each left coset is a ray passing through the point e^{it}.

Exercise 2.3.14. Let G be a group and U be a subset of G such that the family of subsets $\{xU : x \in G\}$ partition G. i.e., G is the disjoint union of the subsets xU. Let x_0 in G such that the identity element e is in x_0U. Show that x_0U is a subgroup of G and xU are the left cosets of the subgroup x_0U.

Let $H = x_0U$ and $h = x_0u$ be an element of H. We will show that the translate $(h^{-1}x_0)U$ has identity element e. We can write $e = (h^{-1}x_0)u = u^{-1}(x_0^{-1}x_0)u$. This means that the two translates $(h^{-1}x_0)U$ and $H = x_0U$ of the set U have identity element in common. Since the translates of U

are disjoint, it follows that $(h^{-1}x_0)U = H$ for every element $h \in H$. Therefore $h^{-1} = h^{-1}e \in h^{-1}H = H$ for every element h in H.

We will now show that if x and y are in H, then xy is also in H. Now

$$\begin{aligned}(xy)H &= x(yH)\\ &= xH\\ &= H.\end{aligned}$$

Therefore $xyH = H$ for all x and y in H. Since e is in H, the element $xy = xy.e$ is in H.

This proves that H is a subgroup of G and the sets xU are left cosets of G.

2.4 Subgroups– continued

In this section we will study some special subgroups of a group G.

Definition 2.4.1 (Normalizer). Let S be a non empty subset of a group G. Then the normalizer of S in G, denoted by $N_G(S)$, is the set defined by $N_G(S) := \{g \in G : gSg^{-1} = S\}$.

Proposition 2.4.2. *For any subset S of a group G, the normalizer $N_G(S)$ is a subgroup of G.*

Proof. It is obvious that the identity element e is in $N_G(S)$.

Let x and y be two elements in $N_G(S)$ and $s \in S$. Let $ysy^{-1} = s'$. Then s' is in S and hence $xys(xy)^{-1} = x(ysy^{-1})x^{-1} = xs'x^{-1}$ is in S.

If $ysy^{-1} = s'$, then $y^{-1}s'y = s$ is also in S. This proves that $N_G(S)$ is a subgroup of G. □

Question 2.4.3. If we only assume that $gSg^{-1} \subseteq S$ for all g in G, can we still conclude that $N_G(S)$ is a subgroup of G?

Exercise 2.4.4. If G is abelian, show that $N_G(S) = G$ for any non-empty subset S of G.

Example 2.4.5.

1. Let $G = S_3$ and $S = \left\{\begin{pmatrix}1 & 2 & 3\\ 2 & 1 & 3\end{pmatrix}\right\}$. Then $N_G(S) = \{Id\}$.

2. Let $G = S_3$ and $S := \left\langle\begin{pmatrix}1 & 2 & 3\\ 2 & 3 & 1\end{pmatrix}\right\rangle$, the cyclic subgroup generated by $\begin{pmatrix}1 & 2 & 3\\ 2 & 3 & 1\end{pmatrix}$. Then $N_G(S) = S_3$.

3. Let $G = S_3$ and $S = \left\{e, \begin{pmatrix}1 & 2 & 3\\ 2 & 1 & 3\end{pmatrix}\right\}$. Then $N_G(S) = S$, a proper subgroup of G.

4. Let $G = \mathrm{GL}(n, \mathbb{R})$ and $S = \mathrm{SL}(n, \mathbb{R})$. Then $N_G(S) = G$. More generally, for $\alpha \in \mathbb{R}^*$, let $S_\alpha := \{A \in \mathrm{GL}(n, \mathbb{R}) : \det(A) = \alpha\}$. Then $N_G(S_\alpha) = G$.

Exercise 2.4.6. 1. Let $G = \mathrm{GL}(n, \mathbb{R}) \ltimes \mathbb{R}^n$. Find $N_G(S)$ if

   (a) $S = \mathbb{R}^n$.

   (b) $S = \mathrm{GL}(n, \mathbb{R})$.

2. Let $G = S_n$ and $S = \{(1\,2\,\ldots, n)\}$. Find $N_G(S)$.

   $N_G(S)$ is the cyclic subgroup generated by the n-cycle $(1\,2\,\ldots, n)$.

3. Let $G = S_n$ and $S = \langle \sigma = (1\,2\,\ldots, n)\rangle$, the cyclic subgroup generated by the n-cycle σ. Find $N_G(S)$.

4. If S is a subgroup of G, then show that $S \subseteq N_G(S)$.

In the examples above we have seen that for some subgroups the normalizer is the whole group and such subgroups are called *normal subgroups.*

Definition 2.4.7. A subgroup H of a group G is said to be normal in G if $N_G(H) = G$. That is ghg^{-1} is in H for all $h \in H$ and for all $g \in G$.

Example 2.4.8. 1. Let $G = \mathbb{Z}$ and H a subgroup of G. Then H is normal in G.

2. Let $G = \mathbb{R}^n$ and $a \in \mathbb{R}^n$ be fixed. Let $H = \{x \in \mathbb{R}^n : \langle x, a\rangle = 0\}$. Then H is a normal subgroup of G.

3. Let $G = \mathbb{C}^*$ and $H = S^1$. Then H is normal in G.

4. Let $G = \mathbb{C}^*$ and $H = G_n$ for $n \geq 2$. More generally, if G is an abelian group then any subgroup of G is normal in G.

5. Let $G = \mathrm{GL}(n, \mathbb{R})$ and $H = \mathrm{SL}(n, \mathbb{R})$. Then H is a normal subgroup of G.

We characterize normal subgroups of a group G in the following proposition.

Proposition 2.4.9. *Let H be a subgroup of a group G. Then the following are equivalent.*

1. *The subgroup H is normal in G.*

2. *$gHg^{-1} \subseteq H$ for all $g \in G$.*

3. *$gH = Hg$ for all $g \in G$.*

Before we start with the proof of the proposition, we need to explain the statements in the proposition. For a subgroup H of a group G, the set $gHg^{-1} = \{ghg^{-1} : h \in H\}$ is a subgroup of G and it is called a conjugate of H.

The equality $gH = Hg$ means that the sets gH and Hg are equal; that is given h in H, there exists h' in H such that $gh = h'g$ and vice versa. We can't and we don't claim that $gh = hg$ for all g in G and h in H!

The following example shows that gh need not be equal to hg. Let $H = \langle \sigma = (1\ \ 2\ \ 3)\rangle$ be the cyclic subgroup generated by the 3-cycle $(1\ \ 2\ \ 3)$ in $G = S_3$. Let $\tau = (1\ \ 2)$. Then

$$\tau\sigma = (1\ \ 2)\,(1\ \ 2\ \ 3) = (2\ \ 3)$$

is not equal to

$$\sigma\tau = (1\ \ 2\ \ 3)\,(1\ \ 2) = (1\ \ 3).$$

Now we will prove the theorem.

Proof. We will prove that (1) ⇒ (2) ⇒ (3) ⇒ (1).

Let H be a normal subgroup of G. By definition $gHg^{-1} = H$ for every g in G. Hence (1) ⇒ (2) is obvious.

Now let us assume that $gHg^{-1} \subseteq H$ for all $g \in G$. If h is an element of H, then, by assumption $h' = ghg^{-1}$ is in H. Therefore $gh = (ghg^{-1})g = h'g$ is in Hg. Hence $gH \subseteq Hg$ and a similar argument shows that $Hg \subseteq gH$. This proves (2) ⇒ (3).

Let H be a subgroup of G such that $gH = Hg$ for all $g \in G$. If h is in H and g is in G, then by hypothesis, $gh = h'g$ for some h' in H. Hence $(gh)g^{-1} = (h'g)g^{-1} = h'$ is in H. This proves that $gHg^{-1} \subseteq H$ and $H \subseteq gHg^{-1}$ for all g in G. □

Exercise 2.4.10. 1. Let G be an abelian group and H be a subgroup of G. Then H is normal in G.

2. Let H be a subgroup of G of index 2. Then H is normal in G.

   Let g be an element not in H. Then $G = H \cup gH = H \cup Hg$.

3. Let $G = S_3$ and $H := \{\sigma \in G : \sigma(1) = 1\}$. Show that H is not normal in G.

We write a vector x in $\mathbb{R}^n$ as a column vector $x = (x_1, x_2, \dots, x_n)^t$. If A is an $n \times n$ matrix then $A(x)$ is the matrix product of a $n \times n$ and $n \times 1$ matrix, which is again a $n \times 1$ matrix. That is Ax is again a vector in $\mathbb{R}^n$.

Exercise 2.4.11. 1. Let $H := \{A \in \mathrm{GL}(n, \mathbb{R}) : A(e_1) = e_1\}$. Is the subgroup H normal in $\mathrm{GL}(n, \mathbb{R})$? First do this for $n = 2$.

2. Let $G = \mathrm{O}(n, \mathbb{R})$ and $H = \mathrm{SO}(n, \mathbb{R})$. Show that H is normal in G. Is $\mathrm{O}(n, \mathbb{R})$ normal in $\mathrm{GL}(n, \mathbb{R})$?

3. Let $G = \mathrm{GL}(n, \mathbb{R}) \ltimes \mathbb{R}^n$. Let $H = \mathbb{R}^n$. Show that H is normal in G. Is $\mathrm{GL}(n, \mathbb{R})$ normal in G?

Exercise 2.4.12. 1. Let $G = S(X)$ be the group of permutations on a set X and $H := \{\sigma \in G : \sigma = Id \text{ outside a finite subset of } X\}$. Show that H is a normal subgroup of G.

   Let σ and τ be two elements of H. Let A be a finite subset of X such that $\sigma(x) = x$ for x in $X \setminus A$ and B be a finite subset of X such that $\tau(x) = x$ for x in $X \setminus B$. If x is in $X \setminus (A \cup B)$, then $\sigma(x) = x$ and $\tau(x) = x$. Therefore $\sigma \circ \tau(x) = \sigma(x) = x$ for all x in $X \setminus (A \cup B)$. Hence $\sigma \circ \tau$ is also in H.

   (a) The identity permutation is in H.

   (b) Associativity law is easily verified: Let σ, τ and θ be in H. Then we have shown above $\sigma \circ (\tau \circ \theta)$ is in H and $(\sigma \circ \tau) \circ \theta$ is also in H. Since the composition of functions is associative, the equality of these two follows.

(c) If σ is identity on $X \setminus A$ for a finite subset A of X, then σ^{-1} is also identity on $X \setminus A$.

This proves that H is a subgroup of G.

We will now show that H is normal in G. Let σ be in H and θ in $S(X)$. There exists a finite subset A of X such that σ is identity on $X \setminus A$. Let x be an element of $X \setminus \theta(A)$ and $y \in X$ be such that $x = \theta(y)$. Then y is not in A. Therefore

$$\begin{aligned} \theta \circ \sigma \circ \theta^{-1}(x) &= \theta \circ \sigma \circ \theta^{-1} \circ \theta(y) \\ &= \theta \circ \sigma(y) \\ &= \theta(y) \\ &= x \quad \text{for } x \in X \setminus \theta(A). \end{aligned}$$

Thus $\theta \circ \sigma \circ \theta^{-1}$ is also in H for every θ in G and hence H is normal in G.

2. Let H be a subgroup of a group G and X be a non-empty set. Let $\mathcal{F}(X, H) := \{f \in \mathcal{F}(X, G) : f(x) \in H \text{ for all } x \in X\}$. When is the subgroup $\mathcal{F}(X, H)$ normal in $\mathcal{F}(X, G)$?

Exercise 2.4.13. 1. Let G be a group and H be a normal subgroup of G. Show by an example that a normal subgroup K of H need not be normal in G.

Let $G := \{x^i y^j : 1 \leq i \leq 4, 1 \leq j \leq 2, x^4 = y^2 = e, yx = x^{-1}y\}$. Then $H := \{e, y, x^2, yx^2 = x^2 y\}$ being a subgroup of index 2 in G is normal in G. Since the element y is of order 2, the set $K = \{e, y\}$ is a subgroup of H and of index 2 in H. Hence K is normal in H. However K is not normal in G: For the element $x^{-1}yx = x^2 y$ is not in K.

We give another example now.
Let $G = A_4$ and $H = \{Id, (1,2)(3,4), (1,3)(2,4), (1,4)(2,3)\}$. It is easy to see that H is normal in G. Let $K = \{Id, (1,2)(3,4)\}$. Then K is normal in H but not in G.
Can you find an example in infinite groups?

2. Let G be a group and H be a normal subgroup of G. Assume that H is also cyclic. Show that every subgroup K of H is normal in G.

Let G be a finite group and H be a normal, cyclic subgroup of G. Let $x \in H$ be a generator of H. That is $H = \langle x \rangle$.

Let K be a subgroup of H. Then K is also cyclic and there exists a natural number d such that $K = \langle x^d \rangle$. We want to show that, for every element g in G, the subgroup $gKg^{-1} = K$.

The order of gxg^{-1} is same as the order of x. Therefore $K = \langle gx^d g^{-1} \rangle = \langle (gxg^{-1})^d \rangle$. (Why?) If $k = (x^d)^r$ is in K, then $gkg^{-1} = g(x^d)^r g^{-1} = (gx^d g^{-1})^r$ is also in K. This proves that K is normal in G.

We will now introduce a special class of normal subgroup of a group G.

Definition 2.4.14. Let G be a group and $S := \{ghg^{-1}h^{-1} : g, h \in G\}$. The smallest subgroup containing the set S is called the commutator subgroup of G and it is denoted by $[G, G]$.

If G is abelian, then $S = \{e\}$ and hence the smallest subgroup containing S is S itself. If S is not abelian, then $[G, G]$ is non-trivial. If G is non-abelian, the subgroup $[G, G]$ may be the full group G itself. We will see examples of such groups later.

Exercise 2.4.15. Show that

1. $[G, G] = \{x_1x_2 \dots x_r : r \geq 0 \text{ and } x_i \in S\}$.

   Let $H := \{x_1x_2 \dots x_r : r \geq 0 \text{ and } x_i \in S\}$.
   We will first show that H is a subgroup of G. Observe that the identity element is in H. If $x_1x_2 \dots x_r$ and $y_1y_2 \dots y_s$ are in H for some natural numbers r and s, then $x_1x_2 \dots x_ry_1y_2 \dots y_s$, being the finite product of elements x_i and y_j in S, is also in H.
   Let $y_1y_2 \dots y_s$ be an element of H and let $y_i = g_ih_ig_i^{-1}h_i^{-1}$. Then $y_i^{-1} = h_ig_ih_i^{-1}g_i^{-1}$ is in S and

   $$y_s^{-1} \dots y_2^{-1}y_1^{-1} = (h_sg_sh_s^{-1}g_s^{-1}) \dots (h_2g_2h_2^{-1}g_2^{-1})(h_1g_1h_1^{-1}g_1^{-1})$$

   in H is the inverse of $y_1y_2 \dots y_s$. This proves that H is a subgroup of G.
   If there is a subgroup K containing S, then it must contain finite product of elements of S. Thus H is contained in K.

2. $[G, G]$ is normal in G.

   Let $x = aba^{-1}b^{-1}$ be an element in S. Then, for every element g in G, the product

   $$\begin{aligned} gxg^{-1} &= gaba^{-1}b^{-1}g^{-1} \\ &= (gag^{-1})(gbg^{-1})(ga^{-1}g^{-1})(gb^{-1}g^{-1}) \\ &= (gag^{-1})(gbg^{-1})(gag^{-1})^{-1}(gbg^{-1})^{-1} \end{aligned}$$

   is also in S.
   Let $x_1x_2 \dots x_r$ be an element of $[G, G]$ and g be an element of G. We need to show that $gx_1x_2 \dots x_rg^{-1}$ is also in $[G, G]$.
   We have shown above that $s_i := gx_ig^{-1}$ is also in S. Therefore the product

   $$\begin{aligned} g(x_1x_2 \dots x_r)g^{-1} &= (gx_1g^{-1})(gx_2g^{-1}) \dots (gx_rg^{-1}) \\ &= s_1s_2 \dots s_r \end{aligned}$$

   is also in $[G, G]$. Thus the subgroup $[G, G]$ is normal in G.

3. if H is a subgroup of G containing $[G, G]$, then H is normal in G.

   Let h be an element of H and g be an element of G. We need to show that ghg^{-1} is in H.
   Let us observe that $ghg^{-1}h^{-1}$ is in $[G, G]$. Since $[G, G]$ is contained in the subgroup H, it follows that $ghg^{-1} = (ghg^{-1}h^{-1})h$ is also in H.

There are groups which do not have any proper normal subgroups and they are called simple groups.

Definition 2.4.16 (Simple Group). If a group G does not have proper normal subgroups, it is called a simple group.

Example 2.4.17. 1. Let p be a prime number and $G = \mathbb{Z}_p$. Then G is simple.

Definition 2.4.18 (Centralizer). Let G be a group and S a subset of G. The set $C_G(S) := \{g \in G : gsg^{-1} = s \text{ for all } s \in S\}$ is called the centralizer of S in G. That is an element g in G is in the centralizer of S iff g commutes with every element of S.

Let $G = \mathbb{Z}$. Since G is abelian, the centralizer $C_G(S) = G$ for any subset S of G.

Example 2.4.19. Let $G = S_3$ and $S = \{(1\ 2)\}$. The centraliser of S is $C_G(S) = \{e, (1\ 2)\}$. (Why?) If $S = \{(1\ 2\ 3)\}$, then the centraliser $C_G(S) = \langle(1\ 2\ 3)\rangle$, the cyclic subgroup generated by the permutation $(1\ 2\ 3)$.

Exercise 2.4.20. Let G be a group. Show that

1. $C_G(S) = N_G(S)$, if $S = \{x\}$, a singleton.

2. $C_G(S) = \cap_{x\in S} C_G(\{x\})$ for any subset S of G.

Proposition 2.4.21. *The centralizer of a subset S in G is a subgroup of G.*

Proof. The identity e is in $C_G(S)$. Let a, b be in $C_G(S)$.

Then $asa^{-1} = s$ implies that $a^{-1}sa = s$ for all s in S.(Why?) Further $abs(ab)^{-1} = absb^{-1}a^{-1} = a(bsb^{-1})a^{-1} = s$ for $s \in S$. Thus the centralizer of a set is a subgroup of G. □

Definition 2.4.22 (Center of a group). Let $S = G$ in Definition 2.4.18. Then the centralizer of G is called the center of G and denoted by $Z(G)$. i.e., It is the set of all elements of G that commute with all the elements of G.

Remark 2.4.23. 1. If G is abelian, then $Z(G) = G$ and conversely.

2. If H is a subgroup of $Z(G)$, then H is normal in G.

Example 2.4.24. Let $G = S(X)$ be the permutation group on a set X with at least three elements. Then the center $Z(G)$ of G is trivial.

> Let X be a set with at least three elements and let $\sigma : X \to X$ be a permutation such that $\sigma \neq Id$, the identity permutation. Then there exists an element x in X such that $y := \sigma(x) \neq x$. Now we choose an element $z \in X \setminus \{x, y\}$ and define

$$\tau(r) = \begin{cases} r & \text{if } r \notin \{x, y, z\} \\ x & \text{if } r = x \\ z & \text{if } r = y \\ y & \text{if } r = z \end{cases}.$$

Then τ is a permutation of X and

$$\tau\sigma(x) = \tau(y) = z \quad \text{and} \quad \sigma\tau(x) = \sigma(x) = y.$$

Therefore $\sigma\tau \neq \tau\sigma$. This shows that the center of $S(X)$ is trivial if X has at least three elements.

Exercise 2.4.25. Show that $Z(\mathrm{GL}(n,\mathbb{R})) = \{\lambda Id : \lambda \in \mathbb{R}^*\}$.

Let e_i's denote the standard basis of $\mathbb{R}^n$. For $1 \leq i \neq j \leq n$, let $A \in \mathrm{GL}(n,\mathbb{R})$ be the matrix such that $A(e_i) = e_j$ and $A(e_j) = e_i$ and $A(e_k) = e_k$ for $k \neq i, j$.

For a matrix B in the center of $\mathrm{GL}(n,\mathbb{R})$, let $B(e_l) = \sum_{j=1}^{n} b_{jl}e_j$ for $1 \leq l \leq n$. Then

$$\begin{aligned}
\sum_{k=1}^{n} b_{kj}e_k &= B(e_j) \\
&= B(A(e_i)) \\
&= A(B(e_i)) \\
&= A(\sum_{k=1}^{n} b_{ki}e_k) \\
&= \sum_{k=1}^{n} b_{ki}A(e_k) \\
&= b_{ii}e_j + b_{ji}e_i + \sum_{k \neq i,j} b_{ki}e_k.
\end{aligned}$$

By comparing the coefficients, we get $b_{ij} = b_{ji}$ for $r \neq k$ and $b_{ii} = b_{jj}$.

Let C be an element of $\mathrm{GL}(n,\mathbb{R})$ such that $C(e_i) = -e_j$ and $C(e_j) = e_i$ and $C(e_k) = e_k$ for $k \neq i, j$. Then proceeding as above we can show that $b_{rk} = -b_{kr}$ for $k \neq r$. Hence $B = \lambda Id$.

The same proof works to show that $Z(\mathrm{GL}(n,\mathbb{C})) = \{\lambda Id : \lambda \in \mathbb{C}^*\}$.

Remark 2.4.26. For any subset S in G, $C_G(S) \subseteq N_G(S)$. (Why?)

2.5 Homomorphisms

In this section we will study maps between groups and see how they help us in understanding the structure of groups.

Definition 2.5.1 (Homomorphism). Let G_1 and G_2 be two groups. A map $f : G_1 \to G_2$ is a homomorphism if $f(xy) = f(x)f(y)$ for all elements x and y in G_1.

Note that xy is the product in the group G_1 and $f(x)f(y)$ is the product in the group G_2.

Examples 2.5.2.

1. Let $f : \mathbb{Z} \to \mathbb{Z}$ be defined by $f(k) = 0$ for all $k \in \mathbb{Z}$. Then f is a homomorphism and it is called the trivial homomorphism.

2. Let G_1 and G_2 be two groups and $f : G_1 \to G_2$ be the map $f(g) = e$, the identity element of G_2. Then f is a homomorphism.

3. Let $l \in \mathbb{Z}$ be fixed. The map $f : \mathbb{Z} \to \mathbb{Z}$ be defined by $f(k) = kl$ for all $k \in \mathbb{Z}$ is a homomorphism.

4. Let $n \geq 2$. The map $f : \mathbb{Z} \to \mathbb{Z}_n$ defined by $f(k) = [k]$ is a homomorphism.

5. Let G be a group and x be an element of G. The map $f : \mathbb{Z} \to G$ defined by $f(k) := x^k$ is a homomorphism.

6. The map $f : (\mathbb{Z}, +) \to (\mathbb{R}^*, \cdot)$ defined by $f(x) = \begin{cases} 1 & \text{if } x \text{ is even} \\ -1 & \text{if } x \text{ is odd} \end{cases}$ is a homomorphism.

   We can directly check this out. It also follows from the earlier example. (Why?).

7. Let $f : \mathrm{GL}(n, \mathbb{R}) \to \mathbb{R}^*$ be defined by $f(A) := \det(A)$. Then f is a homomorphism.

   we have learnt in linear algebra that $\det(AB) = \det(A)\det(B)$ for two matrices of order n. This shows that $A \mapsto \det(A)$ is a homomorphism.

   Let us observe that this map is also onto.

Exercise 2.5.3. 1. Show that each of the maps

   (a) $\varphi : \mathbb{R}^* \to \mathbb{R}^*$ defined by $\varphi(x) = x^2$

   (b) $\varphi : (\mathbb{R}^+, \cdot) \to (\mathbb{R}^+, \cdot)$ defined by $\varphi(x) = \sqrt{x}$

   (c) $\varphi : (\mathbb{R}, +) \to \mathbb{R}^*$ defined by $\varphi(x) := e^x$

   (d) $\varphi : (\mathbb{R}, +) \to \mathbb{R}^*$ defined by $\varphi(x) := a^x$ for $a > 0$ fixed

   (e) $\varphi : (\mathbb{R}^+, \cdot) \to (\mathbb{R}, +)$ defined by $\varphi(x) := \log(x)$

   is a homomorphism.

2. Let $n \geq 2$. For every $\sigma \in S_n$, let $T_\sigma : \mathbb{R}^n \to \mathbb{R}^n$ be the map

$$T_\sigma(x_1, x_2, \ldots, x_n) := (x_{\sigma(1)}, x_{\sigma(2)}, \ldots, x_{\sigma(n)}).$$

   Show that the map $f : S_n \to \mathrm{GL}(n, \mathbb{R})$ defined by $f(\sigma) := T_\sigma$ is a homomorphism.

   Let $B := \{e_i : 1 \leq i \leq n\}$ be the standard basis of $\mathbb{R}^n$.

   First we will show that $\{T_\sigma(e_i) : 1 \leq i \leq n\}$, the image of the standard basis B is a basis of $\mathbb{R}^n$. Let $\sigma(i) = j_i$. Then $T_\sigma(e_i) = e_{j_i}$ and $\{T_\sigma(e_i) : e_i \in B\} = \{e_{j_i}\} = B$. Thus T_σ is non-singular for every σ in S_n.

   It is now easy to show that $T_{\sigma\tau} = T_\sigma \circ T_\tau$ and this shows that the map $f : S_n \to \mathrm{GL}(n, \mathbb{R})$ defined by $\sigma \mapsto T_\sigma$ is a homomorphism.

3. Let $k \in \mathbb{N}$ be fixed and $f : \mathbb{C}^* \to \mathbb{C}^*$ be defined by $f(z) = z^k$. Show that f is a homomorphism.

4. Show that the map $f : M(n, \mathbb{R}) \to \mathbb{R}$ defined by $f(A) := \text{Trace}(A)$ is a homomorphism.

   We have learnt in linear algebra that $A \mapsto \text{Tr}(A)$ is a linear transformation and hence a hornomorphism. This map is also onto.

5. Let $n \geq 2$. Prove that the projection map $f_i : \mathbb{R}^n \to \mathbb{R}$ defined by $f_i(x_1, x_2, \ldots, x_n) := x_i$ is a homomorphism for each $i \leq n$.

   We know that each of these f_i's are linear maps.

6. Let $a := (a_1, a_2, \ldots, a_n)$ be a fixed vector in $\mathbb{R}^n$. Let $f : \mathbb{R}^n \to \mathbb{R}$ be defined by $f(x_1, x_2, \ldots, x_n) := a_1x_1 + a_2x_2 + \cdots + a_nx_n$. Show that f is a homomorphism.

   In linear algebra we have learned that any linear map from $\mathbb{R}^n \to \mathbb{R}$ is of this form.

7. Show that the map $f : C(\mathbb{N}, \mathbb{R}) \to \mathbb{R}$ defined by $f((a_n)) := \lim_n a_n$ is a homomorphism.

   We have seen in analysis that if (a_n) and (b_n) are two convergent sequences, then the sequence $(a_n) + (b_n) = (a_n + b_n)$ is also convergent and $\lim_n a_n + \lim_n b_n = \lim_n(a_n + b_n)$. Hence this map is a homomorphism.

8. If $G := \{(a_n) \in \mathcal{F}(\mathbb{N}, \mathbb{R}) : \sum_n a_n < \infty\}$ show that $f : G \to \mathbb{R}$ defined by $f((a_n)) = \sum_n a_n$ is a homomorphism.

   This again follows from the algebra of limits of convergent series.

9. Let G be a group and g be an element of G. Prove that the map $x \mapsto gxg^{-1}$ is a homomorphism of G to itself.

10. Let $G = M(n, \mathbb{R})$. Show that $X \mapsto X + X^t$ and $X \mapsto X - X^t$ are homomorphisms of $M(n, \mathbb{R})$ to itself.

Given a homomorphism $f : G_1 \to G_2$, we have two subsets naturally associated with it, namely $\ker(f) := \{a \in G_1 : f(a) = e\}$ called the *kernel* of the homomorphism f and $\text{Im}\,(f) := \{f(a) : a \in G_1\}$ called the *image* of the homomorphism f.

In linear algebra, we have seen the kernel and image of a linear transformation and they play an important role in understanding the structure of the vector spaces and also the subspaces.

In particular if $T : V \to W$ is a linear transformation between two finite dimensional vector spaces then $\dim \text{Ker}(T) + \dim \text{Im}(T) = \dim(V)$. We will explore to see similar results for homomorphisms between groups.

Theorem 2.5.4. *Let G_1 and G_2 be two groups and $f : G_1 \to G_2$ be a homomorphism. Then the following holds.*

1. $f(e) = e$.

2. $f(x^k) = (f(x))^k$ *for all x in G_1 and every $k \in \mathbb{Z}$.*

3. *If x is in G_1 and if $|x|$ is finite, then $|f(x)|$ is finite and $|f(x)|$ divides $|x|$.*

4. $f(x^{-1}) = (f(x))^{-1}$ *for all x in G_1.*

5. $\ker(f)$ *is a normal subgroup of G_1.*

6. $\mathrm{Im}\,(f)$ *is a subgroup of G_2.*

Proof. We denote the identity elements of both the groups by e.

For the identity e of G_1, we have $f(e) = f(e.e) = f(e).f(e) = (f(e))^2$. But we know from Proposition 2.1.16 the only element x in a group G such that $x^2 = x$ is identity. Hence $f(e) = e$ and the proof of the first part of the proposition follows.

Let $k \geq 3$ and x be an element of G_1. Then, by generalized associative law, $f(x^k) = f(xx^{k-1}) = f(x)f(x^{k-1}) = f(x)f(x)^{k-1} = f(x)^k$.

If we now show that $f(x^{-1}) = f(x)^{-1}$, then $f(x^k) = f(x)^k$ for any $k \in \mathbb{Z}$ follows from the argument for $k \in \mathbb{N}$.

Let $x \in G_1$. Then $e = f(e) = f(xx^{-1}) = f(x)f(x^{-1})$. This proves that $f(x)^{-1} = f(x^{-1})$.

Let x be an element of finite order, say k. Then $f(x)^k = f(x^k) = f(e) = e$ and by Proposition 2.2.20 $|f(x)|$ divides k.

We will now show that $\ker(f)$ is a normal subgroup of G_1. Let a be in $\ker(f)$. Then $f(a^{-1}) = f(a)^{-1} = e$. That is a^{-1} is in $\ker(f)$, if a is in $\ker(f)$.

Let a and b be two elements in $\ker(f)$. Then

$$f(ab^{-1}) = f(a)f(b^{-1}) = f(a)(f(b))^{-1} = e.e = e.$$

Thus ab^{-1} is also in $\ker(f)$ and by Proposition 2.2.1, $\ker(f)$ is a subgroup of G_1.

Let x be in G_1 and a in $\ker(f)$. Then

$$f(xax^{-1}) = f(x)f(a)f(x^{-1}) = f(x)ef(x^{-1}) = f(xx^{-1}) = f(e) = e$$

and $\ker(f)$ is a normal subgroup of G_1.

We leave it as an exercise to check that $\mathrm{Im}\,(f)$ is a subgroup of G_2. □

Exercise 2.5.5. Let G be a group and $x \neq e$ be an element in G. Then there exists a unique homomorphism $f : \mathbb{Z} \to G$ such that $f(1) = x$.

Define

$$f : \mathbb{Z} \to G \quad \text{by} \quad f(k) = x^k.$$

Then it is easy to check that f is a homomorphism from $\mathbb{Z}$ to G.

Suppose there exists another homomorphism $g : \mathbb{Z} \to G$ such that $g(1) = x$. Then $g(k) = \overbrace{g(1) \cdot g(1) \cdots g(1)}^{k-\text{times}} = x^k = f(k)$. This proves the uniqueness.

Let $\ker(f) = (0)$. This means that $x^k \neq e$ for every nonzero integer k. Hence order of x is infinite.

Conversely suppose order of x is infinite. Then $x^k \neq e$ for every nonzero integer k. Hence $\ker(f) = (0)$. Thus $|x|$ is infinite iff $\ker(f) = (0)$.

We can show that $\ker(f) = d\mathbb{Z}$ for some $d \neq 0$ iff $|x| = d$.

Exercise 2.5.6. 1. Let C be a cyclic group and let $C = \langle a \rangle$. Let G be a group and $x \in G$. Show that the map $f : C \to G$ defined by $f(a^k) := x^k$ is a homomorphism. Find the $\ker(f)$ and relate it with the order of x.

2. Let C be a cyclic group and G be a group. Find all homomorphisms from $C \to G$.

3. Find the kernel and the image of the homomorphisms in Examples 2.5.2.

Let $f : G_1 \to G_2$ be a homomorphism. Let $y(\neq e) \in G_2$. Then by definition $f^{-1}(y) = \{x \in G_1 : f(x) = y\}$. Let us denote this set by S_y.

Let x_1 and x_2 be in S_y. Then $f(x_1) = y = f(x_2)$ and $f(x_1 x_2^{-1}) = e = f(x_1^{-1}x_2)$. This implies that $x_2^{-1}x_1$ is in $\ker(f)$ and we can write this as $x_1 = x_2\, h$ for some $h \in \ker(f)$. Therefore $S_y \subseteq \{x_1 x : x \in \ker(f)\}$, if x_1 is in S_y.

Conversely, if x_1 is in S_y and x is any element in $\ker(f)$, then, since $f(x_1\, x) = f(x_1)f(x) = f(x_1) = y$, it follows that $x_1\, x$ is in S_y. Thus $S_y = x(\ker(f))$ where $f(x) = y$.

Let us observe that S_y is also equal to the right coset $(\ker(f))x$.

A close observation shows the similarities between what we have seen above and the relation between the solution of system of homogeneous linear equations and a system of non-homogeneous linear equations.

Let $A = (a_{ij})$ be a $m \times n$ matrix and let $A : \mathbb{R}^n \to \mathbb{R}^m$ be defined by

$$AX = \left(\sum_{j=1}^{n} a_{1j}x_j, \sum_{j=1}^{n} a_{2j}, \ldots, \sum_{j=1}^{n} a_{mj}x_j \right)^t$$

where $X = (x_1, x_2, \ldots, x_n)^t$ is a vector in $\mathbb{R}^n$. Then A is a linear map from $\mathbb{R}^n$ to $\mathbb{R}^m$. In particular it is a homomorphism between these two groups and the kernel $\{X \in \mathbb{R}^n : AX = 0\}$ is exactly the set of solutions of the homogeneous system of linear equations

$$\sum_{j=1}^{n} a_{ij}x_j = 0 \text{ for } 1 \leq i \leq n.$$

Let Y be a vector in $\mathbb{R}^m$ and $S_Y := \{X \in \mathbb{R}^n : AX = Y\}$ be the set of solutions of the system of non-homogeneous linear equations

$$\sum_{j=1}^{n} a_{ij}x_j = y_i \text{ for } 1 \leq i \leq m.$$

If X_0 is a vector in $\mathbb{R}^n$ such that $AX_0 = Y$, then it follows from what we have seen above that $S_Y = X_0 + \mathrm{Ker}(A)$. That is if we know one particular solution of a system of non-homogeneous linear equations, then any other solution is a sum of this particular solution and a solution of the system of the homogeneous linear equations.

Theorem 2.5.7. *Let G and L be two groups and $f : G \to L$ be a homomorphism. Assume that G is finite. Then*

1. $|\ker(f)|$ *divides the order of* G *and*

2. $|\text{Im}\,(f)|$ *also divides the order of* G.

Proof. We know that $\ker(f)$ is a subgroup of G. Hence it follows from Lagrange's theorem that the order of $\ker(f)$ divides the order of G.

We have seen above that $S_y = x(\ker(f))$ if $y = f(x)$. Further, if y and z are two distinct elements in $\text{Im}\,(f)$, the sets S_y and S_z are disjoint. Therefore $G = \cup_{y \in \text{Im}\,(f)} S_y$ and

$$|G| = |\cup_{y\in \text{Im}\,(f)} S_y| = \sum_{y\in \text{Im}\,(f)} |S_y| = \sum_{y \in \text{Im}\,(f)} |\ker(f)| = |\text{Im}\,(f)||\ker(f)|.$$

Hence $|\text{Im}\,(f)|$ also divides the order of G. □

Theorem 2.5.7 has many applications. Some of the applications are given below.

Exercise 2.5.8. 1. Let p be a prime number and G be a group of order p. Let $f : G \to G'$ be a homomorphism of G in to a group G'. Show that f is either one-one or f is trivial.

Let G be a group of order p. Then $p = |G| = |\text{Im}(f)||\text{Ker}(f)|$. Therefore either $|\text{Ker}(f)| = 1$ or $|\text{Im}(f)| = 1$. This proves that f is one-one or trivial.

2. Let G be a group and $f : G \to \mathbb{Z}_p$ be a non-trivial homomorphism. Show that

   (a) $\text{Im}\,(f) = \mathbb{Z}_p$ and

   (b) p divides the order of G, if G is a finite group.

3. Let G be a finite group and $f : G \to \mathbb{Z}$ be a homomorphism. Show that f is trivial.

Since $\text{Im}\,(f)$ is a proper subgroup of $\mathbb{Z}$, it is of the form $k\mathbb{Z}$ for some $k \in \mathbb{Z}$. If $k \neq 0$, then $\text{Im}\,(f)$ will be infinite, a contradiction. Hence $k = 0$.

4. Let $n, m \geq 2$. Find all the homomorphism from $\mathbb{Z}_n \to \mathbb{Z}_m$.

The answer to this depends on the value of n and m.

Suppose n is a prime number.

Let $f : \mathbb{Z}_n \to \mathbb{Z}_m$ be a non-trivial homomorphism. We know from Theorem 2.5.7 that $|G| = |\text{Im}\,(f)||\ker(f)|$. Hence either $|\ker(f)| = 1$ and $|\text{Im}\,(f)| = p$ or $|\ker(f)| = p$ and $|\text{Im}\,(f)| = 1$.

Since we have assumed f to be non-trivial, $|\text{Im}\,(f)| \neq 1$. Therefore $|\text{Im}\,(f)| = p$ and by Lagrange's theorem $|\text{Im}\,(f)|$ divides m. Now we conclude that if n is prime and if there is a non-trivial homomorphism from $\mathbb{Z}_n \to \mathbb{Z}_m$, then m is a multiple of p.

If m is a multiple of p, indeed there are homomorphisms from $\mathbb{Z}_p \to \mathbb{Z}_m$. Suppose $m = pm_1$. Then the map $[k] \mapsto [km_1]$ is one such.

Now assume that m is a prime. If $f : \mathbb{Z}_n \to \mathbb{Z}_m$ is a nontrivial homomorphism, then $\mathrm{Im}\,(f) = \mathbb{Z}_m$ and since $|\mathrm{Im}\,(f)|$ divides n, we see that n is a multiple of m. We leave it to the readers to construct a non-trivial homomorphism!

5. Let $G = \cup_{n\geq 1} G_n$, where G_n is the group of n-th roots of unity. Let $L = \mathbb{Z}$, $\mathbb{Q}$, $\mathbb{R}$, or $\mathbb{C}$. Show that any homomorphism $f : G \to L$ is trivial.

   Let $f : G \to L$ be a homomorphism. Let a be an element of G of order k. Then order of $f(a)$ divides the order of a. In particular order of $f(a)$ is finite. However the only element of finite order in L is 0. Therefore $f(a) = 0$ for all $a \in G$ and hence f is trivial.

   Alternatively we can also prove as follows:

   Any proper subgroup of the groups mentioned in the list is infinite. Hence the result follows as in the previous example.

6. Let G be a group such that every element in G is of finite order. Show that any homomorphism $f : G \to L$ is trivial, if $L = \mathbb{Z}$, $\mathbb{Q}$, $\mathbb{R}$ etc.

   The proof is same as in previous problem.

7. Let G be a finite abelian group. Does there exist a homomorphism $f : G \to \mathbb{R}^*$ or $\mathbb{C}^*$?

8. Let $G = \cup_{n\geq 1} G_n$. Does there exist a homomorphism $f : G \to \mathbb{R}^*$, or $\mathbb{C}^*$?

9. Let G_1 and G_2 be two finite groups such that $o(G_1)$ and $o(G_2)$ are relatively prime. Show that any homomorphism $f : G_1 \to G_2$ is trivial.

10. Let G be a group and $f : \mathbb{Z} \to G$ be a homomorphism. Show that there exists an element x in G such that $f(k) = x^k$ for all integers k. Find the kernel and image of this homomorphism.

11. Find all homomorphisms $f : \mathbb{Q} \to \mathbb{Z}$.

    Let $f : \mathbb{Q} \to \mathbb{Z}$ be a homomorphism and let $f(1) = k$. Suppose $k > 0$. Then, for every natural number $n \geq 2$, we have

$$\begin{aligned} k &= f(1) \\ &= f(\overbrace{\frac{1}{n} + \frac{1}{n} + \cdots \frac{1}{n}}^{n-\text{ times}}) \\ &= \overbrace{f(\frac{1}{n}) + f(\frac{1}{n}) + \cdots f(\frac{1}{n})}^{n-\text{ times}} \\ &= nf(\frac{1}{n}). \end{aligned}$$

    This shows that every natural number n divides k, a contradiction. Therefore $k = 0$. We conclude that $f(n) = nf(1) = 0$ and $f(\frac{1}{n}) = 0$ for every integer n. Thus the only homomorphism from $\mathbb{Q} \to \mathbb{Z}$ is the trivial homomorphism.

12. Let $f : \mathbb{R}^* \to \mathbb{R}^*$ be a homomorphism. Show that

(a) $f(-1) = \pm 1$.

(b) $f(x) > 0$, if x is positive.

(c) $\mathrm{Im}(f) \subseteq \mathbb{R}_+^*$, if $f(-1) = 1$.

13. Does there exist a non-trivial homomorphism from $\mathbb{R}$ to $\mathbb{Q}$?

    Consider $\mathbb{R}$ as a vector space over $\mathbb{Q}$ and use a $\mathbb{Q}$ basis of $\mathbb{R}$ to construct a homomorphism.

Exercise 2.5.9. 1. Let $f : G \to K$ be a homomorphism between two groups G and K. Let $\mathrm{Graph}(f) := \{(x, f(x)) \in G \times K : x \in G\}$. Show that

(a) $\mathrm{Graph}(f)$ is a subgroup of $G \times K$ and

(b) $x \mapsto (x, f(x))$ is a homomorphism between G and $\mathrm{Graph}(f)$. Further this map is one-one and on-to.

2. Let $f : G \to K$ be a map between two groups. Show that f is a homomorphism iff $\mathrm{Graph}(f) := \{(x, f(x)) \in G \times K : x \in G\}$ is a subgroup of $G \times K$.

Let $f : G \to L$ be a homomorphism. We have seen in Proposition 2.5.4 that image of a subgroup under a homomorphism is a subgroup. Now let us start with a subgroup K in L. We want to know if $f^{-1}(K) := \{x \in G : f(x) \in K\}$ is a subgroup of G.

Since the identity e is in $f^{-1}(K)$, this set is non-empty. Let x and y be two elements of $f^{-1}(K)$, then $f(xy^{-1}) = f(x)f(y)^{-1}$ is in K. Therefore xy^{-1} is in $f^{-1}(K)$ and $f^{-1}(K)$ is a subgroup of G.

Suppose K is normal in L. Let x be an element of $f^{-1}(K)$. Then, since K is normal in L, $f(g)f(x)f(g)^{-1}$ is in K for every $g \in G$. But $f(g)f(x)f(g)^{-1} = f(gxg^{-1})$. This shows that gxg^{-1} is in $f^{-1}(K)$. Hence $f^{-1}(K)$ is normal in G.

We summarize this in the following proposition.

Proposition 2.5.10. *Let $f : G \to L$ be a homomorphism. Then the following holds.*

1. *If K is a subgroup of L, then $f^{-1}(K) := \{x \in G : f(x) \in K\}$ is a subgroup of G.*

2. *If K is a normal subgroup of L, then $f^{-1}(K)$ is a normal subgroup of G.*

3. $\ker(f) \subseteq f^{-1}(K)$ *for every subgroup K of L.*

4. *The restriction $f\,|_{f^{-1}(K)}: f^{-1}(K) \to K$ is a homomorphism with* $\ker(f)$ *as its kernel.*

Remark 2.5.11. If H is a normal subgroup of G. Then $f(H)$ need not be normal in G.

For example let $n \geq 4$ and $f : \mathbb{Z}_4 \to S_n$ be defined by $f([k]) = \sigma^k$ where σ is the 4-cycle $\begin{pmatrix} 1 & 2 & 3 & 4 \\ 2 & 3 & 4 & 1 \end{pmatrix}$. Then $(2\,3)(1\,2\,3\,4)(2\,3) = (1\,3\,2\,4)$ is not in the image of $\mathbb{Z}_4$.

However there is a sufficient condition on f to make $f(H)$ normal in L and we state it as an exercise.

Exercise 2.5.12. Let G and and L be two groups and $f : G \to L$ be a homomorphism. If f is on-to, then $f(H)$ is normal in L for every normal subgroup H of G.

Let G and L be two groups and $f : G \to L$ be an on-to homomorphism. Let H be a normal subgroup of G.

Let y be an element of L. Since f is onto there exists an element x in G such that $f(x) = y$. If h is an element of H, then $yf(h)y^{-1} = f(x)f(h)f(x)^{-1} = f(x)f(h)f(x^{-1}) = f(xhx^{-1})$. Since H is normal in G, it follows that xhx^{-1} is in H. As a consequence $yf(h)y^{-1}$ is in $f(H)$ for every element h in H and y in L.

Exercise 2.5.13. Let $n \geq 2$ and $f : S_n \to \mathrm{GL}(n, \mathbb{R})$ be a homomorphism defined by $f(\sigma) = A_\sigma$ where $A_\sigma : \mathbb{R}^n \to \mathbb{R}^n$ is the linear map

$$f(x_1, x_2, \ldots, x_n) := (x_{\sigma(1)}, x_{\sigma(2)}, \ldots, x_{\sigma(n)}).$$

Show that f is a one-one homomorphism and the image is not normal in $\mathrm{GL}(n, \mathbb{R})$

The image is called the set of permutation matrices.

Let us observe that the image of every element in will be a matrix with exactly one nonzero entry in each and each column and that nonzero entry is also 1.

Let $\sigma = (1\ 2)$ in S_n. The image of σ is the matrix $A_\sigma = \begin{pmatrix} A & B \\ C & I_{n-2} \end{pmatrix}$ where $A = \begin{pmatrix} 0 & 1 \\ 1 & 0 \end{pmatrix}$, B is the zero matrix of size $(n-2) \times 2$, C is the zero matrix of size $2 \times (n-2)$. Let $B = \begin{pmatrix} R_\theta & B \\ C & I_{n-2} \end{pmatrix}$ where $R_\theta = \begin{pmatrix} \cos\theta & -\sin\theta \\ \sin\theta & \cos\theta \end{pmatrix}$ B and C are as above. Then the matrix $BA_\sigma B^{-1} = \begin{pmatrix} R_{2\theta} & B \\ C & I_{n-2} \end{pmatrix}$ and this is not in the image of the group S_n, if we choose $\theta = \frac{\pi}{4}$.

2.5.1 Isomorphisms

Definition 2.5.14. A homomorphism $f : G \to L$ is called an isomorphism if f is one-one and onto. An isomorphism $f : G \to G$ is also called an automorphism.

We say that a group G is isomorphic to another group L if there exists an isomorphism from G onto L.

Remark 2.5.15. Any one-one homomorphism is an isomorphism on to its image.

Before we look at some examples we characterize one-one homomorphisms in the following proposition.

Proposition 2.5.16. *Let G and L be two groups and $f : G \to L$ be a homomorphism. Then the following are equivalent.*

1. *The homomorphism f is one-one.*

2. *For every element $x \in G$, the order of x is same as the order of $f(x)$.*

3. *The subgroup $\ker(f)$ is trivial.*

Proof. Assume that the homomorphism $f : G \to L$ is one-one.

Let x be an element of G. Then $f(x^t) = f(x)^t$ for all $t \in \mathbb{Z}$.

Let $x \neq e$ in G be an element of order k. Then it follows from Theroem 2.5.4 that the order of $f(x)$ and $|f(x)|$ divides k. Assume that the order $f(x)$ is l. Then $f(x^l) = f(x)^l = e$. Since the homomorphism f is one-one, $x^l = e$. This proves that k divides l and the proof of (1) $\Rightarrow$ (2) is complete.

Let us now assume that for every element x in G, the order of x is same as the order of $f(x)$. If x is an element in G such that $f(x) = e$, then $1 = |f(x)| = |x|$. This means that $x = e$ and $\ker(f)$ is trivial.

Now we will prove (3) $\Rightarrow$ (1). Let x and y be two elements of G such that $f(x) = f(y)$. Then $f(xy^{-1}) = f(x)f(y^{-1}) = f(x)f(y)^{-1} = e$. Hence the element xy^{-1} is in the kernel of the homomorphism f. Since $\ker(f)$ is trivial, it follows that $x = y$. Thus f is one-one. □

Now we prove the following theorem using isomorphism.

Theorem 2.5.17. *Let G be a cyclic group. Then G is isomorphic to $\mathbb{Z}$ or $\mathbb{Z}_n$ for some n.*

Proof. Let $G := \langle x \rangle$ be a cyclic group. If G is infinite, then the order of x is infinite and every element y in G is x^k for some $k \in \mathbb{Z}$. Hence there exists an onto map $\varphi : \mathbb{Z} \to G$ defined by $\varphi(k) = x^k$.

We know that the map $k \mapsto x^k$ is a homomorphism from $\mathbb{Z}$ to G. We will now show that φ is a one-one homomorphism.

If k and l are two integers such that $\varphi(k) = \varphi(l)$, then $x^k = x^l$. Therefore $x^{k-l} = e$. Since the element x is of infinite order, it follows that $k = l$. This proves that φ is one-one and G is isomorphic to $\mathbb{Z}$, if G is an infinite cyclic group.

Let us now assume that $G = \langle x \rangle$ is a finite cyclic group of order n. Then $|x| = n$. Let $\varphi : \mathbb{Z}_n \to G$ be the map defined by $\varphi([k]) = x^k$. It is easy to show that φ is a homomorphism. Let $[r]$ and $[s]$ in $\mathbb{Z}_n$ be such that $x^r = x^s$. Then $x^{r-s} = e$ and this means that $r - s = tn$ for some $t \in \mathbb{Z}$. Hence $[r] = [s]$ and the map φ is one-one. Since both the groups are of same order, one-one map is also on-to. Thus $\varphi : \mathbb{Z}_n \to G$ is an isomorphism. □

Examples 2.5.18. 1. Let $f : \mathbb{R} \to \mathbb{R}^*$ be defined by $f(x) = e^x$. Then f is an isomorphism on to $\mathbb{R}^*_+$.

2. Let $f : \mathbb{R}^* \to \mathbb{R}^*$ be the map $x \mapsto \frac{1}{x}$. Then f is an isomorphism.

3. The map $z \mapsto \bar{z}$ is an automorphism of $\mathbb{C}^*$.

4. The map $e^{i\theta} \mapsto \begin{pmatrix} \cos\theta & -\sin\theta \\ \sin\theta & \cos\theta \end{pmatrix}$ is an isomorphism between the groups $\mathbb{T}$ and $SO(2, \mathbb{R})$.

Exercise 2.5.19. 1. Show that any non-trivial homomorphism $f : \mathbb{Z} \to \mathbb{Z}$ is one-one.

2. Show that $(\mathbb{Z}, +)$ is isomorphic to its subgroup $(d\mathbb{Z}, +)$ for every natural number $d \neq 0$.

3. Let $f : \mathbb{Q} \to \mathbb{Q}$ be a homomorphism such that $f(k) = k$ for $k \in \mathbb{Z}$. Show that f is identity.

4. Show that any non-trivial homomorphism $f : \mathbb{Q} \to \mathbb{Q}$ is an isomorphism.

5. Let k be an integer and $f_k : \mathbb{R}^* \to \mathbb{R}^*$ be the map $x \mapsto x^k$. Is the map f_k an isomorphism?

6. Let G be a finite cyclic group of order n. Show that

   (a) for every divisor d of $|G|$, there exists a unique subgroup H of G of order d.

   (b) if H and K are two subgroups of G, then H is contained in K iff $|H|$ divides $|K|$.

7. Let m and n be two natural numbers greater than 1. Show that

   (a) there is a one-one homomorphism $f : \mathbb{Z}_n \to \mathbb{Z}_{mn}$ and an on-to homomorphism $\varphi : \mathbb{Z}_{mn} \to \mathbb{Z}_n$. Further find the kernel of this on-to homomorphism.

   (b) the group $\mathbb{Z}_n \times \mathbb{Z}_m$ is isomorphic to $\mathbb{Z}_{mn}$ iff $\gcd(m, n) = 1$.

8. Let $G := \left\{ A = \begin{pmatrix} 1 & x \\ 0 & 1 \end{pmatrix} : x \in \mathbb{R} \right\} \subseteq GL(2, \mathbb{R})$. Show that the map $\begin{pmatrix} 1 & x \\ 0 & 1 \end{pmatrix} \mapsto x$ is an isomorphism between G and $(\mathbb{R}, +)$.

9. Let $G := \left\{ A \in GL(2, \mathbb{R}) : A = \begin{pmatrix} a & -b \\ b & a \end{pmatrix} \right\}$. Show that $\mathbb{C}^*$ is isomorphic to the group G.

10. Let G be an abelian group of odd order and $f : G \to G$ be defined by $f(x) = x^2$. Show that f is an isomorphism.

    Since G is abelian, the map f is a homomorphism.

    Suppose $f(x) = f(y)$ for x and y in G. That is $x^2 = y^2$. This implies that $(xy^{-1})^2 = e$. Since G is of odd order, the order of xy^{-1} can not be two. This proves that $x = y$.

    One can now finish the proof using the fact that a one-one map from a finite set to itself is a bijection.

    We use number theory and prove that f is onto. Since $|G|$ is odd, the gcd of $|G|$ and 2 is 1. Hence by Bezout's identity, there exist integers

k and l such that $2k + |G|\,l = 1$. Let y be an element of G. Then $y = y^{2k+|G|\,l} = y^{2k}y^{|G|\,l} = (y^k)^2(y^{|G|})^l = (y^k)^2$. Hence $f(y^k) = y$ and the map f is onto.

11. Let G be an abelian group such that every element is of odd order. Show that the map $x \mapsto x^2$ is a one-one and onto homomorphism.

12. Let G be an abelian group and $H_k := \{x \in G : x^k = e\}$ for $k \geq 1$. Let $H := \cup_{k \geq 1} H_{2k+1}$. Show that the map $x \mapsto x^2$ is an isomorphism of the subgroup H.

13. Let G be a finite abelian group. Let $k \in \mathbb{N}$ be relatively prime to $|G|$. Show that the map $f : G \to G$ be defined by $f(x) = x^k$ is an isomorphism.

Exercise 2.5.20. Show that the groups

1. $U(8)$ and $\mathbb{Z}_2 \times \mathbb{Z}_2$ are isomorphic.

2. $U(10)$ and $\mathbb{Z}_2 \times \mathbb{Z}_2$ are isomorphic.

3. $U(7)$ and $\mathbb{Z}_6$ are isomorphic.

4. $U(p)$ and $\mathbb{Z}_{p-1}$ are isomorphic for any prime number p.

Exercise 2.5.21. 1. Let A be an element in $\mathrm{GL}(n, \mathbb{R})$. Show that the map $\varphi_A : \mathrm{SL}(n, \mathbb{R}) \to \mathrm{SL}(n, \mathbb{R})$ defined by $\varphi_A(B) := ABA^{-1}$ is a homomorphism. Is this map an isomorphism?

2. Let A be an element in $\mathrm{GL}(n, \mathbb{R})$ and $\varphi_A : \mathrm{SO}(n, \mathbb{R}) \to \mathrm{SL}(n, \mathbb{R})$ be the map $\varphi_A(B) := ABA^{-1}$. Is this map a homomorphism? Is it one-one? Is it onto? (What happens when $n = 2$?)

Exercise 2.5.22. 1. Show that the groups

(a) $(\mathbb{Q}, +)$ and $(\mathbb{R}, +)$ are not isomorphic.

If r is a rational number, then $f(r)$ is also a rational number!

(b) $\mathbb{Q}^*$ and $\mathbb{R}^*$ are not isomorphic.

(c) $\mathbb{R}^*$ and $\mathbb{C}^*$ are not isomorphic.

Suppose there is an isomorphism $f : \mathbb{R}^* \to \mathbb{C}^*$. Let $x \in \mathbb{R}^*$ be such that $f(x) = i$. Then $f(x)^4 = 1 = f(1)$. Hence $x^4 = 1$ and this means $x = \pm 1$. Since f is an isomorphism and $f(1) = 1$, it follows that $f(-1) = i$. But $-1 = i^2 = f(-1)^2 = f((-1)^2) = f(1)$. This is a contradiction.

(d) $(\mathbb{Q}, +)$ and $(\mathbb{Q}^+ := (0, \infty) \cap \mathbb{Q}, \cdot)$ are not isomorphic.

Let $x\mathbb{Q}$. Then $f(x) = f(x/2 + x/2) = (f(x/2))^2$. Therefore, if $x \in \mathbb{Q}$ is such that $f(x) = 2$, we will have $(f(x/2))^2 = 2$.

(e) $U(8)$ and $\mathbb{Z}_4$ are not isomorphic.

(f) $U(10)$ and $\mathbb{Z}_4$ are not isomorphic.

2. Are the groups $U(8)$ and $U(10)$ isomorphic?

Exercise 2.5.23. 1. Show that

   (a) $\mathrm{SO}(2,\mathbb{R})$ is isomorphic to S^1.

   (b) $\mathrm{SO}(2,\mathbb{R})$ is isomorphic to the group $[0,1)$ in Exercise 2.2.19(5).

   (c) $\mathbb{T}$ is isomorphic to the group $[0,1)$ in Exercise 2.2.19(5).

2. Let $G = [0,1) \cap \mathbb{Q}$ be the subgroup of $[0,1)$ defined as in Exercise 2.2.19(5). Show that G is isomorphic to $\cup_n G_n$, the union of all the roots of unity.

3. Show that the groups $\mathbb{Q}/\mathbb{Z}$ and $G = \cup_{n\geq 1} G_n$ are isomorphic.

Exercise 2.5.24. Let X and Y be two sets. If there is a bijection between the two sets X and Y, then the two permutation groups $S(X)$ and $S(Y)$ are isomorphic.

> Let $\theta : X \to Y$ be a bijection. It is an easy exercise to verify that $\theta \circ \sigma \circ \theta^{-1} : Y \to Y$ is a bijection. Hence the map $\varphi : S(X) \to S(Y)$ defined by $\varphi(\sigma) := \theta \circ \sigma \circ \theta^{-1}$ is an isomorphism.

Now we would like to know if there are non-trivial isomorphism from a group to itself.

Exercise 2.5.25. Let G be a cyclic group and $f : G \to G$ be an automorphism. Show that if x is a generator of G, then $f(x)$ is a generator of G and conversely.

Exercise 2.5.26. 1. Let $G = \mathbb{Z}$. Show that $\mathrm{Aut}(G) = \{\pm Id\}$ where $Id : G \to G$ is the identity map of $\mathbb{Z}$.

2. Let G be an infinite cyclic group. Show that $\mathrm{Aut}(G) = \{\pm Id\}$ where $Id : G \to G$ is the identity map of G.

3. Let $G = G_n$ where G_n is as in Example 2.1.8(18). Find $\mathrm{Aut}(G)$.

4. Let G be a finite cyclic group of order n. Show that $\mathrm{Aut}(G)$ is isomorphic to the group $U_n := \{k \in \mathbb{N} : k < n \text{ and } \gcd(k,n) = 1\}$ where n is the order of the group G.

Exercise 2.5.27. Let $G = \cup_n G_n$ be as in Example 2.1.8(19). Then the map $z \mapsto \overline{z} = z^{-1}$ is a non-trivial automorphism of G.

Let G be a group and $g \in G$ be fixed. Then we have seen that the maps $x \mapsto gxg^{-1}$ is a homomorphism. If the group G is abelian, then this map is identity.

Now let us assume G is non-abelian. Then the center $Z(G)$ is a proper subgroup of G. Therefore, if we choose an element g not in the center of G, this map $x \mapsto gxg^{-1}$ is a one-one and onto homomorphism. Thus if G is non-abelian, we have a non-trivial automorphism of G to itself.

In general we have the following proposition.

Proposition 2.5.28. *Let G be a non-trivial group of odd order. Then it has a non-trivial automorphism.*

Proof. If G is non-abelian, then we have seen above that for $g \notin Z(G)$, the map $\varphi_g : G \to G$ defined by $\varphi_g(x) := gxg^{-1}$ is an automorphism of G. Let us observe that here we don't need the group to be of odd order.

Now assume that G is abelian. Since order of G is odd, the group G does not have elements of order 2. That is $x^2 \neq e$ for all $x \neq e$ in G. Then the map $y \mapsto y^{-1}$ is a nontrivial automorphism of G. □

Remark 2.5.29. If G is abelian and every element is of order 2, still there are non-trivial automorphisms of G. The proof is beyond the scope of this book.

Let G be a group and for every element g in G, let $\varphi_g : G \to G$ be the map $x \mapsto gxg^{-1}$. Then φ_g is an automorphism of G. The automorphism φ_g is called *inner automorphism* of G. Let $I(G)$ denote the set of all inner automorphisms of G. That is $I(G) := \{\varphi_g : g \in G\}$. It is an easy exercise to show that

1. $I(G)$ is a subgroup of $\mathrm{Aut}(G)$, the set of all automorphisms of G.
2. the map $\varphi(g) := \varphi_g$ is a homomorphism from $G \to I(G)$ with kernel $Z(G)$.

Let $\mathrm{Hom}(G, L) := \{f : G \to L : f \text{ is a homomorphism}\}$ be the set of all homomorphism between two groups G and L. For f and g in $\mathrm{Hom}(G, L)$, we define a map $f \cdot g : G \to L$ by $(f \cdot g)(x) := f(x)g(x)$ for all x in G. Is this map a homomorphism?

Let x and y be two elements of G.

Then $(f \cdot g)(xy) = f(xy)g(xy) = f(x)f(y)g(x)g(y)$. For this map to be a homomorphism, we should have, $(f \cdot g)(x)(f \cdot g)(y) = (f \cdot g)(xy) = f(xy)g(xy)$. This is equivalent to $f(x)g(x)f(y)g(y) = f(x)f(y)g(x)g(y)$. This means that $f(y)g(x) = g(x)f(y)$ for all x and y in G. A reasonable assumption to make for this to hold is that L is abelian.

Proposition 2.5.30. *Let G be a group and L be an abelian group. Then $\mathrm{Hom}(G, L)$ is an (abelian) group. In particular $\mathrm{Hom}(G, \mathbb{Z})$ is an abelian group.*

Proof. The discussions above show that the binary operation is well defined.

Let f, g and h be three elements in $\mathrm{Hom}(G, L)$. Then $(f \cdot (g \cdot h))(x) = f(x)(g(x)h(x)) = (f(x)g(x))h(x) = ((f \cdot g)h)(x)$ for all $x \in G$.

Let $\theta : G \to L$ be defined by $\theta(x) = e$ for all $x \in G$. Then $f \cdot \theta = \theta \cdot f = f$ for all $f \in \mathrm{Hom}(G, L)$.

Let f be an element in $\mathrm{Hom}(G, L)$ and $g : G \to L$ be such that $f \cdot g = \theta$. Then $(f \cdot g)(x) = f(x)g(x) = e$ for all $x \in G$. This implies that $g(x) = f(x)^{-1}$ for all $x \in G$. If we show that the map $g : G \to L$ is a homomorphism, then we are through. Let x and y be two elements of G. Then $g(xy) = f(xy)^{-1} = f(y)^{-1}f(x)^{-1}$. Since L is abelian, it follows that $f(y)^{-1}f(x)^{-1} = f(x)^{-1}f(y)^{-1} = g(x)g(y)$. Thus g is a homomorphism from G to L and the proof is complete. □

Exercise 2.5.31.

1. Let G be an abelian group. Then $\mathrm{Hom}(G, G)$ is a group and $\mathrm{Aut}(G)$ is a subgroup of $\mathrm{Hom}(G, G)$.
2. Let G be a group. Then $I(G)$, the set of inner automorphisms, is trivial iff G is abelian.

3. Let G be a group and H be a normal subgroup of G. Let g be an element in G. Show that the conjugation map $\varphi_g : H \to H$ defined by $\varphi_g(h) = ghg^{-1}$ is an automorphism of H. Further show that the map $\varphi : G \to \mathrm{Aut}(H)$ defined by $\varphi(g) := \varphi_g$ is a homomorphism.

Let G be a group and $S(G)$ be the group of permutations of elements of G. That is $S(G)$ is the set of all bijections of G. If g is an element of G, then the map $x \mapsto gx$ is a bijection of G and we denote this map by φ_g. Thus we get a map $\varphi : G \to S(G)$ defined by $\varphi(g) := \varphi_g$. If g and h are two elements of the group G, then $\varphi_{gh}(x) = (gh)x = g(hx) = \varphi_g \circ \varphi_h(x)$ for all elements x in G. This shows that the map $\varphi : G \to S(G)$ is a homomorphism.

If an element g in G is such that $gx = \varphi_g(x) = x$ for all x in G, then $g = e$. Hence the map φ is one-one and we may think of G as a subgroup of $S(G)$, the permutation group on G.

In particular G is a subgroup of S_n, if G is a finite group of order n.

We summarize this discussion as a theorem known as Cayley's Theorem.

Theorem 2.5.32 (Cayley's Theorem). *Every group G is a subgroup of a symmetric group.*

2.6 Product Groups

Let G_1 and G_2 be two groups. The Cartesian product $G := G_1 \times G_2$, of G_1 and G_2, becomes a group under the binary operation

$$(a_1, a_2) \cdot (b_1, b_2) := (a_1 a_2, b_1 b_2).$$

- The identity element of this binary operation is (e_1, e_2) where e_i is the identity element of the group G_i.
- If (a_1, a_2) is in G, then (a_1^{-1}, a_2^{-1}) is the inverse of (a_1, a_2).
- The associativity of this binary operation can be very easily verified.

Definition 2.6.1. The group $G := G_1 \times G_2$ together with the binary operation defined above is called the product group. This product is also called direct product.

Examples 2.6.2. 1. Let $G_1 = G_2 = \mathbb{R}$. Then $G = \mathbb{R}^2$ and the binary operation is the usual addition on $\mathbb{R}^2$.

Remark 2.6.3. We would like to point out that there could be other method to define the structure of a group on the set G. As an example take $G_1 = \mathrm{GL}(n, \mathbb{R})$ and $G_2 = \mathbb{R}^n$. The binary operation is the one defined as in Example 2.1.8(9). Though the underlying set is $\mathrm{GL}(n, \mathbb{R}) \times \mathbb{R}^n$, the binary operation is not product operation. It is to distinguish this difference, we have denoted that group as $\mathrm{GL}(n, \mathbb{R}) \ltimes \mathbb{R}^n$ in Example 2.1.8(9).

Exercise 2.6.4. 1. Let G_1 and G_2 be two groups. Then the product group G is abelian iff the groups G_1 and G_2 are abelian.

2. If G_1 and G_2 are two groups then $G_1 \times \{e\}$ and $\{e\} \times G_2$ are normal in the product group $G_1 \times G_2$.

3. Let G be a group and $\Delta(G) := \{(g, g) : g \in G\}$ be the diagonal subgroup of $G \times G$. Show that $\Delta(G)$ is normal in $G \times G$ iff G is abelian.

   If G is not abelian, there exist elements x and g in G such that $xg \neq gx$. Then $(e, x)(g, g)(e, x^{-1}) = (g, xgx^{-1}) \neq (g, g)$.

Given two groups G_1 and G_2, there are natural homomorphisms between the groups G_1, G_2 and $G := G_1 \times G_2$.

1. The map $x \mapsto (x, e_2)$ from $G_1 \to G_1 \times G_2$ is a one-one homomorphism. This homomorphism is also called the canonical/natural inclusion of G_1 in to $G_1 \times G_2$.

   Similarly the map $y \mapsto (e_1, y)$ from $G_2 \to G_1 \times G_2$ is a one-one homomorphism. This homomorphism is also called the canonical/natural inclusion of G_2 in to $G_1 \times G_2$.

2. The maps $\pi_1 : G \to G_1$ defined by $(x, y) \mapsto x$ and $\pi_2 : G \to G_2$ defined by $(x, y) \mapsto y$ are onto homomorphisms and they are called the natural projection maps.

3. If $G_1 = G_2$, then we have another natural one-one map from $G_1 \to G_1 \times G_1$ defined by $x \mapsto (x, x)$ for all $x \in G_1$.

There is an important property about the product groups called the universal mapping property.

Proposition 2.6.5 (Universal Mapping property of Product Groups). *Let G_1, G_2 and K be groups.*

1. *If $\varphi : K \to G_1 \times G_2$ is a homomorphism, the maps $\varphi_i : K \to G_i$ defined by*

$$\varphi_i(x) := \pi_i(\varphi(x))$$

   are homomorphisms such that $\ker(\varphi) = \ker(\varphi_1) \cap \ker(\varphi_2)$.

2. *Conversely given homomorphisms $\varphi_i : K \to G_i$, there exists a unique homomorphism $\varphi : K \to G_1 \times G_2$ such that* $\ker(\varphi) = \ker(\varphi_1) \cap \ker(\varphi_2)$.

Proof. Since the maps π_i's and φ are homomorphisms, we conclude that φ_i's are homomorphisms.

First we observe that $\varphi(x) = (\varphi_1(x), \varphi_2(x))$ for all $x \in K$. Now an element x is in $\ker(\varphi)$ iff $\varphi(x) = (\varphi_1(x), \varphi_2(x)) = (e_1, e_2)$. This is true iff the element x is in $\ker(\varphi_i)$. Thus $\ker(\varphi) = \ker(\varphi_1) \cap \ker(\varphi_2)$ and completes the proof of the first part.

Conversely given $\varphi_i : K \to G_i$, we define $\varphi : K \to G_1 \times G_2$ by $\varphi(x) = (\varphi_1(x), \varphi_2(x))$ for all x in K. Now the same arguments as above will show that φ is a homomorphism and $\ker(\varphi) = \ker(\varphi_1) \cap \ker(\varphi_2)$. □

It is desirable to write a (finite) group G as a product of two groups G_1 and G_2. For, the groups G_1 and G_2 will be smaller than G and it will be relatively simpler to study the structure of the groups G_i's. Let us look at the following example.

Example 2.6.6. Let $n, m \geq 2$ and $\gcd(n, m) = 1$. Let $G_1 = \mathbb{Z}_n$, $G_2 = \mathbb{Z}_m$ and $K = \mathbb{Z}_{nm}$. Since n and m are relatively prime, there exist integers u and v such that $nu + mv = 1$. If x is an integer, then $x = xnu + xmv$. Therefore $x \equiv xmv \mod n$ and $x \equiv xnu \mod m$. Now we can write $[x]$ in Z_{nm} as $[x] = [xnu + xmv] = [xnu] + [xmv]$. This suggests that we define

$$\varphi_1 : \mathbb{Z}_{nm} \to \mathbb{Z}_n \text{ by } \varphi_1([x]) = [xmv]$$

and

$$\varphi_2 : Z_{nm} \to \mathbb{Z}_m \text{ by } \varphi_2([x]) = [xnu].$$

It can easily be verified that these maps are homomorphisms. Then by universal mapping property of product groups, we have a unique homomorphism

$$\varphi : \mathbb{Z}_{nm} \to \mathbb{Z}_n \times \mathbb{Z}_m \text{defined by} [x] \mapsto ([xmv], [xnu]).$$

Suppose n divides xmv and m divides xnu. Since the pairs (n, m), (n, v), (m, u) and (u, v) are relatively prime it follows that both the integers n and m divide x. Therefore x is a multiple of nm. This proves that $[x]$ is in $\text{Ker}(\varphi) = \text{Ker}(\varphi_1) \cap \text{Ker}(\varphi_2)$. Hence φ is one-one and the map is an isomorphism. Thus we have written $\mathbb{Z}_{nm}$ as product of smaller groups $\mathbb{Z}_n$ and $\mathbb{Z}_m$.

Alternatively we can also do as follows: There is a natural homomorphism

$$\mathbb{Z}_n \times \mathbb{Z}_m \to \mathbb{Z}_{nm} \text{defined by } ([x], [y]) \mapsto [xm + yn].$$

Now arguments similar to the ones above will show that this map is a one-one homomorphism.

Remark 2.6.7. This can't be done if n and m are not relatively prime.

For instance let $n = m = 2$. Then $\mathbb{Z}_4$ is not isomorphic to $\mathbb{Z}_2 \times \mathbb{Z}_2$.

In general we have the following theorem.

Theorem 2.6.8. *Let H and K be two subgroups of a group G. Then the following holds.*

1. *If either H or K is normal in G, then $HK = KH$ and it is a subgroup of G.*

2. *If $H \cap K = \{e\}$, then the natural map $H \times K \to G$ defined by $(h, k) \mapsto hk$ is one-one.*

3. *If H and K are normal in G and $H \cap K = \{e\}$, then the natural map $(h, k) \mapsto hk$ is a one-one homomorphism.*

First we discuss the following example and then prove the theorem.

Example 2.6.9. Let $G = S_3$, $\sigma := \begin{pmatrix} 1 & 2 & 3 \\ 2 & 3 & 1 \end{pmatrix}$ and $\tau = \begin{pmatrix} 1 & 2 \\ 2 & 1 \end{pmatrix}$. Let $H = \langle \sigma \rangle$ and $K = \langle \tau \rangle$, the cyclic subgroups generated by σ and τ respectively.

Then H is normal in G and K is not. Since $|H| = 3$ and $|K| = 2$, the intersection $H \cap K$ is trivial. Therefore the map $(h, k) \mapsto hk$ is one-one. This shows that $HK = G$. For similar reasons $KH = G$.

However this map is not a homomorphism. We will prove this by contradiction. Let us assume that this map is a homomorphism. We write

$$\begin{aligned} Id &= \sigma\sigma^{-1} = \begin{pmatrix} 1 & 2 & 3 \\ 2 & 3 & 1 \end{pmatrix} \begin{pmatrix} 1 & 2 & 3 \\ 3 & 1 & 2 \end{pmatrix} \quad \text{and} \\ Id &= \tau\tau = \begin{pmatrix} 1 & 2 & 3 \\ 2 & 1 & 3 \end{pmatrix} \begin{pmatrix} 1 & 2 & 3 \\ 2 & 1 & 3 \end{pmatrix}. \end{aligned}$$

Since we are assuming this map to be a homomorphism, we should have $Id = \varphi(Id, Id) = \varphi(\sigma\sigma^2, \tau\tau) = \varphi(\sigma, \tau)\varphi(\sigma^2, \tau) = (\sigma\tau)(\sigma^2\tau)$. We will now show that this is not true. Since H is normal in G, the element $\tau\sigma^2\tau$ is in H and we can check that $\tau\sigma^2\tau = \sigma$. Therefore $\sigma(\tau\sigma^2\tau) = \sigma^2 \neq Id$ and φ is not a homomorphism.

Let us also observe that

$$\begin{aligned} \sigma \circ \tau : &= \begin{pmatrix} 1 & 2 & 3 \\ 2 & 3 & 1 \end{pmatrix} \circ \begin{pmatrix} 1 & 2 & 3 \\ 2 & 1 & 3 \end{pmatrix} = \begin{pmatrix} 1 & 2 & 3 \\ 3 & 2 & 1 \end{pmatrix} \quad \text{and} \\ \tau \circ \sigma &:= \begin{pmatrix} 1 & 2 & 3 \\ 2 & 1 & 3 \end{pmatrix} \circ \begin{pmatrix} 1 & 2 & 3 \\ 2 & 3 & 1 \end{pmatrix} = \begin{pmatrix} 1 & 2 & 3 \\ 1 & 3 & 2 \end{pmatrix} \end{aligned}$$

and $\sigma \circ \tau \neq \tau \circ \sigma$.

If $\tau\sigma = \sigma\tau$, then we would have

$$(\sigma\tau)(\sigma^2\tau) = (\tau\sigma)(\sigma^2\tau) = \tau(\sigma^3\tau) = \tau^2 = Id$$

and the map φ would be a homomorphism.

We will now prove the theorem.

Proof. Assume that the group H is normal in G. If h is in H and k is in K, then $hk = k(k^{-1}hk) = kh'$ for some $h' \in H$. This proves that $HK \subseteq KH$. Now starting with kh we show that $kh = (khk^{-1})k = h''k$ is in HK. Thus $HK = KH$.

We will now show that HK is a subgroup of G. Let h_1k_1 and h_2k_2 be two elements of HK. Then $(h_1k_1)(h_2k_2)^{-1} = h_1(k_1k_2^{-1}h_2)$. Since $KH = HK$, we can write $k_1k_2^{-1}h_2 = h_3k_3$ and $(h_1k_1)(h_2k_2)^{-1} = h_1(k_1k_2^{-1}h_2) = h_1h_3k_3$ is also in HK. Hence HK is a subgroup of G and proves the first part of the proposition.

Assume that $H \cap K$ is trivial and $h_1k_1 = h_2k_2$ for h_1, $h_2 \in, H$ and k_1 and k_2 in K. Then $h_2^{-1}h_1 = k_2k_1^{-1}$ is in $H \cap K$. This means that $h_1 = h_2$ and $k_1 = k_2$. Hence the map $(h, k) \mapsto hk$ is one-one.

We will now assume that both the subgroups H and K are normal in G and $H \cap K$ is trivial.

We will first show that $hk = kh$ for all $h \in H$ and $k \in K$. Using the normality of both the groups, we conclude that $hk(kh)^{-1} = h(kh^{-1}k^{-1})$ is in H and $hk(kh)^{-1} = (hkh^{-1})k^{-1}$ is also in K. Since $H \cap K = \{e\}$ we conclude that $hk = kh$.

The map $(h, k) \mapsto hk$ is a homomorphism is easy to verify now.

$$\begin{aligned} \varphi(h_1h_2, k_1k_2) &= h_1h_2k_1k_2 \\ &= (h_1k_1)(h_2k_2) \\ &= \varphi(h_1, k_1)\varphi(h_2, k_2) \end{aligned}$$

for all pairs (h_1, k_1) and (h_2, k_2) in $H \times K$. □

Let $m \geq 2$ and $H_1, H_2, \ldots, H_m$ be normal subgroups of G. Then by induction it follows that the set $H_1H_2 \cdots H_m := \{h_1h_2 \cdots h_m : h_i \in H\}$ is a subgroup of G. Let us denote this subgroup $H_1H_2 \cdots H_m$ of G by H. As in previous Theorem 2.6.8, we have the natural map

$$H_1 \times H_2 \times H_m \to H \quad \text{defined by} \quad (h_1, h_2, \ldots, h_m) \mapsto h_1h_2 \cdots h_m.$$

However this map need not be one-one even if $H_i \cap H_j = \{e\}$ for $i \neq j$. We will see two examples below.

Let $G = \mathbb{R}^2$ and $H_1 := \{(x, 0) : x \in \mathbb{R}\}$, $H_2 := \{(0, y) : y \in \mathbb{R}\}$ and $H_3 := \{x(1, m) : x \in \mathbb{R}\}$ for some $m \neq 0$ in $\mathbb{R}$. Then $H_i \cap H_j = \{(0, 0)\}$ for $i \neq j$. However the map defined above is not one-one. We can write the element $(1, m) := (1, 0) + (0, m)$.

We will give an example in finite groups also.

Let $G = \mathbb{Z}_2 \times \mathbb{Z}_2$. Let $H_1 := \{(0, 0), (1, 0)\}$, $H_2 := \{(0, 0), (0, 1)\}$ and $H_3 := \{(0, 0), (1, 1)\}$. Here again $H_i \cap H_j = \{e\}$. Since we can write $(1, 1) = (1, 0) + (0, 1)$ the map is not one-one.

This maps is one-one if we assume that $H_i \cap (\prod_{j \neq i} H_j) = \{e\}$ for $1 \leq i \leq m$. Let $h_i, k_i \in H_i$ be such that $h_1h_2 \cdots h_m = k_1k_2 \cdots k_m$. This means that $k_1^{-1}h_1 = (k_2k_3 \cdots k_m)(h_2h_3 \cdots h_m)^{-1}$. The element $k_1^{-1}h_1$ is in H_1 and $(k_2k_3 \cdots k_m)(h_2h_3 \cdots h_m)^{-1}$ is in the subgroup $H_2H_3 \cdots H_m$. By assumption $H_1 \cap (H_2H_3 \cdots H_m) = \{e\}$. Therefore $k_1 = h_1$ and $h_2h_3 \cdots h_m = k_2k_3 \cdots k_m$. Arguing as earlier, we conclude that $h_i = k_i$ for all i.

Let $h_1h_2 \cdots h_m$ be an element of the subgroup H and g be an element of G. Then $g(h_1h_2 \cdots h_m)g^{-1} = (gh_1g^{-1})(gh_2g^{-1}) \cdots (gh_mg^{-1})$ is again in H. Thus H is a normal subgroup of G.

We claim that H is also abelian. We fix $1 \leq i \neq j \leq m$. Let x be an element of H_i and y be two element of H_j. Then we write $xyx^{-1}y^{-1}$ as $xyx^{-1}y^{-1} = (xyx^{-1})y^{-1}$. Since H_j is normal in G, it follows that xyx^{-1} is in H_j. As a consequence the product $(xyx^{-1})y^{-1}$ is also in H_j. On the other hand by writing the product $xyx^{-1}y^{-1}$ as $xyx^{-1}y^{-1} = x(yx^{-1}y^{-1})$, we conclude that it is also in H_i. By assumption $H_i \cap H_j = \{e\}$. Therefore $xy = yx$ and this proves that H is an abelian subgroup of G.

We can now use this fact to show that $(h_1, h_2, \ldots, h_m) \mapsto h_1h_2 \cdots h_m$ is a homomorphism between the groups $H_1 \times H_2 \times \ldots \times H_m$ and H.

We summarize this discussion as the following theorem.

Theorem 2.6.10. *Let $m \geq 2$ and $H_1, H_2, \ldots, H_m$ be subgroups of a group G.*

1. *If each H_i is a normal subgroup of G for $1 \leq i \leq m$, then the product $H := \prod_{i=1}^m H_i$ is a subgroup of G.*

2. *Assume that $H_i \cap \left(\prod_{j\neq i} H_j\right) = \{e\}$ for $1 \leq i \leq m$. Then the natural map $\varphi : H_1 \times H_2 \times \cdots \times H_m \to G$ defined by $\varphi(h_1, h_2, \ldots, h_m) = h_1h_2\cdots h_m$ is a one-one homomorphism.*

   Further if the map φ is on-to, the group G is isomorphic to the group $H_1 \times H_2 \times \cdots \times H_m$.

Corollary 2.6.11. *Let $n = p_1^{r_1}p_2\cdots p_k^{r_k}$ where $p_i \neq p_j$ are distinct primes. Then $\mathbb{Z}_n$ is isomorphic to $\prod_{i=1}^k \mathbb{Z}_{p_i^{r_i}}$.*

Proof. Since the group $\mathbb{Z}_n$ is cyclic, for every divisor d of n there is a unique cyclic subgroup of order d.

If $n = p_1^{r_1}p_2\cdots p_k^{r_k}$, then there exist unique cyclic subgroups H_j of order $p_j^{r_j}$ for $1 \leq j \leq k$. Each of these subgroup H_j is isomorphic to $\mathbb{Z}_{p_j^{r_j}}$. Now we apply the theorem to conclude that $\mathbb{Z}_n$ is isomorphic to $\prod_{j=1}^k \mathbb{Z}_{p_j^{r_j}}$. □

Exercise 2.6.12. If $n = kl$ show that the image of the homomorphism $\mathbb{Z}_k \times \mathbb{Z}_l \to \mathbb{Z}_n$ is the subgroup generated by $([m],[m])$ where m is the lcm of k and l and the kernel is the subgroup generated by $[d]$ where $d = \gcd(k,l)$ is the gcd of the integers k and l

2.7 Quotient Groups

Let H be a subgroup of a group G. Let x and y be two elements of H. Then for any element h in H, the product $(xy).h = (x.e)(y.h)$ belongs to the product of the two cosets $(xH)(yH) := \{(xh)(yh') : h, h' \in H\}$. This shows that the coset $xyH \subseteq (xH)(yH)$ for any two elements x and y in G.

Now we want to know when is the product of two cosets $(xH)(yH)$ is equal to the coset xyH.

Suppose $(xH)(yH) = xyH$ for any two elements x and y in G. Then $xy = xh_1yh_2$ for some h_1 and h_2 in H. This means that $y = h_1yh_2$. Hence $yh_2y^{-1} = h_1^{-1}$ must be in H for all y in G. Therefore we need H to be normal in G.

Let $xH = x_1H$ and $yH = y_1H$. Then we can write $x_1 = xh_1$ and $y_1 = yh_2$ for some h_1 and h_2 in H. The left coset $xyH = x_1y_1H$ iff $(xy)^{-1}x_1y_1$ is in H. That is $y^{-1}x^{-1}xh_1yh_2 = y^{-1}h_1yh_2$ is in H. Since h_2 is in H, we see that $y^{-1}h_1y$ mus be in H. This means that H must be normal in G.

Thus, if a subgroup H is normal in G, the product of two left cosets is again a left coset. Hence the map

$$(xH, yH) \mapsto xyH$$

is a binary operation on G/H, the set of all left cosets of H in G.

We will now show that this binary operation is a group structure on G/H. We denote the left cosets xH by $\overline{x}$.

1. If $\overline{x}$, $\overline{y}$ and $\overline{z}$ are in G/H, then it follows from the associative law of G that $\overline{x}(\overline{y}\overline{z}) = (\overline{x}\overline{y})\overline{z}$.

   $\overline{x}(\overline{y}\overline{z}) = xH(yHzH) = xH(yz)H = x(yz)H = (xy)zH = (xyH)zH = (xHyH)zH = (\overline{x}\overline{y})\overline{z}$.

2. If $\overline{x}$ is in G/H, then $\overline{x}\overline{e} = (xH)(eH) = xH = \overline{x}$.

3. If $\overline{x}$ is in G/H, then $\overline{x^{-1}}$ is the inverse of $\overline{x}$.

This proves that G/H is a group under this binary operation.

This group G/H is called the *quotient group* of G by the subgroup H.

Exercise 2.7.1. Let G be a group and H be a normal subgroup of G. Show that the map $\pi : G \to G/H$ defined by $\pi(g) := gH$ is a homomorphism. This map π is called the quotient map.

The following theorem gives us a method to identify the quotient groups.

Theorem 2.7.2 (First Fundamental Theorem of Homomorphism). *Let G and L be two groups and $f : G \to L$ be a homomorphism. Then there exists a unique isomorphism $\widetilde{f} : G/\ker(f) \to \mathrm{Im}\,(f)$ such that $\widetilde{f} \circ \pi = f$.*

Remark 2.7.3. If f is onto, then $G/\ker(f)$ is isomorphic to L.

Proof. Let $f : G \to L$ be a homomorphism and $H = \mathrm{Ker}(f)$. We define $\widetilde{f} : G/\mathrm{Ker}(f) \to \mathrm{Im}\,(f)$ by $\widetilde{f}(aH) = f(a)$.

If $aH = bH$, then $a^{-1}b \in H$. Hence $f(a^{-1}b) = e$ and $f(a) = f(b)$. Thus $\widetilde{f}$ is well defined and also one-one. By definition $\widetilde{f}$ is onto its image $\mathrm{Im}(f)$. □

We apply this result to identify quotient groups.

Applications 2.7.4.

1. Let $n \geq 2$. Then the group $\mathbb{Z}/n\mathbb{Z}$ is isomorphic to $\mathbb{Z}_n$.

   Let $f : \mathbb{Z} \to \mathbb{Z}_n$ be the map defined by $f(k) := [k]$. We have already shown that f is a homomorphism by proving that $[k] + [l] = [k + l]$ for any two integers k and l.

   If $[k]$ is an element in $\mathbb{Z}_n$, then $f(k) = [k]$. Thus this map is on-to.

   An integer k is in the kernel iff $[k] = [0]$. That is k is a multiple of n. As a consequence $\ker(f) = n\mathbb{Z}$ and by fundamental theorem of homomorphism the quotient group $\mathbb{Z}/n\mathbb{Z}$ is isomorphic to $\mathbb{Z}_n$.

2. Let $f : \mathbb{R} \to \mathbb{C}^*$ be defined by $f(t) := e^{2\pi i t}$. Then $\mathrm{Ker}(f) = \mathbb{Z}$ and $\mathrm{Im}(f) = S^1$. Hence by first fundamental theorem of homomorphism, the quotient group $\mathbb{R}/\mathbb{Z}$ is isomorphic to S^1.

3. Let $G = \mathrm{GL}(n, \mathbb{R})$ and $H = \mathrm{SL}(n, \mathbb{R})$. Then G/H is isomorphic to $\mathbb{R}^*$.

   Let $\mathrm{GL}(n, \mathbb{R}) \to \mathbb{R}^*$ be defined by $f(A) := \det(A)$. Then f is a homomorphism and $\mathrm{Ker}(f) = \mathrm{SL}(n, \mathbb{R})$ and $\mathrm{Im}(f) = \mathbb{R}^*$. Therefore $\mathrm{GL}(n, \mathbb{R})/\mathrm{SL}(n, \mathbb{R})$ is isomorphic to $\mathbb{R}^*$.

4. Let $G = \mathbb{C}^*$ and $H = \mathbb{T}$. Then G/H is isomorphic to $\mathbb{R}^*_+$, the group of positive real numbers.

5. Let $G = \mathbb{C}^*$ and $H = \mathbb{R}^*_+$. Then G/H is isomorphic to $\mathbb{T}$.

6. Let $G = \mathbb{C}^*$ and $G_k = \{z \in \mathbb{C}^* : z^k = 1\}$. Then $\mathbb{C}^*/G_k$ is isomorphic to $\mathbb{C}^*$.

   Let $f : \mathbb{C}^* \to \mathbb{C}^*$ be defined by $f(z) = z^k$. Then f is a homomorphism and $\mathrm{Ker}(f) = G_k$. If we now show that G/K is onto then we are through.
   Let $w \in \mathbb{C}^*$ and $\theta = \mathrm{Arg}(w)$ be its principal argument. Then we can write $w = |w|e^{i\theta}$ and $z = |w|^{\frac{1}{k}} e^{i\theta/k}$ is such that $z^k = w$.

7. Let $f : \mathbb{R} \to \mathbb{R}^*$ be defined by $f(x) = e^x$. Then $\mathrm{Ker}(f) = \{0\}$ and $\mathrm{Im}\,(f) = \mathbb{R}^+$, the subgroup of positive reals.

Exercise 2.7.5. 1. Let $G = \mathbb{R}^*$ and $H = \mathbb{R}^+$. Show that G/H is isomorphic to the subgroup $\{\pm 1\}$ of $\mathbb{R}^*$.

2. Let $G = \mathbb{R}^2$ and $H = \{(x, 0) \in G : x \in \mathbb{R}\}$. Show that G/H is isomorphic to $(\mathbb{R}, +)$.

3. Let $G = \mathbb{R}^2$ and H be a line passing through origin in $\mathbb{R}^2$. Show that G/H is isomorphic to $\mathbb{R}$.

4. Let $G = \mathbb{Z}_n$ and $L = G_n$, the group of n-th roots of unity. Show that G is isomorphic to L.

5. Let $G = S_n$ and $H = A_n$. Show that G/H is isomorphic to $\{\pm 1\}$.

6. Let G be an abelian group and $\Delta(G) := \{(g, g) : g \in G\}$ be the diagonal subgroup in $G \times G$. Then $(G \times G)/\Delta(G)$ is isomorphic to G.

7. Let G_1 and G_2 be two groups. Show that $(G_1 \times G_2)/G_1 \times \{e\}$ is isomorphic to G_2.

8. Let $G = \mathbb{Z} \times \mathbb{Z}$ and $H := \mathbb{Z}(2, 2) = \{t(2, 2) : t \in \mathbb{Z}\}$ be the subgroup generated by the element $(2, 2)$. Show that G/H is isomorphic to $\mathbb{Z}_2 \times \mathbb{Z}$.

   Let m be an integer. By division algorithm there exist unique integers $\alpha \in \mathbb{Z}$ and m_1 such that $m = 2\alpha + m_1$ where $0 \leq m_1 \leq 1$.
   We define the map $\varphi : G \to \mathbb{Z}_2 \times \mathbb{Z}$ by

   $$\varphi(m, n) := ([m_1], n - m).$$

   It is an easy exercise to check that this map does the job!

9. Let $G = \mathbb{Z} \times \mathbb{Z}$ and $H := \mathbb{Z}(5, 7) = \{t(5, 7) : t \in \mathbb{Z}\}$ be the subgroup generated by the element $(5, 7)$. Show that G/H is isomorphic to $\mathbb{Z}$.

   Define $\varphi : G \to \mathbb{Z}$ by $\varphi(m, n) = 5n - 7m$. It is easy to show that φ is a homomorphism and the subgroup H is contained in the kernel of φ.
   Since the integers 5 and 7 are relatively prime, there exist integers u and v such that $5u + 7v = 1$. Therefore any integer m can be written as $m = 5um + 7vm$, proving that the map φ is on-to.
   Let (m, n) be an element of G such that $5n - 7m = 0$. This means that there exists an integer α such that $(m, n) = \alpha(5, 7)$ and hence it is in the subgroup H.

10. Let $G = \mathbb{Z} \times \mathbb{Z}$ and $H := \mathbb{Z}(2,4) = \{t(2,4) : t \in \mathbb{Z}\}$ be the subgroup generated by the element $(2,4)$. Find G/H.

11. Let $G = \mathbb{Z} \times \mathbb{Z}$ and $H := \mathbb{Z}(k,l) = \{t(k,l) : t \in \mathbb{Z}\}$ be the subgroup generated by the element (k,l). Show that G/H is isomorphic to $\mathbb{Z}_d \times \mathbb{Z}$, if $d = \gcd(k,l)$.

    Let $d = \gcd(k,l)$. Further, let k_1, l_1, u and v be such that $k = dk_1$, $l = dl_1$ and $k_1u + l_1v = 1$.
    Define $\varphi : G \to \mathbb{Z}_d \times \mathbb{Z}$ by $\varphi(m,n) := ([mu+nv], k_1n - l_1m)$. The map φ is a homomorphism. Let $t(k,l)$ be an element in H. Then $\varphi(t(k,l)) = ([t(ku+lv)], t(k_1l - l_1k))$. We will show that the subgroup H is contained in the kernel of φ by proving $\varphi(t(k,l)) = ([t(ku+lv)], t(k_1l - l_1k)) = ([0], 0)$.
    Now $ku + lv = dk_1u + dl_1v = d(k_1u + l_1v)$. This shows that the element $[t(ku+lv)] = [dt(k_1u + l_1v)]$ is zero in $\mathbb{Z}_d$. Similarly we see that $k_1l - l_1k = dk_1l_1 - dk_1l_1 = 0$.
    Conversely let (m,n) in $\mathbb{Z} \times \mathbb{Z}$ be an element in the kernel of φ. That is $\varphi(m,n) = ([mu+nv], k_1n - l_1m) = ([0], 0]$. Since the integers k_1 and l_1 are relatively prime, the solutions of the Diophantine equation $k_1n - l_1m = 0$ are $\alpha(k_1, l_1)$ where $\alpha \in \mathbb{Z}$. Therefore $(m,n) = \alpha(k_1,l_1)$. If we now show that α is an integer multiple of d, then we are through. Since $[mu+nv] = [0]$, the integer d divides $mu+nv$. Hence there exists an integer t such that $mu + nv = \alpha(k_1u + l_1v) = \alpha = dt$ and $(m,n) = td(k_1,l_1) = t(k,l)$ in the subgroup $H = \mathbb{Z}(k,l)$.

    Observe that

    - $\varphi(k_1,l_1) = ([k_1u + l_1v], 0) = ([1], 0)$ and
    - $\varphi(-v,u) = ([0], k_1u + l_1v) = ([0], 1)$.

    This proves that φ is on-to.(Why?) Therefore G/H is isomorphic to $\mathbb{Z}_d \times \mathbb{Z}$.

Exercise 2.7.6. We know that given a real number x, there exists a unique integer n such that $n \leq x < n+1$. This integer is called the integral part of x and it is denoted by $[x]$.

Let $G = [0,1)$ and let $\oplus : G \times G \to G$ be defined by $x \oplus y = x + y - [x+y]$. We have seen in Exercise 5 that $(G, \oplus)$ is a group and all the rational numbers in G are of finite order. Let $\varphi : \mathbb{R} \to G$ be defined by $\varphi(x) = x - [x]$. Show that φ is an onto homomorphism and the kernel is $\mathbb{Z}$. Hence conclude that $\mathbb{R}/\mathbb{Z}$ is isomorphic to G.

Example 2.7.7. We end this section with the following two examples.

1. Let G be a group. Let H and K be two normal subgroups of G such that H and K are isomorphic. Then G/H need not be isomorphic to G/K.

   Let $G = \mathbb{Z}$. If $m \neq n$ are two natural numbers, then the subgroups $H = m\mathbb{Z}$ and $K = n\mathbb{Z}$ are isomorphic. However the quotient groups $G/H \simeq \mathbb{Z}_m$ and $G/K \simeq \mathbb{Z}_n$ are not isomorphic.

2. Let G be a group. Let H and K be two normal subgroups of G such that G/H and G/K are isomorphic. Then H and K need not be isomorphic.

Let $G = \mathbb{C}^*$ and $H_n = \{z \in \mathbb{C}^* : z^n = 1\}$, the set of n-th roots of unity. Then G/H_n is isomorphic to G for all $n \geq 1$. However H_n is not isomorphic to H_m for $m \neq n$.

Let $n \geq 2$ and $G = \mathbb{Z}_{n^2} \times \mathbb{Z}_n$. If $H = \mathbb{Z}_n \times \mathbb{Z}_n$ and $K = \mathbb{Z}_{n^2} \times \{0\}$, then the groups G/H and G/K are isomorphic to $\mathbb{Z}_n$. However the subgroups H and K are not isomorphic.

Exercise 2.7.8. Let G be a group. Let $H \subseteq Z(G)$ be subgroup of G such that G/H is cyclic. Show that G is abelian. (Most of the books use this result to show that a group of order p^2 is abelian.)

In Section 5.2, we prove an important result in finite groups, namely *Cauchy's Theorem*: Let G be a finite group and p be a prime number such that p divides the order of G. Then there exists an element x in G of order p.

We give an easy proof of this result for finite abelian groups, using quotient groups .

Theorem 2.7.9 (Cauchy's Theorem for Abelian groups). *Let G be a finite abelian group and $p \geq 2$ be a prime number such that p divides the order of G. Then there exists an element x in G of order p.*

Proof. we prove this theorem by induction on the order of G. If order of G is 2, then G is a cyclic group of order 2. In this case the result is true.

Now let $n > 2$ and assume the result to be true for all abelian groups of order less than n.

Let G be an abelian group of order n and p be prime number that divides n. Let $y \neq e$ be an element of G of order k, say.

If $p = k$ then y is of order p and we are done.

If $p \neq k$ and if p divides k, then $x = y^{\frac{k}{p}}$ is an element of order p in G and we are through.

Let us now assume that p does not divide k. Let $H := \langle y \rangle$, the cyclic subgroup generated by y. Then the quotient group G/H is an abelian group of order $m = \frac{n}{k} < n$. Therefore by induction hypothesis, there exists an element z in G such that the order of $\overline{z}$ in G/H is p. This means that z^p is in the subgroup H and let l be the order z^p. If $x = z^l$, then the order of x is p and the proof is complete. □

As an application of quotient group construction, we prove the converse of *Lagrange's Theorem* for finite abelian groups.

Theorem 2.7.10. *Let G be a finite abelian group of order $n \geq 2$ and m be divisor of n. Then there exists a subgroup H of G of order m.*

Proof. We prove this theorem by induction on the order of G. If $|G| = 2$, the result is obvious.

Let $n \geq 3$ and assume the result to be true for abelian groups of order less than n. That is if G is an abelian group of order $l < n$ and k divides l then there is a subgroup H of G of order k.

Let G be an abelian group of order n and $m \geq 2$ divide n. Let p be a prime divisor of m. By Cauchy's theorem for abelian groups (Theorem 2.7.9) there exists an element x in G of order p. Let $K = \langle x \rangle$ be the cyclic subgroup generated by x. The quotient group G/K is again abelian group of order $n/p < n$. Hence by induction hypothesis, there exists a subgroup L of G/K of order m/p. If $\pi : G \to G/K$ is the natural homomorphism, the $H := \pi^{-1}(L)$ is a subgroup of G of order m. □

2.8 Special Class of Groups

In this section we study Symmetric groups in detail in the first subsection. The second subsection is devoted to proving the simplicity of group of even permutations A_n for $n \geq 5$. In the third and last subsection we study Dihedral groups as group of symmetries of a regular n-sided polygon.

The readers may skip the second and third subsections and come back later.

2.8.1 Symmetric Groups

Let X be a non-empty set and $S(X) := \{f : X \to X : f \text{ is a bijection }\}$ be the group of permutations of X. We have seen that, if two sets X and Y are bijective, then the groups $S(X)$ and $S(Y)$ are isomorphic.

For most part we will consider only finite sets. If X is a finite set with n-elements, then $S(X)$ is isomorphic to $S(I_n)$, where $I_n := \{1, 2, \dots, n\}$. We denote the group $S(I_n)$ by S_n.

In this section we study some of the properties and the structure of the group S_n.

Let us fix some conventions. Let σ be an element of S_n. If $\sigma(i) = a_i$ for $1 \leq i \leq n$, then we write this as $\sigma = \begin{pmatrix} 1 & 2 & \cdots & n \\ a_1 & a_2 & \cdots & a_n \end{pmatrix}$.

If σ and τ are two permutations, then the product $\sigma\tau$ is the permutation $\sigma \circ \tau : I_n \to I_n$, the composition of the two maps σ and τ. That is the product $\sigma\tau = \begin{pmatrix} 1 & 2 & \cdots & n \\ \sigma(b_1) & \sigma(b_2) & \cdots & \sigma(b_n) \end{pmatrix}$ where $\tau(i) = b_i$ for $1 \leq i \leq n$.

Definition 2.8.1. An element $\sigma \in S_n$ is called a k-cycle if there exists a subset $\{a_1, a_2, \dots, a_k\}$ of I_n such that

$$\sigma(x) = \begin{cases} a_{i+1} & \text{if } x = a_i \text{ for } 1 \leq i < k \\ a_1 & \text{if } x = a_k \\ x & \text{if } x \neq a_j \text{ for } 1 \leq j \leq k \end{cases}.$$

Remark 2.8.2. 1. If $k = 1$, then σ is identity.

2. If $k = 2$, then it is also called a transposition.

3. If σ is a k-cycle and $k \geq 2$, we denote σ by $(a_1\, a_2 \cdots a_k)$. This means that the a_is are cyclically permuted and other elements of the set I_n are fixed. The number k is also called the length of the cycle, the reason for this must be clear to the readers!

4. There is no unique way of representing a k-cycle. For, the two k-cycles $(a_1\, a_2\, \cdots\, a_k)$ and $(a_2\, a_3\, \cdots a_k\, a_1)$ represent the same permutation. To avoid the possible confusion arising out of writing in different ways, we make the following convention: the natural number a_1 is the smallest in the string $(a_1\, a_2\, \cdots a_k)$.

Examples 2.8.3. 1. Let $X = I_n$ and $\sigma = (1\,2\,3\, \cdots\, n)$. Then σ is an n-cycle.

2. Let $X = I_5$ and $\sigma := (1\,3\,2)$. Then σ is a 3-cycle mapping $1 \mapsto 3 \mapsto 2 \mapsto 1$.

3. Let $X = I_5$ and $\sigma := (1\,4\,2\,5)$. Then σ is a 4-cycle mapping $1 \mapsto 4 \mapsto 2 \mapsto 5 \mapsto 1$ leaving 3 fixed.

Exercise 2.8.4. 1. Let σ be a k-cycle in S_n. Show that the order of σ is k.

2. Let $\sigma := (1\,3\,5\,7\,9\,11\,8\,6)$. What is the order of σ^2, σ^4, σ^5 and σ^6?

3. Let $\sigma := (1\,3\,5\,7\,9\,11\,8) \in S_n$, $n \geq 11$. Find the orders of σ^i for $1 \leq i \leq 7$. Compare it with the earlier exercise and what do you find?

4. Let $\sigma = (a_1\, a_2\, a_3\, \ldots, a_k)$ be a k-cycle. What is σ^{-1}?

Let us start with $k = 2$. Then $\sigma = (a_1\, a_2)$. That is $\sigma(a_1) = a_2$, $\sigma(a_2) = a_1$ and $\sigma(x) = x$ if $x \neq a_1, a_2$. Therefore $\sigma^2(a_1) = \sigma(a_2) = a_1$, $\sigma^2(a_2) = \sigma(a_1) = a_2$ and $\sigma^2(x) = x$ for all $x \neq a_1, a_2$. Thus $\sigma^2 = Id$. Hence $\sigma = \sigma^{-1}$.

Let $k = 3$. Then $\sigma^2 = (a_1\, a_3\, a_2)$. That is $\sigma(a_1) = a_3$, $\sigma^2(a_2) = a_1$ and $\sigma^2(a_3) = a_2$. Now we can see very easily that $\sigma^3 = Id$. Thus $\sigma^{-1} = \sigma^2$. Notice also that the order of σ is three.

Now let $k \geq 4$. Then one can check that $\sigma_j(a_1) = a_{j+1}$ for $1 \leq j \leq k-1$. Hence $\sigma^j \neq Id$ for $j \leq k-1$ and $\sigma^k = Id$. Thus $\sigma^{-1} = \sigma^{k-1}$ for a k-cycle.

Definition 2.8.5. Let $\sigma := (a_1\, a_2\, \cdots\, a_k)$ and $\tau := (b_1\, b_2\, \cdots\, b_l)$ be two cycles of length $k \geq 2$ and $l \geq 2$ respectively. We say that these two cycles are disjoint if $\{a_1, a_2, \ldots, a_k\}$ and $\{b_1, b_2, \ldots, b_l\}$ are disjoint subsets of I_n.

Example 2.8.6. 1. Let $\sigma := (1\,2\,5)$ and $\tau := (3\,4)$. Then σ and τ are disjoint cycles.

2. Let $\sigma := (1\,2\,5)$ and $\tau := (2\,3\,4\,5)$. Then σ and τ are not disjoint.

Remark 2.8.7. Let σ and τ be two disjoint cycles. Then the product is not a cycle in S_n. However if they are not disjoint, then the product may be a cycle.

For example if $\sigma = (1\,2)$ and $\tau = (2\,3)$, then $\sigma\tau = (1\,2)(2\,3)$ is the three cycle $(1\,2\,3)$.

Let $\sigma := (1\,3\,5\,7\,9)$ and $\tau := (3\,4\,5)$. Then $\sigma \circ \tau(5) = 5$ and $\tau \circ \sigma(5) = 7$. This shows that in general two cycles may not commute. However if they are disjoint they commute.

Proposition 2.8.8. *Let $\sigma := (a_1\, a_2\, \cdots\, a_k)$ and $\tau := (b_1\, b_2\, \cdots\, b_l)$ be two disjoint cycles. Then $\sigma \circ \tau = \tau \circ \sigma$.*

We prove a more general result from which we can deduce this result as a corollary.

Theorem 2.8.9. *Let X be a non-empty set and A, B be two disjoint subsets of X. If $f, g : X \to$ are two bijections such that $f(x) = x$ for $x \in X \setminus A$ and $g(x) = x$ for $x \in X \setminus B$, then $f(g(x)) = g(f(x))$ for all x in X.*

Proof. Let x be an element of X. If $x \in X \setminus (A \cup B)$, then $g(x) = x$ and $f(x) = x$. Therefore $f(g(x)) = g(f(x)) = x$.

If x is in A, then x is not in B. Since f is a bijection $f(x)$ is also in A and $g(x) = x$. This implies that $g(f(x)) = f(x)$ and $f(g(x)) = f(x)$.

Similarly we can show that $f(g(x)) = g(f(x))$ if x is in B. □

Exercise 2.8.10. 1. Find the order of the product $\sigma \circ \tau$, if

(a) $\sigma := (1\,3\,5\,7)$ and $\tau := (2\,4)$.

(b) $\sigma := (1\,3\,5\,7\,9)$ and $\tau := (2\,4)$.

(c) $\sigma := (1\,3\,5\,7\,8\,9)$ and $\tau := (2\,4\,6\,10\,11\,12\,13\,14)$.

2. While doing the previous exercises, have you noticed something? Let $\sigma := (a_1\,a_2\,\cdots a_k)$ and $\tau := (b_1\,b_2\,\cdots b_l)$ be two disjoint cycles. Show that the order of $\sigma \circ \tau$ is the lcm of k and l.

3. Let $G = S_n$ for $n \geq 3$. Let σ be the n-cycle $(1\,2\,\cdots\,n)$ and τ be defined by $\tau(1) = 1$ and $\tau(j) = n+2-j$ for $2 \leq j \leq n$. Then τ is of order 2 and $\tau\sigma$ is also of order 2. Think of σ as a rotation of a regular n-sided polygon and τ as the reflection of the polygon about the x-axis, then $\theta = \tau\sigma$ is a reflection. (Can you find the axis of reflection?). Hence it is of order 2. But $\tau\theta = \sigma$ is of order $n \geq 3$.

We encourage the readers to work out some examples.

We will now show that any permutation is a product of disjoint cycles.

Theorem 2.8.11. *Let $\sigma \neq e$ be an element in S_n. Then σ is a product of disjoint cycles. Further this decomposition is unique: That is, if $\sigma = \sigma_1\sigma_2\cdots\sigma_r$ and $\sigma = \tau_1\tau_2\cdots\tau_s$ then $r = s$ and for each $i \leq r$, the cycle $\sigma_i = \tau_{i_j}$ for some i_j.*

Proof. The proof is by induction on n.

If $n = 2$, it is very easy: the only nontrivial element is a transposition $(1\,2)$ which is a two cycle.

Let $n \geq 3$ and assume the result to be true in S_r for $r < n$.

Let $\sigma \neq e$ in S_n be given. Let $a_1 = 1$ and $a_i := \sigma(a_{i-1})$. Since I_n is finite there exists an integer k_1 such that $a_{k_1} = 1$. Then the permutation

$\sigma_1 : S_n \to S_n$ defined by $\sigma_1(x) := \begin{cases} a_i & \text{if } x = a_{i-1} \text{ for } 1 \leq i \leq k_1 - 1 \\ a_1 & \text{if } x = a_{k_1} \\ x & \text{if } x \neq a_i \text{ for } 1 \leq i \leq k_1 \end{cases}$ is a cycle of length k_1. If $k_1 = n$, then $\sigma_1 = \sigma$ is a n-cycle and we are through.

Assume $k_1 < n$. Then $\sigma_1^{-1} \circ \sigma(a_i) = a_i$ for $1 \leq i \leq k_1$. Therefore $\sigma_1^{-1} \circ \sigma$ can be considered as a permutation on $I_n \setminus \{a_1, a_2, \cdots, a_{k_1}\}$ and by induction it can be written as a product $\sigma_2 \circ \sigma_3 \circ \cdots \circ \sigma_r$ such that each σ_i is a cycle of length k_i and they are all disjoint. We can now write $\sigma = \sigma_1 \circ \sigma_2 \circ \cdots \circ \sigma_r$.

If $\sigma_i = (a_{i_1} a_{i_2} \cdots a_{i_{k_i}})$ then $\sigma = \prod_{i=1}^r \sigma_i = \prod_{i=1}^r (a_{i_1}\, a_{i_2}\, \cdots a_{i_{k_i}})$ is a decomposition in to disjoint cycles. □

Let σ be in S_n and we write $\sigma = \sigma_1 \sigma_2 \sigma_3 \cdots \sigma_r$ as the product of disjoint cycles; in this product we list one cycles also. Let length of σ_i be k_i. Then $k_1 + k_2 + \cdots + k_r = n$ and this is useful to write down the sign of a permutation defined below. Since the disjoint cycles commute, by rearranging this product, we may assume that $k_i \geq k_{i+1}$. We call the sequence $(k_1, k_2, \ldots, k_r)$ the cycle structure of the permutation σ.

For example if $\sigma = (1\ 3\ 5\ 7)$ in S_8, then its cycle structure is $(4\,1\,1\,1\,1)$. If $\sigma = (1\ 2)(3\ 4\ 8)$ then the cycle structure is $(3\,2\,1\,1\,1)$.

Before going further we introduce an important homomorphism on a permutation group.

Sign of a permutation

Theorem 2.8.12. *Let $n \geq 2$. Then the map $\varepsilon \colon S_n \to \mathbb{Q}^\times$ given by*

$$\sigma \mapsto \prod_{1 \leq i < j \leq n} \frac{\sigma j - \sigma i}{j - i}$$

is a group homomorphism. Its image is $\{\pm 1\}$.

Proof. Let us show that ε is a group homomorphism. Let X be the collection of two element subsets of $\{1, 2, \ldots, n\}$. For each $\sigma \in S_n$ and every $\{i, j\} \in X$ define

$$\sigma'(\{i, j\}) := \frac{\sigma j - \sigma i}{j - i}\ ;$$

the right hand side is symmetric in i and j and hence σ' is well-defined. We have

$$\varepsilon(\sigma) = \prod_{x \in X} \sigma'(x). \tag{2.1}$$

S_n acts in a natural way on X. For $\sigma, \tau \in S_n$ and $x \in X$, we show that

$$(\sigma\tau)'(x) = \sigma'(\tau x)\ \tau'(x). \tag{2.2}$$

Let $x := \{i, j\}$. We compute

$$\begin{aligned} (\sigma\tau)'(x) &= (\sigma\tau)'(\{i, j\}) \\ &= \frac{\sigma\tau j - \sigma\tau i}{j - i} \\ &= \frac{\sigma\tau j - \sigma\tau i}{\tau j - \tau i}\ \frac{\tau j - \tau i}{j - i} \\ &= \sigma'(\tau\{i, j\})\ \tau'(\{i, j\}) \\ &= \sigma'(\tau x)\ \tau'(x). \end{aligned}$$

In view of Equation 2.1 and Equation 2.2 we obtain

$$\begin{aligned}\varepsilon(\sigma\tau) &= \prod_{x\in X} (\sigma\tau)'(x) \\ &= \prod_{x\in X} (\sigma'(\tau x)\,\tau'(x)) \\ &= \prod_{x\in X} \sigma'(\tau x) \prod_{x\in X} \tau'(x) \\ &= \varepsilon(\sigma)\,\varepsilon(\tau).\end{aligned}$$

If τ is the transposition (1,2) then $\varepsilon(\tau) = -1$. □

Let $A_n := \ker(\varepsilon) = \{\sigma \in S_n : \varepsilon(\sigma) = 1\}$, the set of all even permutations of S_n. Then A_n is a normal subgroup of S_n. It is called the *alternating group.*

We will now continue with the study of the cycles in S_n.

Let $k \geq 2$ and $\sigma := (a_1 a_2 \cdots a_k)$ be a k-cycle. Let $\tau_1 := (a_1 a_2)$ be a transposition. Then $\tau_1 \circ \sigma$ is such that $\tau_1 \circ \sigma(x) = x$ if $x \neq a_i$, $\tau_1 \circ \sigma(a_1) = a_1$, $\tau_1 \circ \sigma(a_i) = a_{i+1}$ for $2 \leq i \leq k-1$ and $\tau_1 \circ \sigma(a_k) = a_2$. Thus $\tau_1 \circ \sigma$ is the $k-1$-cycle $(a_2\, a_3\, \cdots a_k)$.

By proceeding as above if $\tau_i := (a_i\, a_{i+1})$ for $i \geq 2$ then $\tau_i \circ \cdots \circ \tau_1 \circ \sigma$ is the $(k-i)$-cycle $(a_{i+1}\, a_{i+2}\, \cdots\, a_k)$. This proves that $\sigma = \tau_1 \circ \tau_2 \cdots \circ \tau_k$ is a product of k-transpositions. Observe that $\sigma := (a_1 a_2 \cdots a_k) = (a_1 a_k)(a_1 a_{k-1}) \cdots (a_1 a_3)(a_1 a_2)$.

The discussion above has proved one part of the following proposition.

Theorem 2.8.13. *Let σ be a k-cycle. Then*

1. *σ is a product of transpositions.*

2. *If $\sigma = \tau_1 \circ \tau_2 \cdots \circ \tau_l = \sigma_1 \circ \sigma_2 \circ \cdots \circ \sigma_r$ is written as a product of transpositions in two different ways then $l - r$ is an even integer.*

Proof. Let $\sigma = \tau_1 \circ \tau_2 \cdots \circ \tau_l = \sigma_1 \circ \sigma_2 \circ \cdots \circ \sigma_r$ represented in two different ways as a product of transpositions. Since $\varepsilon : S_n \to \{\pm 1\}$ is a homomorphism, and $\varepsilon(\tau) = -1$ for a transposition, it follows that $\varepsilon(\sigma) = \varepsilon(\tau_1 \circ \tau_2 \cdots \circ \tau_l) = \prod_j \varepsilon(\tau_j) = (-1)^l$ and also $\varepsilon(\sigma) = \varepsilon(\sigma_1 \circ \sigma_2 \circ \cdots \circ \sigma_r) = \prod_j \varepsilon(\sigma_j) = (-1)^r$. Therefore $(-1)^l = (-1)^r$ and hence $l - r$ is an even integer. □

Since every permutation σ in S_n is a product of disjoint cycles, we can write every permutation σ as a product of transpositions. Further if $\sigma = \tau_1 \circ \tau_2 \cdots \circ \tau_k = \sigma_1 \circ \sigma_2 \circ \cdots \circ \sigma_r$ then $(k - r)$ is an even integer. We state this result as a corollary of Theorem 2.8.13.

Corollary 2.8.14. *Let σ be a permutation in S_n. Then*

1. *σ is a product of transpositions and if*

2. *$\sigma = \tau_1 \circ \tau_2 \cdots \circ \tau_k = \sigma_1 \circ \sigma_2 \circ \cdots \circ \sigma_r$ is written as a product of transpositions in two different ways then $(k - r)$ is an even integer.*

Proof. The proof is left as an easy exercise. □

The above result shows in particular that the set of all transpositions generate the group S_n. The set of all transpositions is same as the set of all two subsets whose size $n(n-1)/2$ is very big.

We can actually show that the group S_n is generated by just two elements, namely one n-cycle and one transposition.

Let $\sigma = (1\ 2\ 3 \ldots n)$ and $\tau = (1\ 2)$. Then $\sigma\tau = (1\ 2\ 3 \ldots n)(1\ 2) = (1\ 3\ 4 \ldots n)$. Now we multiply this $(n-1)$-cycle, on the left by σ^{-1} to get the transposition $(2\ 3)$. That is $\sigma\tau\sigma^{-1} = (1\ 3\ 4 \ldots n)(1\ 2\ 3 \ldots n)^{-1} = (2\ 3)$.

If we now conjugate σ with the transposition $(2\ 3)$, we get the transposition $(3\ 4)$: $\sigma^2\tau\sigma^{-2} = \sigma(2\ 3)\sigma^{-1} = (3\ 4)$. Thus we get $\sigma^{i-1}\tau\sigma^{-i+1} = (i\ i+1)$ for every $i \leq n-1$.

Then $(i\,, i+1)(i+1\, i+2)(i\, i+1) = (i\ i+2)$ for $i \leq n-2$. If $i < j$, then we can use this relation successively to get the transposition $(i\ j)$.

This proves that we can write every transposition as appropriate conjugates of the two cycle $(1\ 2)$ and the n-cycle $(1\ 2\ \ldots n)$.

We state this as the following theorem.

Theorem 2.8.15. *The group S_n is generated by an n-cycle and a transposition.*

Proof. It suffices to show that the n-cycle $(1\ 2\ 3\ \ldots n)$ and the transposition $(1\ 2)$ generate the group S_n and this has been done above. □

In what follows, we will find the conjugacy classes of the elements of σ.

First let us observe that if two elements are of different order, then they can't be conjugate to each other.

In view of this let us consider two permutations σ and τ and find a necessary and sufficient condition for these two permutations to be conjugate.

Let $\sigma = (a_1\ a_2\ \cdots\ a_k)$. We will show that σ is conjugate to the k-cycle $\tau := (1\ 2\ \cdots\ k)$. This will prove that any two k-cycles in S_n are conjugate.

Let $A = \{a_1, a_2, \ldots, a_k\}$ and $I_k = \{1, 2, \ldots, k\}$. Let $\theta : A \to I_k$ be the bijection defined by $\theta(a_i) = i$. Since $I_n \setminus A$ and $I_n \setminus I_k$ have same number of elements, there is a bijection, which we still denote it by θ and we use these maps to get a bijection θ of I_n. We shall show that the map $\theta : I_n \to I_n$ conjugates σ and τ. Let us choose i in I_n. Then $\theta^{-1}(i) = a_i$ and $\sigma \circ \theta^{-1}(i) = \sigma(a_i) = \begin{cases} a_{i+1} & \text{if } i \leq k-1 \\ a_1 & \text{if } i = k \\ a_i & \text{if } i \geq k+1 \end{cases}$. Now it follows that $\theta \circ \sigma \circ \theta^{-1}(i) = \begin{cases} i+1 & \text{if } i \leq k-1 \\ 1 & \text{if } i = k \\ i & \text{if } i \geq k+1 \end{cases}$.

Hence $\tau = \theta \circ \sigma \circ \theta^{-1}$ proving our claim. As a consequence any two k-cycles are conjugate in S_n.

Suppose $\sigma = (a_1\ a_2\ \cdots\ a_k)$ is a k-cycle and α is in S_n, then $\alpha \circ \sigma \circ \alpha^{-1}$ is the k-cycle $(\alpha(a_1)\ \alpha(a_2)\ \ldots\ \alpha(a_k))$. This proves the following theorem

Theorem 2.8.16. *Let σ be a k-cycle. Then*

1. *any k-cycle is conjugate to σ and*

2. *$\alpha \circ \sigma \circ \alpha^{-1}$ is a k-cycle for every $\alpha \in S_n$.*

Remark 2.8.17. The set of all k-cycles form a single equivalence class under conjugation.

Example 2.8.18. Let $\sigma := (1\,3\,5)$ and $\tau := (2\,4\,6)$. We define a map $\theta : \{1,2,3,4,5,6\} \to \{1,2,3,4,5,6\}$ by $\theta(1) = 1$, $\theta(2) = 4$, $\theta(3) = 2$, $\theta(4) = 6$, $\theta(5) = 3$ and $\theta(6) = 5$. Then $\theta := \{1,2,3,4,5,6\} \to \{1,2,3,4,5,6\}$ conjugates σ and $(1\ 2\ 3)$.

Exercise 2.8.19. 1. Find the number of 3-cycles in S_3

2. Find the number of 3-cycles in S_4 and S_5.

3. Find the number of 4-cycles in S_5 and S_7.
Do you observe any pattern?

We would now like to know when are two permutations conjugate to each other.

Theorem 2.8.16 suggests that two permutations with same cycle structure are conjugate and we prove this result now.

Theorem 2.8.20. *Let σ and τ be two permutations in S_n. Then they are conjugate in S_n iff they have the same cycle structure.*

Proof. Let $\sigma = \sigma_1\sigma_2\cdots\sigma_r$ and $\tau = \tau_1\tau_2\cdots\tau_r$ be two permutations with the same cycle structure, say, $(k_1, k_2, \ldots, k_r)$.

The permutation σ partitions the set I_n: For $1 \leq i \leq r$, we let J_i denote the subset of I_n of size k_i such that $\sigma_i(J_i) = J_i$ and σ_i is identity on $I_n \setminus J_i$. Then the subsets J_i are disjoint and they partition I_n. We write $I_n = \cup_{i=1}^{r} J_i$.

Similarly, the permutation τ partitions I_n in to subsets K_i such that $\tau(K_i) = K_i$ and τ_i is identity on $I_n \setminus K_i$.

Let $\alpha : I_n \to I_n$ be a bijection such that $\alpha(J_i) = K_i$. Now it is easy to see that α conjugates the permutations σ and τ. That is $\alpha\sigma\alpha^{-1} = \tau$.

Conversely if two permutations σ and τ are conjugate, then it is easy to see that they have the same cycle structure. □

Given a natural number $n \geq 2$, let $k_1 \geq k_2 \geq \ldots \geq k_r$ be such that $k_1 + k_2 + \cdots + k_r = n$. Then this sequence $(k_1\ k_2\ \ldots k_r)$ is called a partition of n.

For example, if $n = 2$, then 2 and $(1,1)$ are partitions of 2.

If $n = 3$, then 3, $(2,1)$ and $(1,1,1)$ are partitions of 3.

Exercise 2.8.21. Find the partitions of 4, 7, 9 and 12.

Corollary 2.8.22. *The number of conjugacy classes of S_n is the number of partition of n.*

Proof. Theorem 2.8.20 says that there is exactly one conjugacy class for each partition of n. Hence the number of conjugacy classes in S_n is same as the number of partitions of n. □

Proposition 2.8.23. *The group A_n is generated by 3-cycles.*

Proof. Let σ be an even permutation and let $\sigma = \sigma_1\sigma_2\cdots\sigma_r$ be the product of transpositions. Since σ is even r is even. Therefore it is enough to show that product of two transpositions $(ij)(kl)$ is a product of 3-cycles. This can be written as $(i\ j)(k\ l) = (i\ j\ k)(j\ k\ l)$. This completes the proof. □

Converse of Lagrange's theorem

Lagrange's Theorem (Theorem 2.3.1) tells us that if H is a subgroup of a finite group G, then the order of H divides the order of G. We have shown in Theorem 2.7.10, that the converse of Lagrange's theorem is true for finite abelian groups.

However if the group is non-abelian, the converse need not be true. We give an example below.

Let $G = A_4$, the subgroup of all even permutations in S_4. Then G is a group of order 12. We will now show that G does not have a subgroup of order greater than or equal to 4. This will show that the converse of Lagrange's Theorem need not be true for non-abelian groups.

Exercise 2.8.24. A_4 has subgroups of order 2, 3 and 4.

> The permutations $(1\,2)(3\,4)$, $(1\,3)(2\,4)$ and $(1\,4)(2,3)$ are the only elements which are of order 2 in A_4. Each of the sets $\{Id, (1\,2)(3\,4)\}$, $\{Id, (1\,3)(2\,4)\}$ and $\{Id, (1\,4)(2,3)\}$ are all the subgroups of order 2 in A_4.
>
> Now there are eight 3-cycles in A_4. Any of these three cycles generates a cyclic subgroup of order 3.
>
> Let $H = \{Id, (1\,2)(3\,4), (1\,3)(2\,4), (1\,4)(2,3)\}$. Then it can be easily verified that H is a subgroup of order 4. □

Now we show that A_4 has does not have a proper subgroup of order bigger than 4 in the following proposition.

Proposition 2.8.25. *Let H be a subgroup of order bigger than 4 in A_4. Then $H = A_4$.*

Proof. The group A_n is generated by all 3-cycles. Therefore the group A_4 has all the eight 3 cycles and the product of even number of transpositions.

The product of transpositions $(1\ 2)(3\ 4)$, $(1\ 3)(2\ 4)$ and $(1\ 4)(2\ 3)$ are all the elements of order 2 in A_4.

Let H be a subgroup of order bigger than 4. Then it can not contain just the three elements of order 2. Now we claim that the subgroup H can not contain only 3 cycles. We will in fact show that if H contains two 3-cycles, then it contains a product of transpositions.

Let $\sigma = (i\ j\ k)$ be a 3-cycle. Then any other 3-cycle τ must be of the form $(j\ k\ l)$ or $(i\ k\ l)$ and its powers. (Why?) Hence $\sigma\tau$ is a product of transpositions.

Therefore H must contain at least one 3 cycle and a product of two transpositions.

Let us assume this 3 cycle to be $(1\ 2\ 3)$ and the transposition to be $(1\ 2)(3\ 4)$.

Then $(1\ 2)(3\ 4)(1\ 2\ 3)(1\ 2)(3\ 4) = (1\ 4\ 2)$ is in H. Hence the product $(1\ 2\ 3)(1\ 4\ 2)(1\ 4\ 3)$ is also in H. Since the inverses of $(1\ 2\ 3)$, $(1\ 4\ 2)$ and $(1\ 4\ 3)$ must also be in H, it follows that H must contain at least eight elements. This proves that H must be equal to A_4. □

Exercise 2.8.26. Show that there exists a one-one homomorphism from S_n to S_m if $n < m$.

Given σ in S_n, let $\sigma_m : I_m \to I_m$ defined by $\sigma_m(j) = \begin{cases} \sigma(j) & \text{if } 1 \leq j \leq n \\ j & \text{if } n+1 \leq j \leq m \end{cases}$.
That is we have extended $\sigma \in S_n$ to a map σ_m. One can easily verify that σ_m is in S_m and that the map $\sigma \mapsto \sigma_m$ is a one-one homomorphism from $S_n \to S_m$.

Remark 2.8.27. The above result is not true for the dihedral groups D_n for $n \geq 3$. There does not exist any one-one homomorphism from $D_n \to D_{n+1}$ for $n \geq 3$.

The rotation R_θ by the angle $\theta = \frac{2\pi}{n}$ is an element in D_n of order n. If there is a one-one homomorphism $\varphi : D_n \to D_{n+1}$, then $\varphi(R_\theta)$ must be an element of order n and it must divide the order D_{n+1}. If n divides $2(n+1) = |D_{n+1}|$, then $2(n+1) = nq$ for some $q \in \mathbb{N}$. Therefore $2 = n(q-2)$ and this is possible iff $n = 2$. This proves our assertion.

Exercise 2.8.28. If n divides m, then there exists a one-one homomorphism from $D_n \to D_m$.

Suppose $m = nk$. Define $\varphi : D_n \to D_m$ by

$$\varphi(R_\theta) = R_{\frac{\theta}{k}} \quad \text{and} \quad \varphi(\rho) = \rho.$$

We now show S_n is a subgroup of alternating group A_{n+2}.

Proposition 2.8.29. *S_n is a subgroup of A_{n+2} for $n \geq 3$.*

Proof. Let $\varphi : S_n \to A_{n+2}$ de defined by

$$\varphi(\sigma) = \begin{cases} \sigma & \text{if } \sigma \text{ is even} \\ \sigma \circ (n+1\ n+2) & \text{if } \sigma \text{ is odd} \end{cases}.$$

We leave it to the readers to verify that φ is a one-one homomorphism. □

Corollary 2.8.30. *Let G be a finite subgroup. Then G is a subgroup of A_n for some n.*

Proof. Let G be a finite group of order k. Then by Cayley's theorem G is a subgroup of S_k. By Proposition 2.8.29 the permutation group S_k is a subgroup of A_{k+2} and therefore the group G is a subgroup of A_{k+2}. □

2.8.2 A_n is simple for $n \geq 5$

In this section we show that A_n is simple for $n \geq 5$.

Let us recall the definition of simple groups.

Definition 2.8.31. A group G is called simple if it has no proper normal subgroups.

Example 2.8.32. Let G be a group of prime order. Then G is simple.

Remark 2.8.33. 1. The group A_3 is a group of prime order and hence simple.

2. This result is not true for A_4: First observe that the order of A_4 is 12 and we will show in Section 5.3.3, that a group of order 12 has a normal subgroup of order 3 or 4.

Let us do some simple exercises that will be used in the proof of the main result of this section. We assume that $n \geq 5$ throughout.

Exercise 2.8.34. 1. Let $r \geq 4$ and $\sigma := (a_1\, a_2 \cdots a_r)$ be an r-cycle.

(a) If $\tau := (a_1\, a_2\, a_3)$ then show that $\tau\sigma\tau^{-1}\sigma^{-1} = (a_1\, a_2\, a_4)$.

(b) If $\theta := (a_1\, a_2\, a_r)$ then show that $\theta\sigma\theta^{-1}\sigma^{-1} = (a_1\, a_3\, a_r)$.

2. Let $\sigma := (a_1\, a_2\, a_3)(a_4\, a_5\, a_6)$ be a product of two disjoint cycles and let $\tau := (a_1\, a_2\, a_4)$. Show that $\tau\sigma\tau^{-1}\sigma^{-1}$ is the 5-cycle $(a_1\, a_2\, a_5\, a_3\, a_4)$.

3. Let $\sigma := (a_1\, a_2\, a_3)(a_4\, a_5)$ be a product of two disjoint cycles and let $\tau := (a_1\, a_4\, a_5)$. Show that $\tau\sigma\tau^{-1}\sigma^{-1} = (a_1\, a_4)(a_2\, a_5)$.

4. Let $\sigma := (a_1\, a_2)(a_3\, a_4)$ be a product of two disjoint transpositions. If $a_5 \notin \{a_1, a_2, a_3, a_4\}$ and $\tau := (a_1\, a_3\, a_5)$ show that $\tau\sigma\tau^{-1}\sigma^{-1} = (a_1\, a_3\, a_5\, a_4\, a_2)$.

Now we state the main theorem of this section.

Theorem 2.8.35. *The alternating group A_n is simple for $n \geq 5$.*

We have already seen that the group A_n is generated by 3-cycles. Now we show that any two three cycles are conjugate in A_n, if $n \geq 5$, and complete the proof. This is done in two lemmas.

Lemma 2.8.36. *Let $n \geq 5$. Then every three cycle in A_n is a product of permutation of cycle type $(2, 2)$.*

Proof. Let $\sigma = (a\, b\, c)$ be a 3-cycle and $d, e \notin \{a, b, c\}$. Then we can write $(a\, b\, c) = (a\, b)(a\, c) = (a\, b)(d\, e)(d\, e)(a\, c)$. □

Remark 2.8.37. We know any two cycles of same size are conjugate in S_n. In particular two 3-cycles are conjugate in S_n. However they need not be conjugate in A_n.

The two cycles $(1\,2\,3)$ and $(1\,3\,2)$ are not conjugate in A_4.

If we assume that $n \geq 5$, then any two 3-cycles are conjugate in A_n.

Lemma 2.8.38. *Let σ and $\theta := (1\,2\,3)$ be two 3-cycles in A_n for $n \geq 5$. Then they are conjugate in A_n.*

Proof. Let π be a permutation in S_n such that $\pi\sigma\pi^{-1} = \theta$. If π is an even permutation we are done. Otherwise we choose k and $l \notin \{1, 2, 3\}$. Since the three cycle θ and the transposition $\tau = (k\, l)$ are disjoint, they commute. In view of this $\tau\theta\tau^{-1} = \theta$. Hence $(\pi\tau)\theta(\pi\tau)^{-1} = \pi(\tau\theta\tau^{-1})\pi^{-1} = \pi\theta\pi^{-1} = \sigma$. The permutation $\pi\tau$ is an even permutation and it conjugates σ and θ. This completes the proof. □

Now we start with the proof of Theorem 2.8.35.

Proof. Let $N \neq \{e\}$ be a normal subgroup of A_n. We need to show that $N = A_n$. To prove this, we will show that N contains a 3-cycle. Then from Lemma 2.8.38 we conclude that N contains all the 3-cycles.(Why?). Since A_n is generated by 3-cycles we are through.

Let σ be an element in N and let $\sigma = \sigma_1\sigma_2\cdots\sigma_r$ be the decomposition in to disjoint cycles. Since these cycles commute, by rearranging them, we may assume that $\sigma_1 = (1\,2\ldots k)$.

- Suppose $k \geq 4$. Then the 3-cycle $(1\,2\,3)$ commutes with the remaining σ_i's. Therefore

$$\begin{aligned}\theta\sigma\theta^{-1} &= (1\,2\,3)\sigma_1\sigma_2\cdots\sigma_r(1\,2\,3)^{-1}\\ &= (1\,2\,3)\sigma_1(1\,2\,3)^{-1}\sigma_2\cdots\sigma_r\\ &= (1\,2\,3)\sigma_1(1\,3\,2)\sigma_2\cdots\sigma_r\\ &= (1\,2\,3)\sigma_1(1\,3\,2)\sigma_1^{-1}\sigma_1\sigma_2\cdots\sigma_r\\ &= (1\,2\,3)\sigma_1(1\,3\,2)\sigma_1^{-1}\sigma\\ &= (1\,2\,4)\sigma.\end{aligned}$$

  Since N is normal in A_n, the permutation $(1\,2\,3)\sigma(1\,2\,3)^{-1} = (1\,2\,4)\sigma$ is in N. Hence the 3-cycle $(1\,2\,4) = (1\,2\,3)\sigma(1\,2\,3)^{-1}\sigma^{-1}$ is also in N.

- Suppose all the cycles σ_i's are of length less than or equal to 3. By relabeling, we may assume $\sigma_1 := (1\,2\,3)$ and $\sigma_2 := (4\,5\,6)$. If $\tau := (1\,2\,4)$, then

$$\begin{aligned}\tau\sigma\tau^{-1} &= \tau(\prod\sigma_i)\tau^{-1}\\ &= \tau\sigma_1\sigma_2\tau^{-1}\sigma_2^{-1}\sigma_1^{-1}\sigma\\ &= (1\,2\,5\,3\,4)\sigma.\end{aligned}$$

  Therefore $\tau\sigma\tau^{-1} = (1\,2\,5\,3\,4)\sigma$ is in N and as earlier we conclude that $\eta := (1\,2\,5\,3\,4) = \tau\sigma\tau^{-1}\sigma^{-1}$ is in N. If we now choose $\theta := (1\,2\,5)$, then it follows that $\theta\eta\theta^{-1} = (1\,2\,3)$ is in the subgroup N.

- Assume $\sigma_1 = (1\,2\,3)$ and other cycles are all of length 2. We may assume that $\sigma_2 = (4\,5)$. If we now let $\tau := (1\,4\,5)$, then $\tau\sigma_1\sigma_2\tau^{-1}\sigma_2^{-1}\sigma_1^{-1} = (1\,4)(2\,5)$.

  A similar computation as in the earlier case show that the element $\psi := (1\,4)(2\,5)$ is in the subgroup N. If we now let $\theta = (1\,3\,5)$, then $\theta\psi\theta^{-1}$ is in N and $\theta\psi\theta^{-1}\psi^{-1} = (1\,3\,5\,4\,2)$ is also in N. Again it follows from Exercise 2.8.34(1) that $(1\,3\,5)(1\,3\,5\,4\,2)(1\,5\,3)(1\,3\,5\,4\,2)^{-1} = (1\,3\,4)$ is in N.

- Assume that all the cycles are transpositions and $\sigma_1 = (1\,2)$ and $\sigma_2 = (3\,4)$. Then the product $(1\,3\,5)(1\,2)(3\,4)(1\,5\,3)(3\,4)(1\,2) = (1\,3\,5\,4\,2)$ is also in N. Now we can show as earlier that a 3-cycle is in N.

This completes the proof. □

Corollary 2.8.39. *A_5 is simple.*

Proof. The 5 cycles (1 2 3 4 5) and (1 3 2 4 5) are not conjugate. Hence A_5 is simple. □

2.8.3 Dihedral Groups

Dihedral groups are the group of symmetries of regular n-sided polygons. The reader should study the cases when $n = 3, 4$ while we look at the general case of n-gon.

We orient the regular n-gons as in Figure 2.1 (the left column picture) by locating a vertex at the top, labeled as 1. We number the positions of the other vertices in clockwise order. It is intuitively obvious that a symmetry σ is completely determined (1) by the position of the image, $\sigma(1)$, of the vertex at 1 under the symmetry σ, and (2) by the position of the image, $\sigma(2)$, of the vertex at 2 relative to $\sigma(1)$—whether it is clockwise or anti-clockwise of $\sigma(1)$. There are n choices for the vertex at 1 and for each such choice, there are are two choices for the image of the vertex at 2. Thus there are $2n$ symmetries of the regular n-gon.

Let R denote the clockwise rotation by $2\pi/n$ radians. Let ρ denote the reflection about the vertical diameter through the vertex at 1. See Figures 2.1–2.2.

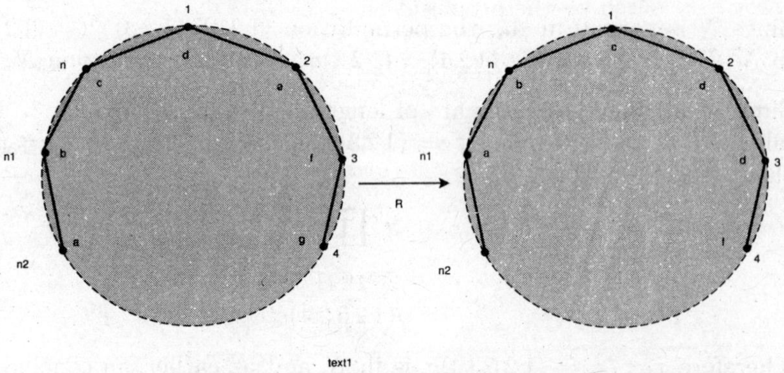

Figure 2.1: Rotation

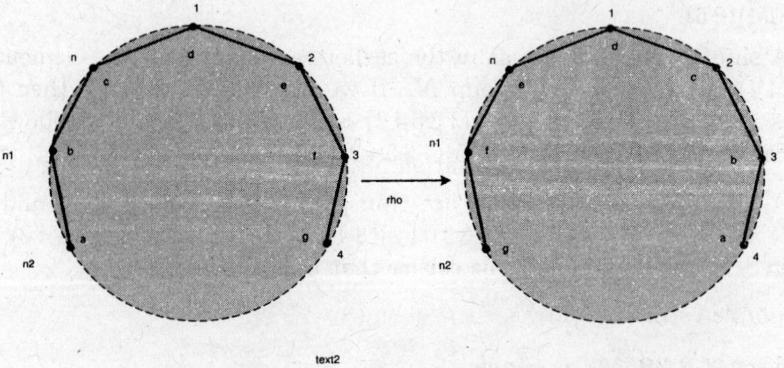

Figure 2.2: Reflection

The following are the consequences of our (intuitively obvious) observation: Any symmetry can be obtained by

1. A rotation of $2\pi k/n$ radians. The corresponding group element is R^k for $0 \leq k < n$.

2. A reflection or no reflection about the vertical diameter. The corresponding group element is ρ^l where $l = 0$ or $l = 1$.

Some typical situations are shown in Figures 2.1–2.2. (Keep in mind that the integers 1,2,3 etc., denote the positions of the vertices, not the vertices themselves.) If we denote the identity symmetry by I, we have the relations

$$R^n = I = \rho^2 \text{ and } R\rho = \rho R^{-1} = \rho R^{n-1}. \tag{2.3}$$

The last relation is visualized in Figure 2.4. See also Observation 2.8.40

We now make three observations.

Observation 2.8.40. The symmetries R and ρ induce permutations of the integers $1, 2, \ldots, n$ in terms of the positions assumed by the vertices. Thus R and ρ are represented by the permutations

$$R = \begin{pmatrix} 1 & 2 & \ldots & n \\ 2 & 3 & \ldots & 1 \end{pmatrix} \text{ and } \rho = \begin{pmatrix} 1 & 2 & 3 & \ldots & n \\ 1 & n & (n-1) & \ldots & 2 \end{pmatrix}. \tag{2.4}$$

Thus, we have $\rho(j) = n + 2 - j$ if $j \neq 1$ and $\rho(1) = 1$. Using this, it is easy to see that $R\rho = \rho R^{-1}$. (Exercise!)

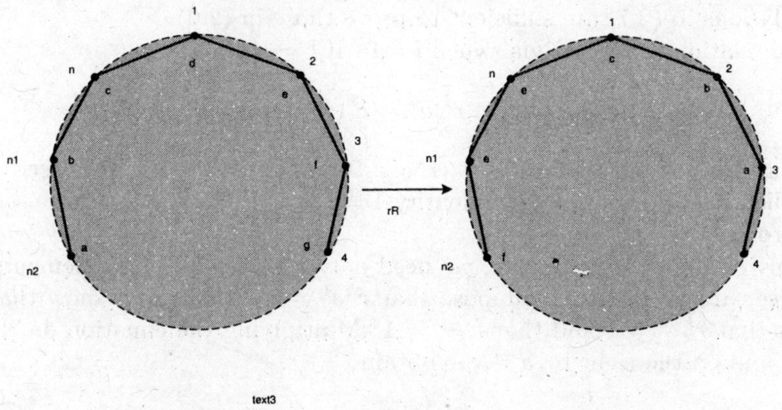

Figure 2.3: Rotation followed by reflection

Observation 2.8.41. From the relation $R\rho = \rho R^{-1}$ we have

$$R^k\rho = \rho R^{-k}, \text{ for } 0 \leq k < n. \tag{2.5}$$

This is seen as follows:

$$R^k\rho = \underbrace{R\cdots R}_{k}\rho = (\underbrace{R\cdots R}_{k-1})(R\rho) = \underbrace{R\cdots R}_{k-1}\rho R^{-1} = \cdots = \rho R^{-k}.$$

Observation 2.8.42. Each element of the group is of the form $R^k\rho^l$ where $0 \le k < n$ and $l = 0, 1$. Thus there are $2n$ elements of this form and we must prove that they are all distinct. If $R^k\rho^l = R^r\rho^s$, then we should have $R^{k-r} = \rho^{l-s}$. In geometric language this means that the rotation R^{k-r} must be a reflection ρ^{l-s}. This is impossible, for example, the rotation will have determinant 1 whereas a reflection will have determinant -1.

While this is perfectly a valid argument, we shall prove this purely algebraically using only the relations (2.3).

Theorem 2.8.43. *Let G be a group. Assume that there exist $r, a \in G$ such that, for some fixed integer $n > 2$*

1. $r^n = 1 = a^2$, $r^k \neq 1$ *if* $0 < k < n$, *and* $a \neq 1$,
2. $ra = ar^{-1}$.

Then G contains at least $2n$ elements of the form $r^k a^l$ where $0 \le k < n$ and $l = 0, 1$. The group multiplication for these elements is given by

$$(r^k a^l)(r^i a^j) = \begin{cases} r^{k+i}a^j, & \text{if } l = 0 \\ r^{k-i}a^{l+j}, & \text{if } l = 1. \end{cases} \tag{2.6}$$

Proof. As shown in Observation 2.8.41, we have

$$r^k a = ar^{-k} \text{ (or equivalently } r^{-k}a = ar^k). \tag{2.7}$$

The relations in (2.7) are sufficient to prove those in (2.6).

The relation (2.6) is trivial when $l = 0$. If $l = 1$, then

$$(r^k a^l)(r^i a^j) = r^k(a^l r^i)a^j = r^k(r^{-i}a^l)a^j = r^{k-i}a^{l+j}.$$

Relation (2.6) shows that the set $\{r^k a^l : 0 \le k < n, l = 0, 1\}$ is closed under multiplication and contains the identity $1 = r^0 a^0$. Since this is a finite set, it is a subgroup of G.

Thus to complete the proof, we need only show that the $2n$ elements $r^k a^l$ of the set are all distinct. Suppose that $r^k a^l = r^i a^j$. We first show that this implies that $r^{k-i} = 1$ and that $a^{l-j} = 1$. Multiplying the equation on the left by r^{-i} and on the right by a^{-j}, we obtain

$$r^{k-i}a^{l-j} = 1.$$

To simplify the notation, let $c = k - i$ and $d = l - j$. We therefore want to show that $r^c a^d = 1$ implies that $r^c = 1$ and $a^d = 1$. If $d = 0$,we are done. If $d = \pm 1$, then, since $a^2 = 1$, $a^d = a$ and hence $r^c a^d = 1$ is the same as saying that $r^c a = 1$ or equivalently $r^c = a$. But then $ra = rr^c = r^c r = ar$. By hypothesis, $ra = ar^{-1}$. Hence we conclude that $r = r^{-1}$ or $r^2 = 1$. This contradicts our hypothesis (1). Thus we infer that $d = 0$.

Finally, if $r^c = 1$ for $0 \le c < n$, then $c = 0$, in view of the hypothesis (1). Therefore we conclude that all elements of the form $r^k a^l$ ($0 \le k < n$, $l = 0, 1$) are distinct. □

Corollary 2.8.44. *For every integer $n > 2$, there is a group of order $2n$ with properties 1 and 2 of the theorem. This group is unique up to isomorphism and will be denoted by D_{2n}.*

Proof. The existence is given by the groups of symmetries of a regular n-gon. The uniqueness is what Theorem 2.8.43 is about. □

Example 2.8.45. Consider D_8. Let the generators by r, s with $r^4 = 1 = s^2$ and $rs = sr^{-1}$ as the relations. The elements $s, b := sr$ are of order 2. Also, the subgroup $\langle s, sr \rangle$ generated by them is D_8, since $r = sb \in \langle s, sr \rangle$. However, not all elements can be written in the form $s^i b^j$. For there are only four such distinct elements whereas $|D_8| = 8$. In particular, the element sbs cannot be brought into the form $s^i b^j$.

Exercise 2.8.46. Let G_n be the subgroup of $\mathrm{GL}(2, \mathbb{C})$ generated by the elements

$$A := \begin{pmatrix} 0 & 1 \\ 1 & 0 \end{pmatrix} \text{ and } B := \begin{pmatrix} e^{\frac{2\pi i}{n}} & 0 \\ 0 & e^{-\frac{2\pi i}{n}} \end{pmatrix}.$$

Show that G_n is isomorphic to D_{2n}.

Exercise 2.8.47. Let G be a finite subgroup of $O(2, \mathbb{R})$. Then G is either cyclic or dihedral.

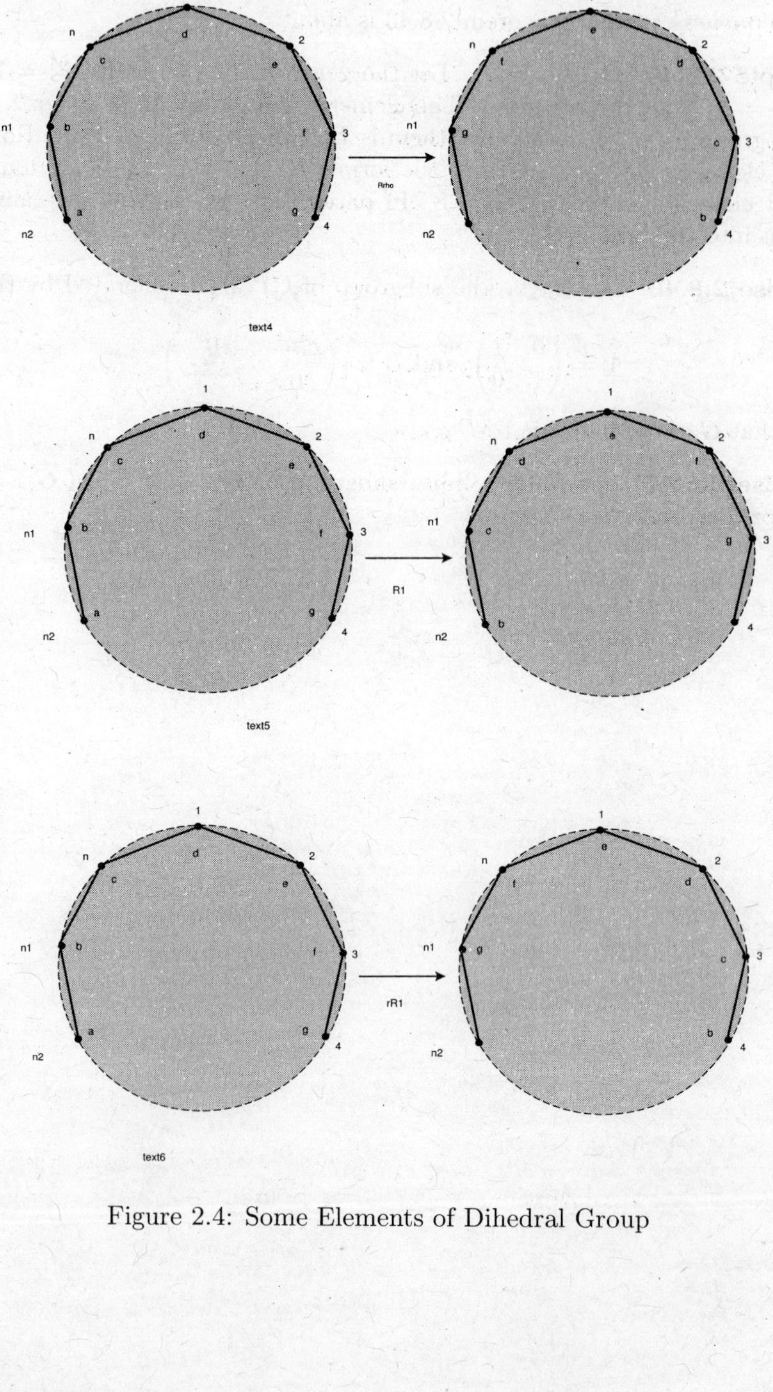

Figure 2.4: Some Elements of Dihedral Group

Chapter 3

Ring Theory

The integers form the model for the concept of a ring. The integers form an abelian group under addition and it is closed under multiplication. The multiplication is associative and it distributes over addition etc.,

The definition of a ring is similar to the above properties of integers.

3.1 Definition and Examples

Definition 3.1.1. 1. A ring R is a set together with two binary operations $+ : R \times R \to R$ called addition and $\cdot : R \times R \to R$ called multiplication satisfying the following axioms:

(a) $(R, +)$ is an abelian group,

(b) $a \cdot (b \cdot c) = (a \cdot b) \cdot c$ for all a, b and c in R,

(c) $a \cdot (b + c) = a \cdot b + a \cdot c$ for all a, b and c in R and

(d) $(a + b) \cdot c = a \cdot c + b \cdot c$ for all a, b and c in R.

2. If $a \cdot b = b \cdot a$ for all a and b in R, the ring R is called a commutative ring.

3. If there exists an element 1 in R such that, $1 \cdot a = a \cdot 1 = a$, for all elements a in R, the ring R is called a ring with multiplicative identity 1.

Remark 3.1.2. 1. If the ring R is commutative, then the axioms (1c) and (1d) are equivalent.

2. In this chapter we consider only commutative rings with multiplicative identity 1 unless specified otherwise.

Examples 3.1.3. 1. $(\mathbb{Z}, +, \cdot)$ is a commutative ring with identity with the usual addition $+$ and usual multiplication $\cdot$ of integers.

2. $(\mathbb{Q}, +, \cdot)$, $(\mathbb{R}, +, \cdot)$ and $(\mathbb{C}, +, \cdot)$ are all commutative rings with identity under usual addition and multiplication.

3. Let $R := \mathbb{Z}[i] := \{a + ib : a, b \in \mathbb{Z}\}$. Then the set R is a subset of $\mathbb{C}$ and the addition and multiplication are binary operations on R as well. With this addition and multiplication $(R, +, \cdot)$ is a commutative ring with identity. This ring is called the ring of *Gaussian Integers*.

   If $a + ib$ and $c + id$ are in the ring R, then the addition

   $$(a + ib) + (c + id) = a + b + i(c + d)$$

   and multiplication

   $$(a + ib)(c + id) = (ac - bd) + i(ad + bc)$$

   of these two complex numbers are again in R. Thus the set R is closed under the addition and multiplication of complex numbers. The remaining axioms follow from Example 2.

4. The example we present now generalizes the earlier one and provides with lot of examples!

   Let D be a nonzero square free integer. If D is positive, then we let $\sqrt{D}$ denote the positive square root of D and if D is negative, then $\sqrt{D}$ is $i\sqrt{|D|}$.

   Let $\mathbb{Z}[\sqrt{D}] := \{a + b\sqrt{D} : a, b \in \mathbb{Z}\}$. Then $\mathbb{Z}[\sqrt{D}]$ is a subset of $\mathbb{C}$ and $\mathbb{Z}[\sqrt{D}]$ is closed under the addition and multiplication as shown below. Let $a + b\sqrt{D}$ and $c + d\sqrt{D}$ be two elements of $\mathbb{Z}[\sqrt{D}]$. Then

   $$(a + b\sqrt{D}) + (c + d\sqrt{D}) := (a + c) + (b + d)\sqrt{D}$$

   and

   $$(a + b\sqrt{D})(c + d\sqrt{D}) := (ac + bd\,D) + (ad + bc)\sqrt{D}$$

   are in $\mathbb{Z}[\sqrt{D}]$. The remaining axioms of a ring can be verified easily as in the previous example.

   The following are very special cases of this example.

   (a) Let $R := \mathbb{Z}[\sqrt{2}] := \{a + b\sqrt{2} : a, b \in \mathbb{Z}\}$. This ring R is contained in $\mathbb{R}$.

   (b) Let $R := \mathbb{Z}[i\sqrt{2}] := \{a + ib\sqrt{2} : a, b \in \mathbb{Z}\}$. This ring is contained in the ring $\mathbb{C}$.

   Similarly $\mathbb{Q}[\sqrt{D}] := \{a + b\sqrt{D} : a, b \in \mathbb{Q}\}$ is also a ring under the natural addition and multiplication.

5. Let $n \geq 2$ and $M(n, \mathbb{R})$ be the set of all $n \times n$ matrices over $\mathbb{R}$. Then the set $M(n, \mathbb{R})$ is a ring under the matrix addition and multiplication. The ring R is not commutative. But it has identity, namely the identity matrix, denoted by I_n.

6. Let $n \geq 2$ and $R := \mathbb{Z}_n$, the set of integers modulo n. Then R is commutative ring with identity under addition and multiplication modulo n.

7. Let D be a square free integer and $R := \left\{ \begin{pmatrix} a & bD \\ b & a \end{pmatrix} : a, b \in \mathbb{Z} \right\}$. Then R is a ring under the addition and multiplication of matrices. This ring is also commutative.

   Let $A := \begin{pmatrix} a & bD \\ b & a \end{pmatrix}$ and $B = \begin{pmatrix} c & dD \\ d & c \end{pmatrix}$ be two elements in R.
   Then their sum $A + B = \begin{pmatrix} a+c & (b+d)D \\ b+d & a+c \end{pmatrix}$ and product $AB = \begin{pmatrix} ac + bd\,D & (ad+bc)D \\ ad+bc & ac+bd\,D \end{pmatrix}$ are again in R and the rest of the properties easily follow from the fact $M(2, \mathbb{R})$ is a ring.
   Does this remind you of something seen earlier?

8. Let R and S be two rings. Then the product $R \times S$ becomes a ring under the addition and multiplication defined as follows:

$$\begin{aligned} (a_1, a_2) + (b_1, b_2) &:= (a_1 + b_1, a_2 + b_2) \text{ and} \\ (a_1, a_2)(b_1, b_2) &:= (a_1 b_1, a_2 b_2). \end{aligned}$$

   It is an easy verification to show that $R \times S$ is a ring. The ring $R \times S$ is a ring with identity, if R and S are rings with identity. This ring $R \times S$ *is called the product ring.*

   In this ring the element $(1, 1)$ is the multiplicative identity.
   Observe that the product $(a, 0)(0, b) = (0, 0)$ for a in R and b in S.
   Thus the product of two nonzero elements in a ring may be zero.

9. Let R be a commutative ring with identity. Then we *complexify* R to get a new ring as follows: Let $R(\mathbb{C}) := R \times R$. We define the addition to be the componentwise addition as in the product ring. Let (a, b) and (c, d) be two elements of $R(\mathbb{C})$. We define the multiplication as

$$(a, b)(c, d) := (ac - bd, ad + bc).$$

   - It is easy to show that $(R(\mathbb{C}), +)$ is an abelian group.
   - Let (a, b), (c, d) and (e, f) be three elements in $R(\mathbb{C})$. Then

$$\begin{aligned} &[(a, b)(c, d)]\,(e, f) \\ &= (ac - bd, ad + bc)(e, f) \\ &= ((ac - bd)e - (ad + bc)f, (ac - bd)f + (ad + bc)e) \\ &= (a(ce - df) - b(cf + de), a(cf + de) + b(ce - df)) \\ &= (a, b)(ce - df, cf + de) \\ &= (a, b)\,[(c, d)(e, f)]\,. \end{aligned}$$

     This proves the associativity of the multiplication.
   - $(1, 0)$ is the multiplicative identity:

$$(a, b)(1, 0) = (a \cdot 1 - b \cdot 0, a \cdot 0 + b \cdot 1) = (a, b)$$

     for all elements (a, b) in $R(\mathbb{C})$.

- If (a,b), (c,d) and (e,f) are in $R(\mathbb{C})$, then

$$\begin{aligned}
&(a,b)\left((c,d)+(e,f)\right)\\
&= (a,b)(c+e,d+f)\\
&= (a(c+e)-b(d+f),a(d+f)+b(c+e))\\
&= ((ac-bd)+(ae-bf),(ad+bc)+(af+be))\\
&= (ac-bd,ad+bc)+(ae-bf,af+be)\\
&= (a,b)(c,d)+(a,b)(e,f)
\end{aligned}$$

Thus $(R(\mathbb{C}),+,\cdot)$ becomes a ring.

In this ring $(0,1)(0,1)=(-1,0)$! That is square of an element is -1. Compare this with the rings $\mathbb{Z}$, $\mathbb{Q}$ and $\mathbb{R}$.

If $R=\mathbb{Z}$, then $R(\mathbb{C})$ is the ring $\mathbb{Z}[i]$ of Gaussian integers.

We leave it as an exercise to find $R(\mathbb{C})$ if $R=\mathbb{Q}$, $\mathbb{R}$ or $\mathbb{C}$. We also urge the readers to find $R(\mathbb{C})$ if $R=\mathbb{Z}[\sqrt{D}]$ or $\mathbb{Q}[\sqrt{D}]$ where D is a square free integer.

Exercise 3.1.4. 1. Let X be any set and $\mathcal{F}(X,\mathbb{R}) := \{f : X \to \mathbb{R} : f \text{ is a function}\}$. On this set define addition and multiplication as

$$(f+g)(x) := f(x)+g(x)$$

and

$$(f\cdot g)(x) := f(x)g(x)$$

for all f and g in R and x in X; here the addition on the right hand side is the usual addition and multiplication of real numbers. Show that $\mathcal{F}(X,\mathbb{R})$ is a ring with these binary operations.

We have seen in Chapter 2 that $\mathcal{F}(X,\mathbb{R})$ is an abelian group under the binary operation defined above.

That the set $\mathcal{F}(X,\mathbb{R})$ satisfies all the axioms of multiplication are seen as follows: If f, g and h are in $\mathcal{F}(X,\mathbb{R})$, then

$$\begin{aligned}
(f(gh))(x) &= f(x)((gh)(x))\\
&= f(x)(g(x)h(x))\\
&= (f(x)g(x))h(x)\\
&= ((fg)h)(x) \qquad \text{and}\\
(f(g+h))(x) &= f(x)((g+h)(x))\\
&= f(x)(g(x)+h(x))\\
&= f(x)g(x)+f(x)h(x)
\end{aligned}$$

for all x in X; all these follow from the fact that for real numbers a, b and c, $a(bc)=(ab)c$ and $a(b+c)=ab+ac$.

The constant function $1 : X \to \mathbb{R}$ defined by $1(x)=1$ for x in X is the multiplicative identity.

This ring R is called the ring of real valued functions.

2. Let R be a ring and $A := \{f : X \to R : f \text{ is a function}\}$. Define addition and multiplication on the set A as in the previous exercise to make A in to a ring. If R is a commutative ring then A is also commutative. Show that the ring A has multiplicative identity, the constant function 1, if R has multiplicative identity. This ring is called the ring of R valued functions.

3. Let $X = \mathbb{N}$ and $C(X, \mathbb{R}) := \{(a_n) : \lim_n a_n \text{ exists}\}$. Then using the algebra of convergent sequences, show that $C(\mathbb{N}, \mathbb{R})$ is a ring under addition and multiplication of sequences.

4. Let J be an interval in $\mathbb{R}$ and let $C(J, \mathbb{R}) := \{f : J \to \mathbb{R} : f \text{ is continuous}\}$. Show that $C(J, \mathbb{R})$ is a ring under addition and multiplication of functions.

5. Let
$$R := \{A := (a_{ij}) \in M(n, \mathbb{R}) : a_{ij} = 0 \text{ for } j < i\}$$
be the set of all *upper triangular* matrices. Show that R is a ring. Further show that this ring is not commutative.

6. Let $n \geq 3$ be a natural number and $\omega (\neq 1)$ in $\mathbb{C}$ be a primitive n-th root of unity. That is $\omega^n = 1$ and $\omega^j \neq 1$ for $0 < j < n$. Let
$$\mathbb{Z}[\omega] := \left\{a_0 + a_1\omega + a_2\omega^2 + \cdots + a_{n-1}\omega^{n-1} : a_i \in \mathbb{Z}\right\}.$$
Show that the restriction of addition and multiplication on $\mathbb{C}$ is a binary operation on $\mathbb{Z}[\omega]$ and it is a ring.

7. Let $n \geq 3$ be a natural number and $\omega \neq 1$ be a primitive n-th root of unity. Show that the set
$$\mathbb{Q}[\omega] := \left\{a_0 + a_1\omega + a_2\omega^2 + \cdots + a_{n-1}\omega^{n-1} : a_i \in \mathbb{Q}\right\}$$
is a ring under the addition and multiplication of complex numbers as in the last example.

The following are some examples of non-commutative rings!

Example 3.1.5. 1. Let $\mathbb{H} := \{a + b\,i + c\,j + d\,k : a, b, c, d \in \mathbb{R}\}$ with the relations $i^2 = j^2 = k^2 = -1$ and $ij = k = -ji$. This set with addition and multiplication defined by the above rules is a ring. This ring is called the ring of *quaternions*. This is a non-commutative ring with identity. This ring has another property, namely, given nonzero element $x = a + b\,i + c\,j + d\,k$ in $\mathbb{H}$, there exists an element $y = \alpha + \beta\,i + \gamma\,j + \delta\,k$ in $\mathbb{H}$ such that $xy = 1$. The element y is given by $y = \frac{a - b\,i - c\,j - d\,k}{a^2 + b^2 + c^2 + d^2}$.
We can also show that $\mathbb{H}(\mathbb{Q}) := \{a + b\,i + c\,j + d\,k : a, b, c, d \in \mathbb{Q}\}$ with the relations $i^2 = j^2 = k^2 = -1$ and $ij = k = -ji$ is also a ring. In this ring too every nonzero element has multiplicative inverse.

2. One can also define the ring $\mathbb{H}(\mathbb{Z})$. However in this ring not all nonzero elements have multiplicative inverse. For example, the elements $i + j$, $j + k$, $i + k$ and $1 + i + j + k$ do not have multiplicative inverses.

Now we give few important examples of commutative rings without identity.

Examples 3.1.6. 1. Let $n \geq 2$ and $R := n\mathbb{Z}$. Then R is a commutative ring without identity.

2. Let $C_c(\mathbb{R}, \mathbb{R}) := \{f \in C(\mathbb{R}, \mathbb{R}) : \text{supp } (f) \text{ is compact}\}$. Then $C_c(\mathbb{R}, \mathbb{R})$ is a commutative ring without multiplicative identity.

   Let us recall that if $f : \mathbb{R} \to \mathbb{R}$ is a continuous function, the support of f is the closure of the set $\{x \in \mathbb{R} : f(x) \neq 0\}$.

   Suppose there is a function $g \in C_c(\mathbb{R}, \mathbb{R})$ such that $f \cdot g = f$ for all $f \in C_c(\mathbb{R}, \mathbb{R})$. That is $f(x)g(x) = f(x)$ for all $x \in \mathbb{R}$. Since the support of the function g is compact, there exists real numbers a and b such that supp $(g) \subseteq [a, b]$. Therefore, if $x \notin [a, b]$, then $f(x) = f(x)g(x) = f(x) \cdot 0 = 0$. This means that supp $(f) \subseteq [a, b]$ for all $f \in C_c(\mathbb{R}, \mathbb{R})$ and this is a contradiction.

   Hence $C_c(\mathbb{R}, \mathbb{R})$ is a commutative ring without multiplicative identity.

3. Let $R = \{(a_n) \in C(\mathbb{N}, \mathbb{R}) : \lim_n a_n = 0\}$. Then R is a ring under the usual addition and multiplication of sequences and R does not have multiplicative identity.

We will prove some simple properties of a ring R.

Proposition 3.1.7. *Let R be a ring with multiplicative identity* 1. *Then* 1 *is unique.*

Proof. Let $1'$ be another identity. Using the property of multiplicative identity, we see that $1' = 1' \cdot 1 = 1$. □

Let R be a ring with multiplicative identity 1. Then $(1 + 1)(a + b) = 1 \cdot (a+b) + 1 \cdot (a+b) = a+b+a+b$. On the other hand we expand $(1+1)(a+b)$ as $(1 + 1)(a + b) = (1 + 1) \cdot a + (1 + 1) \cdot b = a + a + b + b$. Equating the two expressions, we get $a + a + b + b = a + b + a + b$. This proves that $a + b = b + a$ for all a and b in R. Thus if a ring R has multiplicative identity, then the group $(R, +)$ is abelian is a consequence of distributive properties of addition and multiplication.

Example 3.1.8. Let $(R, +)$ be an abelian group. Define $a \cdot b = 0$ for all a and b in R. Then $(R, +, \cdot)$ is a ring. This ring is called the *trivial ring*.

Definition 3.1.9. If $R = \{0\}$, then R is called the *zero ring*.

Proposition 3.1.10. *Let R be a ring. Then*

1. $a \cdot 0 = 0 \cdot a = 0$ *for all a in R.*
2. $(-a) \cdot b = a \cdot (-b) = -(a \cdot b)$ *for all a and b in R.*
3. $-(-a) = a$ *for all a in R.*
4. $(-a) \cdot (-b) = a \cdot b$ *for all a and b in R.*
5. $-a = (-1) \cdot a$ *for all a in R if R is a ring with multiplicative identity.*

Proof. Let a be an element of R. Then,

$$a \cdot 0 = a \cdot (0 + 0) = a \cdot 0 + a \cdot 0.$$

Adding $-(a \cdot 0)$ to the equation $a \cdot 0 = a \cdot (0 + 0) = a \cdot 0 + a \cdot 0$, we get

$$\begin{aligned} 0 &= -(a \cdot 0) + a \cdot 0 \\ &= -(a \cdot 0) + (a \cdot 0 + a \cdot 0) \\ &= (-(a \cdot 0) + a \cdot 0) + a \cdot 0 \\ &= 0 + a \cdot 0 \\ &= a \cdot 0 \end{aligned}$$

This proves $a \cdot 0 = 0$ for all $a \in R$.

Let a and b two elements in R. Then

$$0 = (a + (-a)) \cdot b = a \cdot b + (-a) \cdot b.$$

Hence $(-a) \cdot b = -(a \cdot b)$. Similarly we can show that $a \cdot (-b) = -(a \cdot b)$.

Let a be an element in R. Then

$$-(-a) = -(-a) + 0 = -(-a) + (-a + a) = (-(-a) + (-a)) + a = 0 + a = a.$$

Let a and b be two elements in R. Then

$$0 = (-a) \cdot 0 = (-a) \cdot (b + (-b)) = (-a) \cdot b + (-a) \cdot (-b) = -(a \cdot b) + (-a) \cdot (-b).$$

Hence

$$(-a)(-b) = -(-(a \cdot b)) = a \cdot b$$

for all elements a and b in R.

Let us now assume that the ring R has multiplicative identity 1. Then

$$0 = (1 + (-1)) \cdot a = 1 \cdot a + (-1) \cdot a.$$

This means that $(-1) \cdot a$ is the additive inverse of a. Hence $-a = (-1).a$ for all $a \in R$. □

Exercise 3.1.11. Let R be a commutative ring and $a_1, a_2, \ldots, a_k$ be elements in R for $k \geq 3$. We denote the product $a_1(a_2(a_3 \cdots (a_{k-1}a_k)))$ by $a_1a_2 \cdots a_k$. Show that the product $a_1 \cdot a_2 \cdots a_k = (a_1 \cdot a_2 \cdots a_r)(a_{r+1} \cdot a_{r+2} \cdots a_k)$ for any $r < k$. This is called the *generalized associative law for multiplication.*

We will now prove an important result in a commutative ring called the binomial theorem.

Theorem 3.1.12 (Binomial Theorem)**.** *Let R be a commutative ring. Then, for any two elements a, b in R and any natural number $n \geq 2$,*

$$(a + b)^n = \sum_{r=0}^{n} \binom{n}{r} a^r b^{n-r}.$$

Proof. The proof is by induction on n. Let $n = 2$. Then

$$(a+b)^2 = (a+b)(a+b) = a(a+b)+b(a+b) = a^2+ab+ba+b^2 = a^2+2\,ab+b^2.$$

Let us assume the result to be true for all natural numbers $m < n$. That is, if $m < n$, then $(a+b)^m = \sum_{r=0}^{m} \binom{m}{r} a^r b^{m-r}$.

Now we write $(a+b)^n = (a+b)(a+b)^{n-1}$ by generalized associative law of multiplication. By induction hypothesis, we have

$$(a+b)^{n-1} = \sum_{r=0}^{n-1} \binom{n-1}{r} a^r b^{n-1-r}.$$

Hence

$$\begin{aligned}
(a+b)^n &= (a+b)(a+b)^{n-1} \\
&= (a+b)\left(\sum_{r=0}^{n-1} \binom{n-1}{r} a^{n-1-r} b^r\right) \\
&= a\left[a^{n-1} + \sum_{r=1}^{n-1} \binom{n-1}{r} a^{n-r-1} b^r\right] \\
&\quad +b\left[\sum_{r=0}^{n-2} \binom{n-1}{r} a^{n-r-1} b^{r+1} + b^{n-1}\right] \\
&= a^n + \sum_{r=1}^{n-1} \binom{n-1}{r} a^{n-r} b^r + \sum_{r=1}^{n-2} \binom{n-1}{r} a^{n-1-r} b^r + b^{n-1} \\
&= a^n + \sum_{r=1}^{n-1} \binom{n-1}{r} a^{n-r} b^r + \sum_{r=1}^{n-1} \binom{n-1}{r-1} a^{n-r} b^r + b^n \\
&= a^n + \sum_{r=1}^{n-1} \left\{\binom{n-1}{r} + \binom{n-1}{r-1}\right\} a^{n-r} b^r + b^n
\end{aligned}$$

It follows from Pascal's Lemma 3.1.13 that

$$\binom{n-1}{k} + \binom{n-1}{k-1} = \binom{n}{k}$$

Hence

$$\begin{aligned}
(a+b)^n &= a^n + \sum_{r=1}^{n-1} \left\{\binom{n-1}{r} + \binom{n-1}{r-1}\right\} a^{n-r} b^r + b^n \\
&= a^n + \sum_{k=1}^{n-1} \binom{n}{k} a^{n-k} b^k + b^n \\
&= \sum_{k=0}^{n} \binom{n}{k} a^{n-k} b^k
\end{aligned}$$

□

We will now prove Pascal's lemma.

Lemma 3.1.13 (Pascal's Lemma). *Let $n \geq 2$ be an integer and $k \leq n-1$. Then*

$$\binom{n-1}{k} + \binom{n-1}{k-1} = \binom{n}{k}.$$

Proof. Let $k \leq n-1$. Then

$$\begin{aligned}
\binom{n-1}{k} + \binom{n-1}{k-1} &= \frac{(n-1)!}{(n-1-k)!k!} + \frac{(n-1)!}{(n-1-(k-1))!(k-1)!} \\
&= \frac{(n-1)!}{(n-1-k)!k!} + \frac{(n-1)!}{(n-k)!(k-1)!} \\
&= \frac{(n-1)!}{(k-1)!(n-k-1)!}\left[\frac{1}{k} + \frac{1}{n-k}\right] \\
&= \frac{(n-1)!}{(k-1)!(n-k-1)!}\frac{n}{k(n-k)} \\
&= \frac{n!}{(n-k)!k!} \\
&= \binom{n}{k}.
\end{aligned}$$

□

Another important result is the following theorem.

Theorem 3.1.14. *Let R be a commutative ring. If a and b are two elements in R, then*

$$a^n - b^n = (a-b)\left(a^{n-1}b + a^{n-2}b^2 + \cdots + b^{n-1}\right)$$

for natural numbers $n \geq 2$.

Proof. The proof follows by multiplying out the expression on the right hand side and observe cancellation:

$$\begin{aligned}
(a-b)\left(a^{n-1} + a^{n-2}b + \cdots + b^{n-1}\right) &= a^n - a^{n-1}b + a^{n-1}b - a^{n-2}b^2 \\
&\quad + \cdots + ab^{n-1} - b^n \\
&= a^n - b^n.
\end{aligned}$$

□

Corollary 3.1.15. *Let R be a commutative ring. If a and b are two elements in R, then $a-b$ divides $a^n - b^n$ for every $n \geq 1$.*

Remark 3.1.16. If R is a field and if $a \neq 0$, then we can write $a^n - b^n = a^n(1-(a^{-1}b)^n)$ and expand $(1-(a^{-1}b)^n)$ using geometric series to prove Theorem 3.1.14.

3.2 Divisibility in rings

In Chapter 1 on Elementary Number Theory, we learnt the divisibility properties of integers. In this section we will study the analogous properties in a ring.

Definition 3.2.1. An element u in a ring R with identity is called a unit in R if there exists an element v in R such that $uv = 1$. The element v is called an inverse of u and it is denoted by u^{-1} or $\frac{1}{u}$.

This is not to be confused with the additive inverse of u, which we denote by $-u$.

Definition 3.2.2. A ring R is called a field if every nonzero element in R is a unit in R.

Example 3.2.3.

1. In $\mathbb{Z}$, the integers ± 1 are the only units.
2. In $\mathbb{Q}$, $\mathbb{R}$ and $\mathbb{C}$, all nonzero elements are units.
3. Let $n \geq 2$ and $R = \mathbb{Z}_n$ and $\overline{k}$ be an element in R. Then $\overline{k}$ is a unit in R iff k and n are relatively prime.

   In particular, if p is a prime number, the ring $\mathbb{Z}_p$ is a field.

   Let $\overline{k}$ be unit in $\mathbb{Z}_n$. Then there exists an element $\overline{l}$ in $\mathbb{Z}_n$ such that $\overline{kl} = 1$ in $\mathbb{Z}_n$. This means the integer $kl \equiv 1 \pmod{n}$ and there exists an integer r such that $kl - nr = 1$. Thus k and n are relatively prime integers.

   Conversely let $k \in \mathbb{Z}$ be relatively prime to n. By Bezout's identity there exist integers l and r such that $kl + nr = 1$. Hence $\overline{kl} = 1$ in $\mathbb{Z}_n$.

4. Let $R = M(n, \mathbb{R})$. Then an element $A \in M(n, \mathbb{R})$ is a unit iff $\det(A) \neq 0$. Therefore $\mathrm{GL}(n, \mathbb{R})$ is the set of all units in $M(n, \mathbb{R})$.
5. Let $R := \left\{ \begin{pmatrix} a & -b \\ b & a \end{pmatrix} : a, b \in \mathbb{R} \right\}$. Then R is a field under the matrix addition and multiplication.

   To show that this set R is a ring under the usual matrix addition and multiplication, we need only show that, if A and B two matrices in R, then their sum and multiplication is also in R. (Why?) We will check this now.

   Let $A = \begin{pmatrix} a_1 & -a_2 \\ a_2 & a_1 \end{pmatrix}$ and $B = \begin{pmatrix} b_1 & -b_2 \\ b_2 & b_1 \end{pmatrix}$ be two elements of R.

   Then the matrices

   $$A + B = \begin{pmatrix} a_1 + b_1 & -(a_2 + b_2) \\ a_2 + b_2 & a_1 + b_1 \end{pmatrix}$$

   and

   $$AB = \begin{pmatrix} a_1b_1 + a_2b_2 & -(a_1b_2 + a_2b_1) \\ a_1b_2 + a_2b_1 & a_1b_1 + a_2b_2 \end{pmatrix}$$

   are also in R. Thus the usual matrix addition and multiplication define two binary operations on R. This ring is a field.

If

$$A = \begin{pmatrix} a_1 & -a_2 \\ a_2 & a_1 \end{pmatrix}$$

is nonzero, then the matrix

$$B = \begin{pmatrix} \frac{a_1}{a_1^2+a_2^2} & \frac{a_2}{a_1^2+a_2^2} \\ -\frac{a_2}{a_1^2+a_2^2} & \frac{a_1}{a_1^2+a_2^2} \end{pmatrix}$$

is the inverse of the matrix A.

We will show later that this ring R is the field of *complex numbers.*

6. Let D be a square free integer and $R := \left\{ \begin{pmatrix} a & bD \\ b & a \end{pmatrix} : a, b \in \mathbb{Q} \right\}$. Then R is a field under the addition and multiplication of matrices.

Let

$$A := \begin{pmatrix} a & bD \\ b & a \end{pmatrix} \text{ and } B = \begin{pmatrix} c & dD \\ d & c \end{pmatrix}.$$

be in R. Then their sum $A + B = \begin{pmatrix} a+c & (b+d)D \\ b+d & a+c \end{pmatrix}$ and the product $AB = \begin{pmatrix} ac+bd\,D & (ad+bc)D \\ ad+bc & ac+bd\,D \end{pmatrix}$ are again in R and the rest of the properties are easy to verify.

We will now show that every nonzero element in R is a unit. Let

$$A = \begin{pmatrix} a & bD \\ b & a \end{pmatrix}.$$

If

$$B = \begin{pmatrix} c & dD \\ d & c \end{pmatrix}$$

is such that $AB = Id$, then

$$ac + bd\,D = 1 \quad \text{and } ad + bc = 0.$$

Solving these two equations we get

$$c = \frac{a}{a^2 - Db^2} \quad \text{and} \quad d = -\frac{b}{a^2 - Db^2}.$$

(Observe that $a^2 - Db^2 \neq 0$ by Exercise 1.5.3).

That is the matrix

$$B = \begin{pmatrix} \frac{a}{a^2-Db^2} & -\left(\frac{b}{a^2-Db^2}\right) D \\ -\left(\frac{b}{a^2-Db^2}\right) & \frac{a}{a^2-Db^2} \end{pmatrix}$$

is the inverse of the matrix A. Thus R is a field.

Definition 3.2.4 (Norm Function)**.** Let $R = \mathbb{Z}[\sqrt{D}]$ where D is a square free integer. Let $N : R \to \mathbb{N} \cup \{0\}$ be defined by $N(a + b\sqrt{D}) := |a^2 - Db^2|$. This function N is called the norm function.

If D is negative, then $N(a + b\sqrt{D})$ is the square of the absolute value of the complex number $a + b\sqrt{D}$ and $N(a + b\sqrt{D})$ is also equal to the product $(a + ib\sqrt{-D})(a - ib\sqrt{-D})$.

It is easy to show that $N(\alpha\beta) = N(\alpha)N(\beta)$ for α and β in $\mathbb{Z}[\sqrt{D}]$.

Exercise 3.2.5. Let D be a positive square free integer and $R = \mathbb{Z}[\sqrt{D}]$. Let $a + b\sqrt{D}$ be a nonzero element in R. Show that $N(a + b\sqrt{D}) \neq 0$. (Observe that this exercise is a recast of Exercise 1.5.3.)

Let $a + b\sqrt{D}$ be a nonzero element in R. Then $N(a + b\sqrt{D}) = a^2 - Db^2$. If $D < 0$, then $a^2 - Db^2 > 0$ if either a or b nonzero.

Now suppose $D > 0$ and $N(a + b\sqrt{D}) = 0$. i.e., $a^2 = Db^2$. In this case both a and b are nonzero. Let d be the gcd of a and b. Let $a = da_1$ and $b = db_1$ for some integers a_1 and b_1. Then the equation $a^2 - Db^2 = d^2(a_1 - Db_1^2) = 0$. Hence $a_1^2 - Db_1^2 = 0$.

If a prime number p divides D, then it also divides a_1^2 and hence a_1. Let $a_1 = p^k a_2$ for some $k \geq 1$. Then p^{2k} divides $a_1^2 = Db_1^2$. Since a_1 and b_1 are relatively prime, it follows that p^{2k} divides D and p^{2k+1} does not divide D. Hence we can write $D = p_1^{2k_1} p_2^{2k_2} \cdots p_t^{2k_t}$, a contradiction to our assumption that D is square free.

Norm function is very useful as we will see throughout this chapter. Here we use norm function to find units in some rings.

Example 3.2.6. 1. Let $R = \mathbb{Z}[\sqrt{2}]$. Then $\{(1 \pm \sqrt{2})^k : k \in \mathbb{N} \cup \{0\}\}$ are the units in R.

Let $a + b\sqrt{2}$ be a unit in R and $c + d\sqrt{2}$ in R be its inverse. That is $(a + b\sqrt{2})(c + d\sqrt{2}) = (ac + 2bd) + (ad + bc)\sqrt{2} = 1$. Therefore $1 = N(1) = N((a + b\sqrt{2})(c + d\sqrt{2})) = |a^2 - 2b^2||c^2 - 2d^2|$. As a consequence, an element $a + b\sqrt{2}$ in R is a unit in R iff $a^2 - 2b^2 = \pm 1$. The equation $a^2 - 2b^2 = \pm 1$ is called *Brahmagupta- Bhaskara-Pell's* Equation and all the integer solution are also known. The solutions are $(1 \pm \sqrt{2})^k$ for some natural number k. (See [9] for a proof). Thus the only units in $\mathbb{Z}[\sqrt{2}]$ are $\{(1 \pm \sqrt{2})^k : k \in \mathbb{N} \setminus \{0\}\}$.

2. More generally, if D is an integer and $R = \mathbb{Z}[\sqrt{D}]$, then an element $a + b\sqrt{D}$ is a unit iff $|a^2 - Db^2| = 1$.

3. Let $R = \mathbb{Z}[i\sqrt{2}]$. The only units in R are ± 1.

An element $a + i\sqrt{2}$ is a unit in R iff $a^2 + 2b^2 = 1$. Hence $a = \pm 1$ are the only units in R.

Exercise 3.2.7. 1. Let $D \geq 2$ be a positive square free integer. Show that the integers ± 1 are the only units in $\mathbb{Z}[\sqrt{-D}]$.

Let $a + b\sqrt{-D}$ be a unit in R. Then there exists $c + d\sqrt{-D}$ in R such that $(a + b\sqrt{-D})(c + d\sqrt{-D}) = 1$. Since the norm function is multiplicative, it follows that

$$1 = N(1) = N((a + b\sqrt{-D})(c + d\sqrt{-D})) = (a^2 + Db^2)(c^2 + Dd^2).$$

Therefore $a^2 + Db^2 = 1$ and $c^2 + Dd^2 = 1$ and this is possible iff $(a, b) = (\pm 1, 0)$, $(c, d) = (\pm 1, 0)$.

2. Let D be a square free integer and $R := \left\{ \begin{pmatrix} a & bD \\ b & a \end{pmatrix} : a, b \in \mathbb{Z} \right\}$. What are the units in R?

A matrix
$$A = \begin{pmatrix} a & b\,D \\ b & a \end{pmatrix}$$
be a unit. Then there exists a matrix
$$B = \begin{pmatrix} c & d\,D \\ d & c \end{pmatrix}$$
such that
$$AB = \begin{pmatrix} ac + bd\,D & (ad+bc)\,D \\ (ad+bc) & ac + bd\,D \end{pmatrix} = \pm \begin{pmatrix} 1 & 0 \\ 0 & 1 \end{pmatrix}.$$
This gives rise to two equations
$$ac + D\,bd = \pm 1 \text{ and } ad + bc = 0.$$
Solving these two equations as we have done earlier, we conclude that a matrix
$$A = \begin{pmatrix} a & b\,D \\ b & a \end{pmatrix}$$
is a unit iff $a^2 - D\,b^2 = \pm 1$.

For example, if we let $D = 3$, then the pairs $(a,b) = (\pm 2, \pm 1)$ are solutions to the equation $a^2 - 3b^2 = 1$. Hence the matrices
$$\begin{pmatrix} \pm 2 & \pm 1 \\ \pm 3 & \pm 2 \end{pmatrix}^k$$
for $k \in \mathbb{N}$ are all units in the ring R.

Let us now take $D = 5$. Then again $a = \pm 2$ and $b = \pm 1$ are solutions to the equation $a^2 - 5b^2 = -2$. Hence we can write down a sequence of units in this ring too! The readers may try for small primes like 7, 11 and square free integers like product of two distinct primes! For example if $D = 35$, then a solution to the equation $a^2 - 35b^2 = \pm 1$ is the pair $(a,b) = (\pm 6, \pm 1)$!

If D is negative then $\pm \begin{pmatrix} 1 & 0 \\ 0 & 1 \end{pmatrix}$ are the only units in R.

3. Let $R = \mathbb{Z}[\sqrt{7}]$. Show that $(8 + 3\sqrt{7})^n$ is a unit in R for every $n \geq 0$.

4. Let p be a prime number. Find the units in the ring $\mathbb{Q}[\sqrt{p}]$.

5. Let D be a square free integer. Find the units in the ring $\mathbb{Q}[\sqrt{D}]$.

Exercise 3.2.8. 1. Let X be a non-empty set and $R = \mathcal{F}(X, \mathbb{R})$. Find the units in R.

Let f be a unit in R. Then there exists g in R such that $fg = 1$, the constant function $1(x) = 1$ for all x in X. That is $f(x)g(x) = 1$ for all x in X. This means that $f(x) \neq 0$ for all $x \in X$ and the inverse $g(x) = \frac{1}{f(x)}$ for x in X. We denote g by $\frac{1}{f}$.

Thus a function $f : X \to \mathbb{R}$ is a unit in R iff $f(x) \neq 0$ for all $x \in X$.

2. Let X be a nonempty set and R be a ring. What are the units in $\mathcal{F}(X, R)$.

The units are those functions $f : X \to R$ such that $f(x)$ is a unit in R for every x in X.

Proposition 3.2.9. *Let R be a commutative ring with 1 and $U(R) := \{a \in R : a \text{ is a unit}\}$. Then $U(R)$ is a group under multiplication.*

Proof. Let a and b be two elements in $U(R)$. Then there exists two elements u and v in R such that $au = ua = 1 = bv = vb$. Now

$$(ab)(vu) = a(bv)u = 1, \ \ (vu)(ab) = v(ua)b = 1$$

and

$$(ba)(uv) = b(au)v = bv = 1, \ \ (uv)(ba) = u(vb)a = ua = 1.$$

This proves that ab is a unit in R.

If a, b and c are in $U(R)$, then from what we have seen above it follows that $a(bc)$ and $(ab)c$ are units. Since multiplication is associative in R, we get that $a(bc) = (ab)c$ in $U(R)$.

If a in R is a unit then its inverse a^{-1} is also a unit. Thus $U(R)$ is a group under multiplication. □

Definition 3.2.10 (Division in rings)**.** Let R be a ring and a, b in R. We say that a divides b, if there exists c in R such that $b = ac$. If a divides b, we also write it by $a \mid b$.

An element $a \neq 0$ is called a zero divisor, if there exists $c \neq 0$ such that $ac = 0$. That is, if a divides zero, it is called a zero divisor.

A ring R without zero divisors is called an integral domain. In other words, if $ab = 0$ for two elements a and b in R, then either $a = 0$ or $b = 0$.

Remark 3.2.11.

1. If the ring R is not commutative, the element c is called a *right zero divisor* of a. If there exists an element b in R such that $ba = 0$, then b is called a *left zero divisor* of a. If a has both right and left zero divisors then a is called a *zero divisor*.

2. Let a be an element of a ring R and let b be a divisor of a. If u is a unit in R then bu is also a divisor of a.

   Let a be an element of R and b be a divisor of a. Then there exists an element c in R such that $a = bc$.

   If u is a unit in R, then $a = (bu)(u^{-1}c)$. This means that bu is also a divisor of a.

3. Let R be a ring and u be a unit. Then u divides every element a of the ring R.

   Let a be an element of the ring R. Then we can write $a = u(u^{-1}a)$ for every unit u in R. Thus every unit u in R is a divisor of a in R.

Definition 3.2.12.

1. Let R be a ring. Two elements a and b are called associates in R if they are not units in R and there exists a unit u such that $a = bu$.

2. Let R be a ring and a be an element in R. A divisor b in R is called a proper divisor of a if b is not a unit and b is not an associate of a.

   If an element a in R is such that its only divisors are units and its associates, we say that a does not have proper divisors in R.

Examples 3.2.13. 1. Let $R = \mathbb{Z}$ and let $n = ab$ be a composite number. Then a divides n.

2. Let R be a commutative ring. If a and b are two elements in R, then $a-b$ divides $a^n - b^n$ for every integer $n \geq 2$.

3. Let R be a field. If a and b are in R and $a \neq 0$, then a divides b. If both a and b are nonzero, then a divides b and b divides a. In particular this holds for the rings $\mathbb{Q}$, $\mathbb{R}$ or $\mathbb{C}$.

   If $a \neq 0$, then $b = a\frac{b}{a}$. If both a and b nonzero then $a = b\frac{a}{b}$ and $b = a\frac{b}{a}$.

4. Let $R = \left\{ \begin{pmatrix} a & -b \\ b & a \end{pmatrix} : a, b \in \mathbb{R} \right\}$. If A and B are in R and if A is nonzero, then A divides B.

   This is a special case of earlier example. We have already shown that R is a field.

5. Let $R = \mathbb{Z}_{12}$ and $a = [4]$ and $b = [8]$. Then $b = ac$ where $c = [2]$ or $[5]$ or $[8]$. That is $[8] = [2][4] = [5][4] = [8][4]$.

   This example shows that c appearing in Definition 3.2.10 need not be unique. One can find many such examples and we encourage the readers to find some.

Exercise 3.2.14. 1. Let $R = \mathbb{Z}[i]$. If $a + ib \neq 0$ in R, then show that $a + ib$ divides the integer $a^2 + b^2$ in R.

   For we can write $a^2 + b^2 = (a + ib)(a - ib)$. For instance $1 + i$ divides 2: $2 = (1 + i)(1 - i)$, $2 + i$ divides 5: $5 = (2 + i)(2 - i)$.

2. Let $R = \mathbb{Z}[\sqrt{2}]$. Show the nonzero element $a + b\sqrt{2} \neq 0$ divides the integer $a^2 - 2b^2$ in R. Further show that, for every k in $\mathbb{N}$, the element $(a + b\sqrt{2})(1 + \sqrt{2})^k$ also divides $a^2 - 2b^2$ in R.

3. Let $R = \mathbb{Z}[\sqrt{D}]$. Let a and b two elements in R. If a divides b in R, show that $N(a)$ divides $N(b)$ in $\mathbb{Z}$, the ring of integers. Is the converse true? Substantiate your claim.

4. Let $R = \mathbb{Z}[i]$. Show that the integer 7 does not have proper divisors in R.

   Suppose $\alpha = a + ib$ divides 7. Then $N(a + ib) = a^2 + b^2$ divides $N(7) = 49$. The proper divisors of 49 are 7 and 49. Therefore $a^2 + b^2 = 7$ or $a^2 + b^2 = 49$.

   Suppose $a^2 + b^2 = 7$. Then $a^2 \leq 7$ and $b^2 \leq 7$ and it follows that both $|a|, |b| \leq 2$. Now it is easy to check the equation $a^2 + b^2 = 7$ does not have integer solutions.

Similarly we can show that the integer solutions of equation $a^2+b^2 = 49$ are $a = \pm 7$ and $b = 0$ or $a = 0$ and $b = \pm 7i$. Hence 7 does not have proper divisor in $\mathbb{Z}[i]$.

5. The integer 3 does not have a proper divisor in $\mathbb{Z}[\sqrt{2}]$.

   Suppose $a + b\sqrt{2}$ is a proper divisor of 3. Let $c + d\sqrt{2}$ be a nonzero and nonunit in $\mathbb{Z}[\sqrt{2}]$ such that $3 = (a + b\sqrt{2})(c + d\sqrt{2})$. Therefore

$$9 = N(3) = N(a + b\sqrt{2})N(c + d\sqrt{2}) = |a^2 - 2b^2||c^2 - 2d^2|.$$

   The only possibilities are $a^2 - 2b^2 = \pm 3$ and $c^2 - 2d^2 = \pm 3$. Suppose $a^2 - 2b^2 = 3$. Then $a^2 = 2b^2 + 3$. Hence a is odd , say $a = 2t + 1$ for some $t \geq 1$ and

$$\begin{aligned} 3 &= a^2 - 2b^2 \\ &= (2t+1)^2 - 2b^2 \\ &= 4t(t+1) + 1 - 2b^2. \end{aligned}$$

   This shows that $2t(t+1) = b^2 + 1$. If we now write $b = 2l + 1$, then we get $t(t+1) = 2l(l+1) + 1$, a contradiction. (Why?). Similarly, we can show that $a^2 - 2b^2 = -3$ is not possible either. Thus 3 does not have a proper divisor in $\mathbb{Z}[\sqrt{2}]$.

6. Let $R = \mathbb{Z}[i]$. Then 3 does not have proper divisor in $\mathbb{Z}[i]$.

   Suppose 3 is divisible by an element $a + i\,b$ in $\mathbb{Z}[i]$. Then there exists a nonzero and nonunit element $c + i\,d$ in $\mathbb{Z}[i]$ such that $3 = (a + i\,b)(c + i\,d)$. Hence

$$9 = N(3) = N(a + ib)N(c + id) = (a^2 + b^2)(c^2 + d^2).$$

   We can easily prove that $(a^2 + b^2)(c^2 + d^2) = 9$ iff $a + ib = \pm 3$ or $\pm 3i$. Thus the only divisors of 3 are its associates.

7. Let a be a nonzero element in a ring R such that a divides every nonzero element of R. Show that a is a unit in R.

We will now give some example of rings with zero divisors.

Examples 3.2.15. 1. Let $R = \mathbb{Z}_n$. Let $Z(R)$ be the set of zero divisors in $\mathbb{Z}_n$. Then

$$Z(R) = \{[k] \in \mathbb{Z}_n : \gcd(k, n) = d \text{ for some } d \neq 1\}.$$

Let k be an integer and $d = \gcd(k, n) \geq 2$. If we now write $k = k_1\,d$ and $n = n_1\,d$, then $k \cdot n_1 = k_1(d \cdot n_1) = k_1 \cdot n$. Let us also note that $[n_1] \neq 0$ in $\mathbb{Z}_n$. Hence

$$[k][n_1] = [kn_1] = [k_1(dn_1)] = [k_1 n] = [k_1][n] = 0.$$

Converse follows easily. Thus

$$Z(R) = \{[k] \in \mathbb{Z}_n : (k, n) = d \text{ for some } d \neq 1\}.$$

2. Let $n = k\,l \geq 2$ be a composite number and $R = \mathbb{Z}_n$. Then the elements $[k]$ and $[l]$ are zero divisors.

3. Let $R = \mathbb{Z}_{12}$ and $a = [4]$. Then $[0] = [4][3] = [4][6]$. That is if $ab = ac$ in a ring we can not conclude that $b = c$.

4. Let $R = M(n, \mathbb{R})$. If A is not invertible, then it is a zero divisor.

   Let us write a matrix A in $M(n, \mathbb{R})$ as $A = [C_1, C_2, \ldots, C_n]$, where C_i denotes the i-th column vector. If A is not invertible, then these column vectors are linearly dependent. Hence there exists scalars x_i not all zero such that $\sum_{i=1}^{n} x_i C_i = 0$. Let $X = (x_1, x_2, \ldots, x_n)^t$ and $B = [\overbrace{X, X, \cdots, X}^{n-\text{times}}]$. Then $AB = 0$, the zero matrix.

   If we write A as $A = \begin{pmatrix} R_1 \\ R_2 \\ \vdots \\ R_n \end{pmatrix}$ as column vectors, then there exists scalars y_i not all zero such that $\sum_{i=1}^{n} y_i R_i = 0$. Let us now take $Y = (y_1, y_2, \ldots, y_n)$ and $C = \begin{pmatrix} Y \\ Y \\ \vdots \\ Y \end{pmatrix}$. Then CA is zero matrix. Thus every non-invertible element in $M(n, \mathbb{R})$ is a zero divisor in $M(n, \mathbb{R})$.

   Another Method: Let A in $M(n, \mathbb{R})$ be such that $\det(A) = 0$. Then we know that there exists a nonzero column vector $X := (x_1, x_2, \ldots, x_n)^t$ in $\mathbb{R}^n$ such that $AX = (0, 0, \ldots, 0)^t$. If we now let $B = [\overbrace{X, X, \cdots, X}^{n-\text{times}}]$, then $AB = 0$, the zero matrix. Since $\det(A^t) = \det(A) = 0$, there exists a nonzero column vector $Y = (y_1, y_2, \cdots, y_n)^t$ such $A^t Y = 0$. Let $C = [\overbrace{Y, Y, \cdots, Y}^{n-\text{times}}]$. Then $A^t C = 0$ and hence $C^t A = (A^t C)^t = 0$. Thus a non-invertible matrix in $M(n, \mathbb{R})$ is a zero-divisor.

5. Let $C(\mathbb{R}, \mathbb{R})$ be the ring of all continuous functions on $\mathbb{R}$. Let

$$f(x) = \begin{cases} 0 & \text{if } x \leq 0 \\ x & \text{if } x > 0 \end{cases} \text{ and } g(x) = \begin{cases} x & \text{if } x \leq 0 \\ 0 & \text{if } x > 0 \end{cases}.$$

   Then the functions f and g are nonzero. However the product $f \cdot g$ is identically zero function. Hence the functions f and g are zero divisors in R. Can you find more zero divisors in R?

6. Let $R = \mathcal{F}(X, \mathbb{R})$. If a function f in $\mathcal{F}(X, \mathbb{R})$ is such that $f(x_0) = 0$ for some point x_0 in X, then it is a zero divisor.

   Let $f : X \to \mathbb{R}$ be a function and x_0 in X be such that $f(x_0) = 0$. Let $Z(f) := \{x \in X : f(x) = 0\}$ and define the function $g : X \to \mathbb{R}$ by

$$g(x) = \begin{cases} 1 & \text{if } x \in Z(f) \\ 0 & \text{if } x \notin Z(f) \end{cases}.$$

Then $f(x)g(x) = 0$ for all x in X.

In the examples above all the nonzero and nonunits are zero divisors. The following are some examples of rings R in which nonzero and nonunits that need not be not zero divisors in R.

Examples 3.2.16. 1. If $R := C(J, \mathbb{R})$ is the ring of continuous functions on an interval J, there are functions which are neither units nor zero divisors.

Let x_0 be a point in the interval J. Let $f : J \to \mathbb{R}$ be defined by $f(x) = x - x_0$. The function f is a continuous on J. Since $f(x_0) = 0$, it is not unit in R. Suppose there exists a continuous function $g : J \to \mathbb{R}$ such that $f(x)g(x) = 0$ for all points x in J. Since $f(x) \neq 0$ for $x \neq x_0$, it follows that $g(x) = 0$ for all $x \neq x_0$. Now we use the continuity of the function g at the point x_0 to conclude that $g(x_0) = \lim_{x \to x_0} g(x) = 0$. We already know that $g(x) = 0$ for $x \neq x_0$. Thus $g \equiv 0$, the zero function. This shows that f is neither a unit nor a zero divisor.

One can find lot of continuous functions which are neither units nor zero divisors.

2. Let $X = \mathbb{N}$ and $C(\mathbb{N}, \mathbb{R}) := \{(a_n) : \lim_n a_n \text{ exists}\}$. Then we know that $C(\mathbb{N}, \mathbb{R})$ is a ring under addition and multiplication of sequences. In this ring let $a_n = \frac{1}{n}$. Then (a_n) is a neither a unit nor a zero divisor.

Let $a_n = \begin{cases} \frac{1}{n} & n \text{ is even} \\ 0 & n \text{ is odd} \end{cases}$. The sequence (a_n) is a zero divisor.

Exercise 3.2.17. 1. Let R be a ring and let $A := \mathcal{F}(X, R)$. What are the zero divisors in the ring A? If the ring R is an integral domain, what are the zero divisors in A?

2. Show that the ring $C_c(\mathbb{R}, \mathbb{R})$ does not have units.

3. Let R be a ring with 1 and a be a zero divisor. Show that a is not a unit.

Proposition 3.2.18. *Let R be a ring. If an element a in R is a unit then its inverse is unique and it is not a zero divisor.*

Proof. Let a be a unit. Let b and c be inverses of a. Then

$$b = b \cdot 1 = b \cdot (ac) = (ba) \cdot c = c.$$

This proves the uniqueness of the multiplicative inverse of a unit.

Let d in R be such that $ad = 0$. Then

$$0 = b \cdot 0 = b(ad) = (ba) \cdot d = 1 \cdot d = d.$$

Hence a unit in ring is not a zero divisor. □

Proposition 3.2.19. *Let R be an integral domain. Then the following holds.*

1. *If a and b are two nonzero elements in R such that a divides b and b divides a, then a and b are associates in R.*

2. *Let a, b and c be three elements in R such that $ab = ac$. If $a \neq 0$, then show that $b = c$. This is called the* cancellation law for multiplication *in R.*

   Further show that this assertion is not valid if R is not an integral domain.

Proof. Let a and b be two nonzero elements dividing each other. Then there exists u and v in R such that $b = au$ and $a = bv$. Therefore $a = bv = auv$ and $a(1 - uv) = 0$. Since a is nonzero, it follows that $uv = 1$. Thus a and b are associates in R.

Let a, b and c in R be such that $ab = ac$. Then $a(b - c) = 0$. Since R is an integral domain, either $a = 0$ or $b = c$. If $a \neq 0$, then $b = c$ necessarily. □

Remark 3.2.20. Let us now contrast the cancellation law for multiplication with the cancellation law in a group G.

Let G be a group. If a, b and c are three elements in G such that $ab = ac$, then $b = c$. This is done by using the inverse of a: $b = a^{-1}(ab) = a^{-1}(ac) = c$. However if R is an integral domain, we don't need a to be a unit.

Proposition 3.2.21. *Let R be a field. If a and b are two nonzero elements in R, then they are associates in R.*

Proof. In a field every nonzero element is a unit. Thus, if a and b are two nonzero elements in R, we can write $a = b(b^{-1}a)$. □

Example 3.2.22. 1. The ring of integers $\mathbb{Z}$ is an integral domain.

2. Let p be a prime number. Then $\mathbb{Z}_p$ is an integral domain.

   Let $[k]$ and $[l]$ in $\mathbb{Z}_p$ such that $[kl] = [0]$ in $\mathbb{Z}_p$. This means that the integer $kl \equiv 0 \pmod{p}$. Since p is a prime number p divides k or l, which means that $[k] = 0$ or $[l] = 0$ in $\mathbb{Z}_p$.

3. Let D be a square free integer. Then the ring $\mathbb{Z}[\sqrt{D}]$ is an integral domain.

   Let $a + b\sqrt{D}$ be a nonzero element in $\mathbb{Z}[\sqrt{D}]$. If there exists an element $c + d\sqrt{D}$ in R such that $(a + b\sqrt{D})(c + d\sqrt{D}) = 0$, then

$$\begin{aligned} N((a + b\sqrt{D})(c + d\sqrt{D})) &= N(a + b\sqrt{D})N(c + d\sqrt{D}) \\ &= |a^2 - Db^2||c^2 - Dd^2| = 0. \end{aligned}$$

   We have seen that $N(a + b\sqrt{D}) \neq 0$ iff $a + b\sqrt{D} \neq 0$. Hence $c + d\sqrt{D} = 0$ and this shows that $\mathbb{Z}[\sqrt{D}]$ is an integral domain.

Proposition 3.2.23. *A finite integral domain is a field.*

Proof. Let R be a finite integral domain and $R \setminus \{0\} := \{a_1, a_2, \ldots, a_n\}$.

Let a be a nonzero element in R. Then using cancellation law for multiplication in R, we conclude that $aa_i \neq 0$ for all $i = 1, 2, \cdots, n$ and $aa_i \neq aa_j$ if $i \neq j$. Therefore $\{a_1, a_2, \ldots, a_n\} = \{aa_1, aa_2, \ldots, aa_n\}$. Since $1 \neq 0$ in R, there exists an integer j such that $aa_j = 1$. This proves that a is a unit in R. Hence R is a field. □

Remark 3.2.24. This result need not be true, if R is infinite. For example the ring $\mathbb{Z}$ is an integral domain. But it is not a field.

Exercise 3.2.25. Let R be a field. Show that R is an integral domain.

Definition 3.2.26. Let R be a ring. An element x in R is called nilpotent if there exists $n \geq 2$ in $\mathbb{N}$ such that $x^n = 0$.

Remark 3.2.27. Let R be a ring and x be a nonzero nilpotent element in R. Then x is a zero divisor in R.

Examples 3.2.28. 1. Let $R = M(n, \mathbb{R})$ and $A = (a_{ij})$ with $a_{ij} = 0$ for $1 \leq j \leq i \leq n$. Then A is nilpotent.

Let A be a matrix in R such that $A = (a_{ij})$ with $a_{ij} = 0$ for $1 \leq j \leq i \leq n$.
Let us write A as $A = [C_1, C_2, \ldots, C_n]$ where C_j is the j-th column of A. First we observe that the first column C_1 is zero. The matrix A^2 will have first two columns as zero and we can show by successive multiplication that the first i-columns of the matrix A^i will be zero. Thus A^n will be an identically zero matrix.

2. Let R be an integral domain. Then R does not have any nilpotent elements.

3. The rings $\mathcal{F}(X, \mathbb{R})$ and $C(I, \mathbb{R})$ do not have nilpotent elements.

   More generally if R is an integral domain, then the ring $A = \mathcal{F}(X, R)$ does not have nilpotent elements. However this ring A is not an integral domain.

   Suppose f in $\mathcal{F}(X, \mathbb{R})$ is nilpotent. This means that there exists a natural number k such that $f^k \equiv 0$, the zero function. That is $f^k(x) = (f(x))^k = 0$ for every point x in X. If $f(x)$ is a nonzero real number then every power of $f(x)$ is nonzero. Hence $f(x) = 0$ for every point x in X and it proves that $\mathcal{F}(X, \mathbb{R})$ has no nilpotent elements.

Exercise 3.2.29. Let $n = p_1^{r_1} p_2^{r_2} \cdots p_k^{r_k}$. Show that the elements $[p_i^{r_i}]$ are not nilpotent in $\mathbb{Z}_n$. What are the nilpotent elements in $\mathbb{Z}_n$?

Proposition 3.2.30. *Let R be a commutative ring.*

1. *If x is a nilpotent element in R, then $1 \pm x$ is a unit in R. Further if y is the inverse of $1 + x$, then $y - 1$ is nilpotent in R.*

2. *If x and y are nilpotent in R then $x \pm y$ are also nilpotent in R.*

3. *If x is nilpotent then rx is nilpotent for every element r in the ring R.*

4. *The set $\{1 + x : x \text{ is nilpotent in } R\}$ is a subgroup of group of units in R.*

Proof. Let x be nilpotent and $n \geq 2$ be such that $x^n = 0$. Observe that $-x$ is also nilpotent in R. (Why?) Then

$$1 = 1 - x^n = (1 - x)(1 + x + x^2 + \cdots + x^{n-1}).$$

This proves that $1-x$ is a unit and its inverse is $y = 1+x+x^2+\cdots+x^{n-1}$. Now $y-1 = x(1+x+x^2+\cdots+x^{n-2})$. Since $x^n = 0$, it follows that $(y-1)^n = x^n(1+x+x^2+\cdots+x^{n-2})^n = 0$. Hence $y-1$ is nilpotent.

Similarly we can show that $1+x$ is also unit. Thus $1 \pm x$ are units in R.

Let x and y be two nilpotent elements in R such that $x^n = 0 = y^k$ for some integers $k, n \geq 2$. Then

$$(x \pm y)^{nk} = \sum_{r=0}^{kn} \binom{kn}{r} x^r y^{kn-r}.$$

In this sum, if $r \leq n$, then $kn - r \geq kn - n = (k-1)n \geq n$. Hence $y^{kn-r} = 0$. If $r = n+t$ for some $t \neq 0$, then $x^r = x^{n+t} = x^n x^t = 0$. Therefore $x^r y^{nk-r} = 0$ for every $r \leq kn$ and we conclude that $(x \pm y)^{nk} = 0$. This proves that $x \pm y$ is also nilpotent in R.

If x is nilpotent and r in R, then $(rx)^k = r^k x^k$ for all $k \in \mathbb{N}$. Therefore, if $x^n = 0$, then $(rx)^n = 0$. □

Definition 3.2.31 (Greatest Common Divisor). Let R be a ring. Let a and b be two elements in R. We say an element d in R is a greatest common divisor- written shortly as gcd- of the elements a and b

1. if d divides a and b in R, and

2. if an element e in R divides a and b in R, then e divides d.

Definition 3.2.32 (Least Common Multiple). Let R be a ring. Let a and b be two elements in R. We say an element l in R is a least common multiple- written shortly as lcm- of the elements a and b

1. if l is a multiple of a and b in R, and

2. if k in R is a multiple of a and b in R, then k is a multiple of l in R.

Example 3.2.33. Let $R = \mathbb{Z}$. Let $a = 4$ and $b = 6$. Then, by our definition, the gcd of two integers 4 and 6 are ± 2 and the lcm are ± 12. But in the ring $\mathbb{Z}$, conventionally, the gcd and lcm are taken to be positive. Therefore 2 is the gcd of the integers is 2 and 6 and 12 is the lcm.

However in a general ring R, we may not be able to do as in integers. Let R be a ring. Let d be a gcd of two nonzero elements a and b. If u is a unit in R, then du is also a gcd of the two elements a and b. As a consequence there may be many (possibly infinite) gcds for two elements in a ring R. We will give example of such a ring now.

Example 3.2.34. Let $R = \mathbb{Z}[\sqrt{2}]$. Let $a = 4$ and $b = 6$ and d be a proper divisor of a and b. Then $N(d)$ divides $N(4) = 16$ and $N(6) = 36$. Hence the integer $N(d)$ divides the integer $4 = \gcd(16, 36)$ in $\mathbb{Z}$. Therefore $N(d) = 1, 2$ or 4. Since d is not a unit, it follows that $N(d) = 2$ or 4. Suppose $d = d_1 + d_2\sqrt{2}$. Then $N(d) = |d_1^2 - 2d_2^2| = 2$ or 4.

Suppose $|d_1^2 - 2d_2^2| = 2$. If $d_1 \neq 0$, then the integer d_1 is even. If we write $d_1 = 2u_1$, then $2d_2^2 = 2 - 4u_1^2 = 2(1 - 2u_1^2) \leq 0$. This is a contradiction. Thus $d_1 = 0$ and hence $d = \pm\sqrt{2}$.

Let us now assume that $|d_1^2 - 2d_2^2| = 4$. In this case $d_1 \neq 0$.(why?) Let $d_1 = 2u_1$. Then $2d_2^2 = 4(u_1^2 - 1) \geq 0$. If $d_2 \neq 0$, then it is even and we write $d_2 = 2u_2$ to conclude that $d = 2(u_1 - u_2\sqrt{2})$, where $u_1^2 - 2u_2^2 = \pm 1$. Thus $d = 2u$ where u is a unit. Hence the integer 2 and all the unit multiples $2u$ are gcds of a and b.

We have seen above an example of a ring in which two nonzero elements have infinitely many gcds. We will now give an example of a ring R and two nonzero elements a and b in R such that they do not have gcd d in R.

Example 3.2.35. Let $R = \mathbb{Z}[\sqrt{-5}]$. Let $a = 6$ and $b = 2(1+\sqrt{-5})$. These two elements are nonzero and nonunits in R. First observe that a does not divide b and b does not divide a.(Why?)

We claim that a and b do not have a gcd in R. Suppose they have a gcd and let d be a gcd of a and b. Let $a = d\alpha$ and $b = d\beta$ for some α and β in R. Then $36 = N(6) = N(d)N(\alpha)$ and $24 = N(b) = N(d)N(\beta)$. This shows that $N(d)$ divides the integers 24 and 36. Therefore 2, 3, 4, 6 or 12 are the possible values for $N(d)$.

We write $6 = 2 \times 3 = (1+\sqrt{-5})(1-\sqrt{-5})$. This shows that the integer 2 and the element $1+\sqrt{-5}$ divide both a and b in R. Therefore, by definition of gcd, the integer 2 and $1+\sqrt{-5}$ divide d in R. As a consequence the integers $4 = N(2)$ and $6 = N(1+\sqrt{-5})$ divide $N(d)$ in $\mathbb{Z}$. Hence $N(d) = 12$. If we write $d = d_1 + d_2\sqrt{-5}$, then $N(d) = d_1^2 + 5d_2^2 = 12$. It can be shown easily that the equation $d_1^2 + 5d_2^2 = 12$ does not have integer solution. Therefore the elements a and b do not have gcd in R.

Definition 3.2.36.

1. An element $a \neq 0$ in a ring R is called reducible in R, if there exist elements b and c in R, both nonzero and nonunit such that $a = bc$. In other words, the element a has a proper divisor in R.

   The elements b and c are called proper divisors of a.

2. An element $a \neq 0$ in a ring R is called irreducible if a is not a unit and if $a = bc$ for some elements b and c in R, then either b or c is a unit in R. Equivalently only divisors of a in R are units or the associates of a in R.

3. An element p in a ring R is called prime in R if p is a nonunit and if p divides the product ab of two elements a and b in R, then p divides either a or b in R.

Example 3.2.37.

1. All prime numbers are irreducible in $\mathbb{Z}$. We have shown this in first chapter and we recall it here.

   Let $p \geq 2$ be a prime number and let $p = ab$ for some integers a and b. We need to show that either a or b is a unit. Since p is a prime number, it follows that p divides either a or b. If p divides a and if $a = pc$ for some integer c, then we get $p = pcb$. This means that $p(1 - cb) = 0$. Hence $cb = 1$ proving that b is a unit in $\mathbb{Z}$.

2. Let R be an integral domain. Then every prime element in R is irreducible in R.

   The proof is same as above.

3. The integers 2 is irreducible in $\mathbb{Z}$ but reducible in $\mathbb{Z}[i]$. The integer 3 is irreducible in $\mathbb{Z}$ and also in $\mathbb{Z}[i]$.

The integers 2 and 3 are irreducible in $\mathbb{Z}$. We will show that 2 is reducible in $\mathbb{Z}[i]$ and 3 is irreducible in $\mathbb{Z}[i]$.

We know that ± 1 and $\pm i$ are the only units in $\mathbb{Z}[i]$. But we can write 2 as the product of two nonunits $1 \pm i$. That is $2 = (1+i)(1-i)$.

We will now show that 3 is irreducible in $\mathbb{Z}[i]$. Let us assume that $3 = (a+ib)(c+id)$. Then $9 = N(3) = N((a+ib)(c+id)) = (a^2+b^2)(c^2+d^2)$. Since the equation $a^2+b^2 = 3$ does not have integer solutions, the only possibilities are $a^2+b^2 = 1$ and $c^2+d^2 = 9$ or $a^2+b^2 = 9$ and $c^2+d^2 = 1$.

The solutions for the equation $a^2+b^2 = 9$ are $(a, b) = (\pm 3, 0)$ or $(0, \pm 3)$.

This proves that the integer 3 is irreducible in $\mathbb{Z}[i]$.

4. We have shown that in an integral domain a prime element is irreducible. This example shows that the converse is not true.

The integers 2 and 3 are irreducible in $\mathbb{Z}[\sqrt{-5}]$ but they are not prime in $\mathbb{Z}[\sqrt{-5}]$.

First we show that 2 is irreducible in $\mathbb{Z}[\sqrt{-5}]$. Assume that 2 is reducible in $\mathbb{Z}[\sqrt{-5}]$. Let $a+b\sqrt{-5}$ and $c+d\sqrt{-5}$ in $\mathbb{Z}[\sqrt{-5}]$ be two nonzero and nonunit such that $2 = (a+b\sqrt{-5})(c+d\sqrt{-5})$. Therefore $4 = N(2) = N(a+b\sqrt{-5})N(c+d\sqrt{-5}) = (a^2+5b^2)(c^2+5d^2)$. This leads us to two inequalities $a^2+5b^2 \leq 4$ and $c^2+5d^2 \leq 4$. This is possible iff $b = d = 0$ and $(a, c) = (\pm 2, \pm 1)$ or $(a, c) = (\pm 1, \pm 2)$. That is $a+b\sqrt{-5} = \pm 2$ and $c+d\sqrt{-5} = \pm 1$ or $a+b\sqrt{-5} = \pm 1$ and $c+d\sqrt{-5} = \pm 2$. This proves that 2 is irreducible in $\mathbb{Z}[\sqrt{-5}]$. Using similar arguments one can show that 3 is also irreducible in $\mathbb{Z}[\sqrt{-5}]$.

We will now show that 3 is not prime in $\mathbb{Z}[\sqrt{-5}]$.

We write $6 = 2 \times 3 = (1+\sqrt{-5})(1-\sqrt{-5})$. Thus 3 divides $6 = (1+\sqrt{-5})(1-\sqrt{-5})$. If 3 were prime, then 3 divides either $1+\sqrt{-5}$ or $1-\sqrt{-5}$. Let us assume that 3 divides $1+\sqrt{-5}$. Then there exists $a+b\sqrt{-5}$ such that $1+\sqrt{-5} = 3(a+b\sqrt{-5})$. Therefore $6 = N(1+\sqrt{-5}) = N(3)N(a+b\sqrt{-5}) = 9N(a+b\sqrt{-5}) \geq 9$, a contradiction. Hence 3 is not a prime in $\mathbb{Z}[\sqrt{-5}]$.

Similarly we can show that 2 is not prime either.

Exercise 3.2.38. 1. Show that the integer 29 is reducible in $\mathbb{Z}[\sqrt{7}]$.

2. Show that the integer 7 is irreducible in $\mathbb{Z}$, but reducible in $\mathbb{Z}[\sqrt{2}]$.

For $7 = (2\sqrt{2}+1)(2\sqrt{2}-1)$. we need to show that $2\sqrt{2} \pm 1$ are not units in $\mathbb{Z}[\sqrt{2}]$. We have seen in an earlier example that the units in $\mathbb{Z}[\sqrt{2}]$ are of the form $(1 \pm \sqrt{2})^n$ for $n \in \mathbb{Z}$ and these two elements are not of that form.(Why?)

Alternately we can use the norm function to prove this and we leave it as an exercise.

We have also shown that 7 and $1 \pm 2\sqrt{2}$ are not associates in $\mathbb{Z}[\sqrt{2}]$.

3. Show that the integers 2 and 3 are irreducible but not primes in the ring $R = \mathbb{Z}[\sqrt{7}]$.

4. Let I be an interval and $R = C(I, \mathbb{R})$, the ring of continuous functions.

   (a) Let y be a point in I. Is the function $\varphi(x) := x - y$ irreducible in R?

   (b) If $s \neq t$ are two points in I, then show that the function $f(x) := (x - t)(x - s)$ is reducible in R. In general, if t_i for $1 \leq i \leq k$ are in I, then the function $f(x) = \prod_{i=1}^{k}(x - t_i)$ is reducible.

Exercise 3.2.: 9. 1. Let R be an integral domain. If $a_1, a_2 \ldots a_n$ are irreducible elements in R, show that the product $a_1 a_2 \cdots a_n$ is not a unit in R.

2. Let R be an integral domain. If an element a in R divides all the nonzero elements in R, then show that a is a unit in R.

   The element a divides 1. Hence there exists q in R such that $1 = aq$.

3.3 Polynomials, Power Series, Laurent Polynomials and Laurent Series

3.3.1 Polynomial Rings

We are already familiar with polynomials over real and complex numbers and we know how to work with them. Now we will formally define a polynomial ring with coefficients in a ring R.

Let R be a ring. A polynomial over the ring R is a formal expression of the form $a_0 + a_1X + a_2X^2 + \cdots + a_nX^n$ where $a_i \in R$ called the coefficients of the polynomial.

In this description X is just a symbol and the powers X^k are called monomials.

We say that two polynomials $f(X) = a_0 + a_1X + a_2X^2 + \cdots + a_nX^n$ and $g(X) = b_0 + b_1X + b_2X^2 + \cdots + b_mX^m$ are same if $a_i = b_i$ for $i = 0, 1, 2, \ldots$.

We define the *degree* of a polynomial to be the largest integer n such that the coefficient a_n of the monomial X^n is nonzero. The coefficient a_n is also called the *leading coefficient* of the polynomial $a_0 + a_1X + a_2X^2 + \cdots + a_nX^n$.

When we work with the polynomials some of the coefficients being zero creates a problem.

For example the polynomial $1 + X^2 + X^3$ can also be written as $1 + 0.X + X^2 + X^3$ and also as $1 + 0.X + X^2 + X^3 + 0.X^4$. To avoid this we may insist that we write only nonzero coefficients of a polynomial. But even this creates problem. For example if $f(X) = 1 + X + X^2$ and $g(X) = 1 - X + X^2$, then $f(X) + g(X) = 2 + (1-1).X + 2X^2$ produces a zero coefficient. To avoid this we may opt to write a polynomial with the leading coefficient and list all the preceding terms. This also has problem.

If

$$\begin{aligned} f(X) &= X^n - X^{n-1} + 0.X^{n-2} + \cdots + 0.X + 0 \text{ and} \\ g(X) &= -X^n + X^{n-1} + 0.X^{n-2} + \cdots + 0.X + 0 \text{ then} \\ f(X) + g(X) &= 0.X^n + 0.X^{n-1} + 0.X^{n-2} + \cdots + 0.X + 0. \end{aligned}$$

We get around these by defining a polynomial to be a formal expression

$$a_0 + a_1X + a_2X^2 + \cdots$$

with the convention that only finitely many a_i are nonzero.

That is

$$R[X] := \left\{ \sum_{k=0}^{\infty} a_k X^k : a_k \in R \text{ and } a_k \neq 0 \text{ only for finitely many } k \right\}.$$

Let $f(X) := a_0 + a_1X + a_2X^2 + \cdots$ and $g(X) = b_0 + b_1X + b_2X^2 + \cdots$. If $a_i = 0$ for $i \geq n_1$ and $b_j = 0$ for $j \geq n_2$, then $a_i + b_i = 0$ for $i \geq \max\{n_1, n_2\}$ and $a_ib_j = 0$ if $i \geq n_1$ or $j \geq n_2$. Thus the sum $f(X) + g(X) := \sum_{k\geq 0}(a_k + b_k)X^k$ and the product $f(x)g(x) := \sum_{k\geq 0}\left(\sum_{i+j=k} a_ib_j\right) X^k$ are again in $R[X]$.

It is now easy to check that $R[X]$ is a ring under the addition and multiplication defined above. This is called the *polynomial ring* over R.

There is another definition of a polynomial ring and we will describe it now.

Let (a_n) be a sequence such that the terms a_i are from the ring R. We say that the sequence (a_n) is *eventually zero* if there exists a non-negative integer $k \geq 0$ such that $a_n = 0$ for all $n \geq k$. Given an eventually zero sequence, we associate the polynomial

$$a_0 + a_1X + \cdots + a_rX^r + \cdots.$$

Conversely given a polynomial

$$a_0 + a_1X + a_2X^2 + \cdots$$

we get an eventually zero sequence $(a_0, a_1, \cdots, a_k, \cdots)$ of terms from the ring R. Thus the set of polynomials is in one-one correspondence with the set of all eventually zero sequences in R.

On this set of eventually zero sequences, if we now define the addition and multiplication as

$$(a_n) + (b_n) := (a_n + b_n)$$

and

$$(a_n)(b_n) := \sum_{n\geq 0} \sum_{i+j=n} a_ib_j$$

for eventually zero sequences (a_n) and (b_n), then the sum $(a_n) + (b_n)$ and the product $(a_n)(b_n)$ are also eventually zero sequences. Further it agrees with the addition and multiplication of polynomials.

Definition 3.3.1. Let $f(X) := a_0 + a_1X + a_2X^2 + \cdots + a_kX^k + \cdots$ be a polynomial in $R[X]$. Let n be a non-negative integer such that $a_j = 0$ for $j \geq n+1$ and $a_n \neq 0$. Then n is called the degree of the polynomial $f(X)$.

If $a_i = 0$ for all i, the degree is defined to be -1.

A polynomial ring $R[X]$ has following simple properties.

Proposition 3.3.2. *Let R be a ring. Let $f(X) := a_0 + a_1X + a_2X^2 \cdots$ and $g(X) = b_0 + b_1X + b_2X^2 + \cdots$ be two polynomials in $R[X]$. Then*

1. $\deg(f(X) + g(X)) \leq \max\{\deg f(X), \deg g(X)\}$,
2. $\deg(f(X)g(X)) \leq \deg f(X) + \deg g(X)$, *and*
3. $\deg(f(X)g(X)) = \deg f(X) + \deg g(X)$ *if R is an integral domain and both $f(X)$ and $g(X)$ are nonzero.*

Proof. Let $f(X) = a_0 + a_1X + a_2X^2 \cdots$ and $g(X) = b_0 + b_1X + b_2X^2 + \cdots$ be two polynomials in $R[X]$. Let n and m be the degree of the polynomials f and g respectively.

If $n < m$, then

$$f(X) + g(X) = a_0 + b_0 + (a_1 + b_1)X + \cdots + (a_n + b_n)X^n + \cdots + b_mX^m$$

is a polynomial of degree $m = \deg g(X)$. Similarly if $m < n$, then $f(X) + g(X)$ is a polynomial of degree $n = \deg f(X)$. If $n = m$, then $f(X) + g(X) = a_0 + b_0 + (a_1 + b_1)X + \cdots + (a_n + b_n)X^n$. Further $f(X) + g(X)$ is a polynomial of degree less than n if $a_n + b_n = 0$ and equal to n if $a_n + b_n \neq 0$. In all possible cases we see that $\deg(f(X) + g(X)) \leq \max\{\deg f(X), \deg g(X)\}$.

The product $f(X)g(X) = a_0b_0 + (a_0b_1 + a_1b_0)X + \cdots + a_nb_mX^{m+n} + \cdots$. This shows that the degree of the polynomial $f(X)g(X)$ is less than $n+m$ if $a_nb_m = 0$ and equal to $n+m$ if $a_nb_m \neq 0$. Hence $\deg(f(X)g(X) \leq \deg f(X) + \deg g(X)$. If R is an integral domain, then $a_nb_m \neq 0$ and in this case $\deg(f(X)g(X) = \deg(f(X)) + \deg(g(X))$. □

Corollary 3.3.3. *If R is an integral domain, then $R[X]$ is an integral domain.*

Proof. We will prove this result by showing that the product of two nonzero polynomials is a nonzero polynomial.

Let $f(X) = a_0 + a_1X + a_2X^2 + \cdots$ be a nonzero polynomial of degree n and $g(X) = b_0 + b_1X + b_2X^2 + \cdots$ be a nonzero polynomial of degree m. Therefore $a_n \neq 0$ and $b_m \neq 0$. Since R is an integral domain, the product a_nb_m is also nonzero. As a consequence, product $f(X)g(X)$ is nonzero in $R[X]$. □

Definition 3.3.4. Let R be a ring and $f(X) = a_0 + a_1X + \cdots + a_nX^n$ be a polynomial of degree n with coefficients a_i in R. We say that $f(X)$ is a monic polynomial in $R[X]$ if the leading coefficient a_n is a unit in R.

Example 3.3.5.
1. The polynomial $f(X) = 1 + X$ is a monic polynomial in $\mathbb{Z}[X]$.
2. The polynomial $f(X) = 1 + X + 2X^2$ is not a monic polynomial in $\mathbb{Z}[X]$. However it is a monic polynomial in $\mathbb{Q}[X]$.

3. If R is a field, then every nonzero polynomial $f(X)$ is a monic polynomial in $R[X]$.

Exercise 3.3.6. 1. Let $R = \mathbb{Z}[X]$, the polynomial ring over integers. What are the units in R?

Let $f(X) := a_0 + a_1X + a_2X^2 + \cdots + a_nX^n$ be a unit and $g(X) := b_0 + b_1X + b_2X^2 + \cdots + b_mX^m$ be its inverse. Then

$$f(X)g(X) := 1.$$

Since the degree of a constant polynomial is zero, we see that

$$0 = \deg(1) = \deg(f(X)g(X)) = \deg f(X) + \deg g(X).$$

Therefore both the polynomials are constant, $f(X) = a_0$ and $g(X) = b_0$. This proves that the units in $\mathbb{Z}[X]$ are the same as the units in $\mathbb{Z}$.

2. Let R be an integral domain. Find the units in $R[X]$.

One of the most important result in polynomial rings is the *division algorithm*.

Theorem 3.3.7 (Division Algorithm). *Let $\mathbb{F}$ be a field and $R = \mathbb{F}[X]$, the polynomial ring over $\mathbb{F}$. Let $f(X)$ be a nonconstant polynomial in R. Let $g(X)$ be a polynomial in $\mathbb{F}[X]$. Then there exist unique polynomials $q(X)$ and $r(X)$ in $\mathbb{F}[X]$ such that $g(X) = f(X)q(X) + r(X)$ such that either $r(X) \equiv 0$ or $0 \leq \deg r(X) < \deg f(X)$.*

The polynomial $q(X)$ is called the quotient and $r(X)$ the remainder.

First we will prove the following lemma.

Lemma 3.3.8. *Let R be a ring. Let $f(X)$ and $g(X)$ be two polynomials in $R[X]$. Assume $f(X)$ is monic and $\deg g(X) \geq \deg f(X)$. Then there exists a polynomial $q(X)$ in $R[X]$ such that $\deg(g(X) - q(X)f(X)) < \deg g(X)$.*

Proof. Let $f(X) = a_0 + a_1X + a_2X^2 + \cdots + a_nX^n$ where a_n is a unit and $g(X) = b_0 + b_1X + b_2X^2 + \cdots + b_mX^m$ where $m = n + k$ for some $k \geq 0$.

We multiply the polynomial $f(X)$ by the polynomial $q(X) = \frac{b_m}{a_n}X^k$ and subtract from the polynomial $g(X)$ to get

$$\begin{aligned}
&g(X) - \left(\frac{b_m}{a_n}\right) X^k f(X) \\
&= b_0 + b_1X + b_2X^2 + \cdots + b_{k-1}X^{k-1} + \left(b_k - \frac{b_m}{a_n}a_0\right) X^k \\
&\quad + \left(b_{k+1} - \frac{b_m}{a_n}a_1\right) X^{k+1} + \cdots + \left(b_m - \frac{b_m}{a_n}a_n\right) X^m \\
&= b_0 + b_1X + b_2X^2 + \cdots + b_{k-1}X^{k-1} + \left(b_k - \frac{b_m}{a_n}a_0\right) X^k \\
&\quad + \left(b_{k+1} - \frac{b_m}{a_n}a_1\right) X^{k+1} \cdots + \left(b_{m-1} - \frac{b_m}{a_n}a_{n-1}\right) X^{m-1} \\
&= c_0 + c_1X + c_2X^2 + \cdots + c_{m-1}X^{m-1}
\end{aligned}$$

where $c_i = b_i$ for $1 \leq i \leq k-1$ and $c_i = \left(b_i - \frac{b_m}{a_n} a_i\right)$ for $k \leq i \leq m-1$. This completes the proof. $\square$

We will now prove Theorem 3.3.7.

Proof. Let $f(X) = a_0 + a_1 X + a_2 X^2 + \cdots + a_n X^n \in \mathbb{F}[X]$ be given. Assume that $n \geq 1$.

Let $g(X)$ be a polynomial in $\mathbb{F}[X]$. If $\deg g(X) < \deg f(X)$, we take $q(X) \equiv 0$ and $r(X) = g(X)$. Then $g(X) = 0 \cdot f(X) + g(X)$.

Now we assume that $\deg g(X) \geq \deg f(X) = n$. We prove the result by the induction on the degree of the polynomial $g(X)$. Let $\deg g(X) = n + k$ for some $k \geq 0$.

Assume $k = 0$. Then $g(X) = b_0 + b_1 X + b_2 X^2 + \cdots + b_n X^n$. It follows from Lemma 3.3.8 that there exists a polynomial $q(X)$ in $\mathbb{F}[X]$ such that $g(X) - q(X)f(X)$ is a polynomial of degree less than n. Let $r(X) = g(X) - q(X)f(X)$. Then $g(X) = q(X)f(X) + r(X)$.

Now assume the result to be true for all polynomials of degree less than or equal to $n + k - 1$.

Let $g(X) = b_0 + b_1 X + b_2 X^2 + \cdots + b_n X^n + \cdots + b_{n+k} X^{n+k}$. As earlier, by Lemma 3.3.8, there exists a polynomial $q_1(X)$ in $\mathbb{F}[X]$ such that $g_1(X) = g(X) - q_1(X)f(X)$ is a polynomial of degree less than $n + k$. Therefore by induction hypothesis, there exist polynomials $q_2(X)$ and $r_2(X)$ in $\mathbb{F}[X]$ such that $g_1(X) = q_2(X)f(X) + r_2(X)$ where $r_2(X) \equiv 0$ or $0 \leq \deg r_2(X) < \deg f(X)$.

Since $g(X) = g_1(X) + q_1(X)f(X)$, we get that

$$\begin{aligned} g(X) &= g_1(X) + q_1(X)f(X) \\ &= q_2(X)f(X) + r_2(X) + q_1(X)f(X) \\ &= (q_1(X) + q_2(X))f(X) + r_2(X). \end{aligned}$$

This proves the existence of the polynomials $q(X)$ and $r_2(X)$ such that $g(X) = q(X)f(X) + r_2(X)$.

We will now prove the uniqueness of the polynomials $q(X)$ and $r(X)$.

If possible let $q(X)$, $q'(X)$, $r(X)$ and $r'(X)$ be polynomials such that

$$g(X) = q(X)f(X) + r(X) = q'(X)f(X) + r'(X).$$

Then

$$(q(X) - q'(X))\, f(X) = r'(X) - r(X).$$

Since the degree of the polynomials $r(X)$ and $r'(X)$ is less than $\deg f(X)$, we see that $\deg\left[(q(X) - q'(X))\, f(X)\right] = \deg(r'(X) - r(X)) < \deg f(X) = n$.

If $q(X) - q'(X)$ is a nonzero polynomial, then it follows from Proposition 3.3.2(3) that

$$\begin{aligned} n &= \deg f(X) \\ &> \deg(r'(X) - r(X)) \\ &= \deg\,(q(X) - q'(X))\, f(X) \\ &= \deg\,(q(X) - q'(X)) + \deg f(X) \\ &\geq n. \end{aligned}$$

This is a contradiction.

Hence $q(X) = q'(X)$ and $r(X) = r'(X)$ and the uniqueness follows. □

In the proof of the theorem, we have used only the fact that the leading coefficient of a_n is a unit to prove division algorithm. This leads us to the following Corollary.

Corollary 3.3.9. *Let R be a commutative ring and $f(X) := a_0 + a_1X + a_2X^2 + \cdots + a_nX^n$ be a nonconstant monic polynomial in $R[X]$. If $g(X)$ is a polynomial in $R[X]$, then there exist unique polynomials $q(X)$ and $r(X)$ in $R[X]$ such that $g(X) = q(X)f(X) + r(X)$ where $r(X) \equiv 0$ or $0 \leq \deg(r(X)) < n$.*

Proof. The proof is verbatim same as in Theorem 3.3.7. □

Division algorithm is an important result with many applications and some of them are discussed here.

3.3.2 Applications of Division Algorithm

Definition 3.3.10. Let R be a commutative ring and $f(X) = a_0 + a_1X + a_2X^2 + \cdots + a_nX^n$ be a polynomial in $R[X]$. An element α is called a root of the polynomial $f(X)$, if $f(\alpha) = a_0 + a_1\alpha + a_2\alpha^2 + \cdots + a_n\alpha^n = 0$.

Example 3.3.11.

1. Let R be a commutative ring and a be an element in R. Then a is a root of the polynomial of $f(X) = X - a$.

2. Let $f(X) = 1 + 2X + X^2$ in $\mathbb{R}[X]$. Then -1 is the root of polynomial $f(X)$.

3. Let $f(X) = 1 + X^2$ in $\mathbb{Z}[X]$. Then $f(X)$ does not have a root in $\mathbb{Z}$ and $\pm i$ are the roots of $f(X)$, when considered as a polynomial over $\mathbb{C}$.

Proposition 3.3.12. *Let R be a ring and $f(X) = aX + b$ be a polynomial of degree 1. Then the following holds.*

1. *The polynomial $f(X)$ has a root in R iff a divides b in R.*
2. *If R is an integral domain, then the polynomial $f(X)$ has at most one root in R.*

Proof. Let $f(X) = aX + b$ be a polynomial of degree 1.

Suppose α is a root of $f(X)$. Then $f(\alpha) = a\alpha + b = 0$ and hence $a\alpha = -b$. This shows that a divides b.

Conversely suppose a divides b in R and $b = ac$ for some c in R. Then $f(X) = aX + b = aX + ac = a(X + c)$ and we conclude that $-c$ is a root of $f(X)$. Hence the polynomial $f(X)$ has a root in R iff a divides b and this proves the first part of the proposition.

Let us now assume that the ring R is an integral domain. Suppose the polynomial $f(X) = aX + b$ has at least two roots, say α_1, α_2 in R. Then $a\alpha_1 + b = a\alpha_2 + b = 0$. Hence $a(\alpha_1 - \alpha_2) = 0$. Since R is an integral domain and $a \neq 0$, it follows that $\alpha_1 - \alpha_2 = 0$. This proves that the polynomial $f(X)$ has at most one root in R. □

Proposition 3.3.13. *Let R be an integral domain and $f(X)$ be a polynomial in $R[X]$. If α in R is a root of the polynomial $f(X)$, then there exists a unique polynomial $q(X)$ of degree $(n-1)$ in $R[X]$ such that $f(X) = (X-\alpha)q(X)$.*

Proof. Let $f(X)$ and α be as in the statement of theorem. Since the polynomial $(X-\alpha)$ is a monic polynomial, there exist unique polynomials $q(X)$ and $r(X)$ such that

$$f(X) = (X-\alpha)q(X) + r(X)$$

where $0 \leq \deg r(X) < 1$. Then r is a constant polynomial. i.e., $r \in R$. If we now substitute α in the above equation, we get

$$0 = f(\alpha) = (\alpha - \alpha)q(\alpha) + r = r.$$

Therefore $r = 0$ and this completes the proof. □

Theorem 3.3.14. *Let R be an integral domain and $f(X) = a_0 + a_1X + a_2X^2 + \cdots + a_nX^n$ be a polynomial in $R[X]$. Then $f(X)$ has atmost n-roots in R.*

Proof. We prove the result by induction on the degree of the polynomial $f(X)$.

Let $f(X) = aX + b \in R[X]$ be a polynomial of degree 1. Then it follows from Proposition 3.3.12, that the polynomial has at most one root in R.

Now let $n \geq 2$ and assume that any polynomial of degree $k \leq n-1$ has at most k roots in R.

Let $f(X)$ be a polynomial of degree n. If the polynomial $f(X)$ has no root in R, the result is true.

If the polynomial $f(X)$ has a root α in R, then by Proposition 3.3.13 there exists a polynomial $q(X)$ in $R[X]$ such that $f(X) = (X-\alpha)q(X)$ where $\deg(q(X)) = n-1$.

Suppose β in R is root of $f(X)$. Then $0 = f(\beta) = (\beta-\alpha)q(\beta) = 0$. Since R is an integral domain either $\beta - \alpha = 0$ or $q(\beta) = 0$. If $\beta \neq \alpha$, then $q(\beta) = 0$.

Since the degree of the polynomial $q(X)$ is $n-1$, by induction hypothesis, $q(X)$ has at most $n-1$ roots in R. This proves that the polynomial $f(X)$ has at most n roots in R. □

Remark 3.3.15. The above result about the number of roots of a polynomials is no longer true, if the ring R is not an integral domain.

For example let $R = \mathbb{Z}_{12}$ and $f(X) = X^2 - 1$. Then the polynomial $f(X)$ has four roots and the roots are 1, 5, 7 and 11.

Exercise 3.3.16.

1. Let $f(X) = 2X + 4$ in $\mathbb{Z}_6[X]$. Find all the roots of $f(X)$ in $\mathbb{Z}_6$.
2. Let $f(X) = X^2 + 1 \in \mathbb{Z}_8[X]$. Find all the roots of $f(X)$ in $\mathbb{Z}_8[X]$.
3. Find all the roots of the polynomial $f(X) = X^2 - 1$ in $\mathbb{Z}_{15}[X]$.
4. Let $f(X) = X^3 + 1 \in \mathbb{Z}_9[X]$. Find all the roots of $f(X)$ in $\mathbb{Z}_9[X]$.

Example 3.3.17. Let $f(X) := (X^2+1)(X^2-2)$. The polynomial $f(X)$

1. has no root in $\mathbb{Q}$,

2. only two roots in $\mathbb{Z}_{10}$,

3. two roots in $\mathbb{R}$ and

4. four roots in $\mathbb{C}$.

Exercise 3.3.18. Let $f(X) = a_0 + a_1X + a_2X^2 + \cdots + X^n$ be a polynomial in $\mathbb{Z}[X]$ such that $f(\sqrt{2}) = 0$. Show that there exists a polynomial $q(X) \in \mathbb{Z}[X]$ such that $f(X) = (X^2 - 2)q(X)$ for a polynomial $q(X)$ in $\mathbb{Z}[X]$.

Let $\varphi(X) = X^2 - 2$ in $\mathbb{R}[X]$. Then the polynomial $\varphi(X)$ has a root $\sqrt{2}$ in $\mathbb{R}$. Since $\varphi(X)$ is in $\mathbb{Z}[X]$ and it is a monic polynomial, we can apply division algorithm to the pair $(f(X), \varphi(X))$. Hence there exist unique polynomials $q(X)$ and $r(X)$ in $\mathbb{Z}[X]$ such that $f(X) = (X^2 - 2)q(X) + r(X)$ where $r(X) \equiv 0$ or $r(X)$ is a polynomial of degree less than 2.

Now we interpret this decomposition in $\mathbb{R}[X]$. Since $\sqrt{2}$ is a root of $f(X)$, we have $0 = f(\sqrt{2}) = 0 + r(\sqrt{2})$. That is $r(\sqrt{2}) = 0$. If $r(X) = aX + b$ in $\mathbb{Z}[X]$, then $r(\sqrt{2}) = 0$ means that $a\sqrt{2} + b = 0$. This leads to a contradiction unless $a = 0 = b$. Thus we have shown that $f(X) = (X^2 - 2)q(X)$ for a polynomial $q(X)$ in $\mathbb{Z}[X]$.

Exercise 3.3.19. Let $f(X) \in \mathbb{R}[X]$ be such that $f(\alpha) = 0$ for some α in $\mathbb{C}$. Show that

1. $\overline{\alpha}$ is also a root of $f(X)$,

2. the polynomial $(X - \alpha)$ divides the polynomial $f(X)$ if $\alpha \in \mathbb{R}$ and

3. the polynomial $X^2 - 2\text{Re}(\alpha)X + |\alpha|^2$ divides the polynomial f if α is a complex number.

Let $f(X) = a_0 + a_1X + \ldots + a_nX^n$ be a nonconstant polynomial in $\mathbb{R}[X]$.

Let $\alpha \in \mathbb{C}$ be a root of $f(X)$. Then $f(\alpha) = a_0 + a_1\alpha + \ldots + a_n\alpha^n = 0$. If α is a real number the $\overline{\alpha} = \alpha$ and $f(\overline{\alpha}) = f(\alpha) = 0$. We will now assume that $\text{Im}(\alpha \neq 0)$. Since the coefficients a_is are real numbers, it follows that

$$\begin{aligned} f(\overline{\alpha}) &= a_0 + a_1\overline{\alpha} + \ldots + a_n\overline{\alpha}^n \\ &= \overline{a_0 + a_1\alpha + \ldots + a_n\alpha^n} \\ &= \overline{f(\alpha)} = 0. \end{aligned}$$

If we consider $f(X)$ as a polynomial over $\mathbb{C}$, then there exists a a polynomial $q(X)$ in $\mathbb{C}[X]$ such that $f(X) = (X - \alpha)q(X)$. Now $f(\overline{\alpha}) = (\overline{\alpha} - \alpha)q(\overline{\alpha})$. The root $\overline{\alpha}$ is not equal to α. Hence $q(\overline{\alpha}) = 0$ and we can write $q(X) = (X - \overline{\alpha})q_1(X)$ for a polynomial $q_1(X)$ in $\mathbb{C}[X]$. Therefore we can write

$$\begin{aligned} f(X) &= (X - \alpha)(X - \overline{\alpha})q_1(X) \\ &= \left(X^2 - 2\text{Re}(\alpha)X + |\alpha|^2\right) q_1(X). \end{aligned}$$

If we now show that $q_1(X)$ is in $R[X]$, then we are through. This again follows by comparing the coefficients of the two expressions of $f(X)$.

The Proposition 3.3.13 can also be proved without appealing to Division algorithm as shown below.

Proposition 3.3.20. *Let R be a commutative ring and $f(X)$ be a polynomial in $R[X]$. If a is an element in R, then the polynomial $X-a$ divides the polynomial $f(X)-f(a)$.*

Proof. Let $f(X) = a_0+a_1X+a_2X^2+\cdots+a_nX^n$ be a polynomial in $R[X]$. If a is in R, then $f(a) = a_0+a_1a+\cdots+a_na^n$. Let us recall Theorem 3.1.14 and write $X^k - a^k = (X-a)\sum_{r=0}^{k-1} X^r a^{k-r}$. Let us denote $\sum_{r=0}^{k-1} X^r a^{k-r} = f_{k-1}(X)$. Therefore

$$\begin{aligned} f(X)-f(a) &= a_1(X-a)+a_2(X^2-a^2)+\cdots+a_n(X^n-a^n) \\ &= a_1(X-a)+a_2(X-a)(X+a)+\cdots \\ &\quad +a_n(X-a)\left(\sum_{r=0}^{n-1} a^r X^{n-r}\right) \\ &= (X-a)\left(a_1+a_2f_1(X)+\cdots+a_nf_{n-1}(X)\right). \end{aligned}$$

Hence the polynomial $X-a$ divides the polynomial $f(X)-f(a)$. □

Theorem 3.3.21. *Let R be a commutative ring and $f(X)$ be a polynomial in $R[X]$. If a in R is a root of the polynomial $f(X)$ then $X-a$ divides the polynomial $f(X)$.*

Proof. We have shown in Proposition 3.3.20 that $X-a$ divides the polynomial $f(X)-f(a)$ for every a in R. Since a is a root of the polynomial $f(X)$, we get $f(a)=0$. Hence $X-a$ divides $f(X)$. □

Exercise 3.3.22.

1. Let R be a ring. Show that the polynomial $X-a$ is irreducible in $R[X]$ for every a in R.

2. Let $f(X)$ be a polynomial in $R[X]$. If $f(a)=0$ fo some a in R and degree$(f(X)) \geq 2$, then show that $f(X)$ is reducible i1 $R[X]$.

3.3.3 Power series rings

Let R be a ring. A formal power series with coefficients in the ring R is a formal expression $\sum_{k\geq 0} a_kX^k$ with a_k in R for $k=0,1,2,\ldots$. Observe that here we don't insist on only finitely many coefficients being nonzero.

The power series ring over a ring R denoted by $R[\![X]\!]$ is the set

$$\left\{\sum_{i\geq 0} a_iX^i : a_i \in R,\right\}.$$

On this set addition and multiplication are defined as follows:

$$\left(\sum_{k=0}^{\infty} a_kX^k\right)+\left(\sum_{k=0}^{\infty} b_kX^k\right) := (a_0+b_0)+(a_1+b_1)X+(a_2+b_2)X^2+\cdots$$

and

$$(a_0 + a_1X + a_2X^2 + \cdots) \cdot (b_0 + b_1X + b_2X^2 + \cdots) := \sum_{k\geq 0}\left(\sum_{i+j=k} a_i\, b_j)\right) X^k.$$

Let us observe that, since $\sum_{i+j=k} a_i\, b_j$ is a finite sum it makes sense in R.

With these two binary operations $(R[\![X]\!], +, \cdot)$ becomes a ring. It is very easy to check all the axioms. We just check the associativity of the multiplication.

If $\sum_{k\geq 0} a_k X^k$, $g(X) = \sum_{j\geq 0} b_j X^j$ and $h(X) := \sum_{l\geq 0} c_l X^l$ are three elements in $R[\![X]\!]$, then

$$\begin{aligned}
(f(X)g(X))\,h(X) &= \left(\sum_{k\geq 0}\left(\sum_{i+j=k} a_ib_j\right)X^k\right)\left(\sum_{l\geq 0} c_lX^l\right)\\
&= \sum_{n\geq 0}\left(\sum_{k+l=n}\left(\sum_{i+j=k} a_ib_j\right)c_l\right)X^n\\
&= \sum_{n\geq 0}\sum_{i+j+l=n} a_ib_jc_lX^n\\
&= \sum_{n\geq 0}\sum_{i+t=n} a_i\left(\sum_{j+l=t} b_jc_l\right)X^n\\
&= \left(\sum_{i\geq 0} a_iX^i\right)\left(\sum_{t\geq 0}\left(\sum_{j+l=t} b_jc_l\right)X^t\right)\\
&= f(X)\,(g(X)h(X))\,.
\end{aligned}$$

The identity 1 in R is the multiplicative identity of $R[\![X]\!]$.

We will now find the units in $R[\![X]\!]$.

Exercise 3.3.23. 1. Let R be a ring. Show that the element $1-X$ is a unit in $R[\![X]\!]$.

We have seen in analysis that the numerical series $\sum_{n=0}^{\infty} a^n$ converges to $\frac{1}{1-a}$ if $|a| < 1$.

This tells us that the formal series $\sum_{n=0}^{\infty} X^n$ is the inverse of $1 - X$. One can prove that this is indeed the inverse of $1 - X$ and we leave it to the readers to work out the details.

2. Let $\mathbb{F}$ be a field. What are the units in $\mathbb{F}[\![X]\!]$.

Let $p(X) = a_0 + a_1X + a_2X^2 + \cdots$ be a unit in $\mathbb{F}[\![X]\!]$ and $q(X) = b_0 + b_1X + b_2X^2 + \cdots$ be its inverse. Then

$$1 = p(X)q(X) = a_0b_0 + \sum_{k\geq 1}\left(\sum_{i+j=k} a_ib_j\right)X^k.$$

This shows that $a_0b_0 = 1$ and $\sum_{i+j=k} a_ib_j = 0$ for all $k \geq 1$.

A calculation now shows the following.

We multiply the equation $a_1b_0+a_0b_1 = 0$ by b_0, to get $b_1a_0b_0 = -a_1b_0^2$. Hence $b_1 = -a_1b_0^2$. Let us now assume that we have found b_j for $j \leq k$ for some k. We multiply the equation $a_0b_{k+1} + \sum_{i+j=k} a_ib_j + a_{k+1}b_0 = 0$ by b_0 and get $b_{k+1} = -a_{k+1}b_0^2 - \left(\sum_{i+j=k} a_ib_j\right) b_0$. Thus inductively, we can find all the coefficients b_j. This proves that a power series $p(x) = a_0 + a_1X + a_2X^2 + \cdots$ is a unit in the power series ring $\mathbb{F}[\![X]\!]$ iff the constant term a_0 is a unit in $\mathbb{F}$. Since $\mathbb{F}$ is a field, every non zero element in $\mathbb{F}$ is a unit. Thus $\{\sum_{n=0}^{\infty} a_nX^n \in \mathbb{F}[\![X]\!] : a_0 \neq 0\}$ is the set of all units in $\mathbb{F}[\![X]\!]$.

Why doesn't this proof work in polynomial rings?

3. Let R be a ring. Find the units in $R[\![X]\!]$.

A power series $\sum_{k\geq 0} a_kX^k$ is a unit in $R[\![X]\!]$ iff a_0 is a unit in R.

Theorem 3.3.24. *Let R be an integral domain and $R[\![X]\!]$ be the power series ring over R. Then $R[\![X]\!]$ is an integral domain.*

Proof. Let $\sum_{n\geq 0} a_nX^n$ and $\sum_{n\geq 0} b_nX^n$ be two nonzero elements in $R[\![X]\!]$. Let a_r be the first nonzero coefficient in $\sum_{n\geq 0} a_nX^n$ and b_s be the first nonzero coefficient in $\sum_{n\geq 0} b_nX^n$. Then

$$\begin{aligned}\left(\sum_{n\geq 0} a_nX^n\right)\left(\sum_{n\geq 0} b_nX^n\right) &= \left(\sum_{n\geq r} a_nX^n\right)\left(\sum_{n\geq s} b_nX^n\right)\\ &= a_rb_sX^{r+s} + \sum_{n\geq r+s+1}\left(\sum_{i+j=n} a_ib_j\right)X^n\end{aligned}$$

is a power series whose first coefficient a_rb_s is nonzero. Hence the product is nonzero and this proves that $R[\![X]\!]$ is an integral domain. □

3.3.4 Laurent Polynomials

Let R be a commutative ring. A formal expression $\sum_{n\in\mathbb{Z}} a_nX^n = \cdots + a_{-2}X^{-2} + a_{-1}X^{-1} + a_0 + a_1X + a_2X^2 + \cdots$ is called a *Laurent polynomial* if $a_n \neq 0$ only for finitely many integers. We write X^{-k} as $\frac{1}{X^k}$ for $k \in \mathbb{N}$.

For example $\frac{1}{X} + X$ is a Laurent polynomial. But $\sum_{k\geq 0} X^k$ is a not a Laurent polynomial.

Let $f(X) := \sum_{n\in\mathbb{Z}} a_nX^n$ and $g(X) = \sum_{n\in\mathbb{Z}} b_nX^n$ be two Laurent polynomials. If $a_n = 0$ for $|n| \geq n_1$ and $b_n = 0$ for $|n| \geq n_2$, then $a_n + b_n = 0$ if $n \geq \max\{n_1, n_2\}$.

Since $\{a_n : a_n \neq 0\}$ and $\{b_n : b_n \neq 0\}$ are finite sets, it follows that

$\{a_ib_j : a_i, b_j \neq 0\}$ is also a finite set. Therefore, if we define

$$f(X) + g(X) := \sum_{n\in\mathbb{Z}} (a_n + b_n)\, X^n \text{ and}$$

$$f(X)g(X) := \sum_{n\in\mathbb{Z}} \left(\sum_{i+j=n} a_ib_j \right) X^n$$

then $f(X) + g(X)$ and $f(X)g(X)$ are also Laurent polynomials.

Now we can check, as we did in the case of polynomials that the set of Laurent polynomials also becomes a ring under this addition and multiplication. That is

$$\left\{ \sum_{n\in\mathbb{Z}} a_nX^n : a_n \neq 0 \text{ only for finitely many } n \right\}$$

is a ring.

Exercise 3.3.25. Let R be an integral domain. Show that the ring of Laurent polynomials over R is also an integral domain.

3.3.5 Laurent Series

Let R be a commutative ring. A formal expression

$$\cdots + a_kX^k + \cdots + a_{-2}X^{-2} + a_{-1}X^{-1} + a_0 + a_1X + a_2X^2 + \cdots$$

with $a_n \in R$ and $a_n \neq 0$ only for finitely many negative integers is called a Laurent series with coefficients in R.

Let $\sum_{k\in\mathbb{Z}} a_kX^k$ and $\sum_{k\in\mathbb{Z}} b_kX^k$ be two Laurent series. Then there exists integers n_1 and n_2 such that $a_n = 0$ if $n < n_1$ and $b_n = 0$ if $n < n_2$. Hence the product $a_ib_j = 0$, if $i < n_1$ or $j < n_2$ and if $r < n_1 + n_2$, then $\sum_{i+j=r} a_ib_j = 0$.

In view of this we define the addition and multiplication on the set

$$A := \left\{ \sum_{n=N}^{\infty} a_nX^n : a_i \in \mathbb{R}, N \in \mathbb{Z} \right\}$$

of Laurent series by

$$\sum_{k\in\mathbb{Z}} a_kX^k + \sum_{k\in\mathbb{Z}} b_kX^k := \sum_{k\in\mathbb{Z}} (a_k + b_k)\, X^k$$

and

$$\left(\sum_{k\in\mathbb{Z}} a_kX^k \right) \left(\sum_{j\in\mathbb{Z}} b_jX^j \right) := \sum_{k\in\mathbb{Z}} \left(\sum_{i+j=k} a_ib_j \right) X^k$$

for $\sum_{k\in\mathbb{Z}} a_kX^k$ and $\sum_{k\in\mathbb{Z}} b_kX^k$ in A.

Though it is little unpleasant to check the details, one can show that A is a commutative ring with identity.

Remark 3.3.26. A power series is a Laurent series with $a_k = 0$ for $k < 0$.

If the ring R has more structure then A also has.

Theorem 3.3.27. *Let $A = \{\sum_{n=N}^{\infty} a_n X^n : a_i \in R, N \in \mathbb{Z}\}$ be the ring of Laurent series over a commutative ring R with identity. Then*

1. *A is an integral domain if R is an integral domain and*
2. *A is a field if R is a field.*

Proof. It is easy to show that A is an integral domain if R is an integral domain. We will show that, if R is a field then A is also a field.

Let $\sum_{n\in\mathbb{Z}} a_n X^n$ in A be nonzero and let $n_1 := \min\{k \in \mathbb{Z} : a_k \neq 0\}$. Then $a_n = 0$ if $n < n_1$ and $a_{n_1} \neq 0$. We write $\sum_{n\in\mathbb{Z}} a_n X^n$ as

$$\begin{aligned}\sum_{n\in\mathbb{Z}} a_n X^n &= X^{n_1}\left(a_{n_1} + a_{n_1+1}X + a_{n_1+2}X^2 + \cdots\right)\\ &= X^{n_1}\sum_{k\geq 0} C_k X^k\end{aligned}$$

where $C_k := a_{n_1+k}$ for $k \geq 0$.

Since $a_{n_1} \neq 0$ it is a unit in R and hence the power series $\sum_{k\geq 0} C_k X^k$ is a unit in the power series ring $R[\![X]\!]$. Let $\sum_{r\geq 0} d_r X^r$ be the inverse of $\sum_{k\geq 0} c_k X^k$. Then the Laurent series $\sum_{r\geq 0} d_r X^{r-n_1}$ is the inverse of $\sum_{k\in\mathbb{Z}} a_k X^k$ and this proves that A is a field. □

3.4 Subrings and Ideals

Let S be a subset of a ring R. If we restrict the addition and multiplication to the set S, we get

$$+ : S \times S \to R$$

and

$$\cdot : S \times S \to R.$$

If the set S is closed under these binary operations, then $(S, +)$ becomes a subgroup of $(R, +)$ and since S is closed under multiplication, the associativity of multiplication and the distributive properties of multiplication over addition follows easily.

For example, if $k \geq 2$, then $S = k\mathbb{Z}$ is closed under addition and multiplication of integers and it satisfies all the axioms of a ring except the one on multiplicative identity.

If there is an element 1_S in S such that $a.1_S = 1_S.a = a$ for all elements a in R, then $(S, +, \cdot)$ becomes a ring. However we can't conclude that $1_S = 1_R$ as we did in the case of subgroups!(Why?) Therefore our definition of subring is the following.

Definition 3.4.1 (Subring). Let $(R, +, \cdot)$ be a ring and S be a subset of R. We call S a subring of S, if

1. $(S, +)$ is a subgroup of $(R, +)$,

2. the multiplication is a binary operations on S, and
3. the multiplicative identity 1_R is in S.

Examples 3.4.2. 1. Consider the inclusions $\mathbb{Z} \subseteq \mathbb{Q} \subseteq \mathbb{R} \subseteq \mathbb{C}$ with usual addition and multiplication. Then each of them is a subring in which they are contained in.

2. Let $R = \mathbb{Z}[i]$ the ring of Gaussian integers. Then R is a subring of $\mathbb{C}$. In general $\mathbb{Z}[\sqrt{D}]$ is a subring of $\mathbb{C}$ for every integer D. If D is positive, then R is a subring of $\mathbb{R}$ too.

3. Let $R = \mathbb{Z}[\sqrt{2}] \subseteq \mathbb{Q}[\sqrt{2}] \subseteq \mathbb{R}$. Each of them are subrings of the bigger rings containing them.

4. Let $R = M(2, \mathbb{R})$ and
$$S := \left\{ \begin{pmatrix} a & -b \\ b & a \end{pmatrix} : a, b \in \mathbb{R} \right\}.$$
Then S is a subring of $M(2, \mathbb{R})$.

5. Let J be an interval in $\mathbb{R}$. Let
$$C(J, \mathbb{R}) := \{f : J \to \mathbb{R} : f \text{ is continuous}\}$$
and
$$D(J, \mathbb{R}) := \{f \in C(J, \mathbb{R}) : f \text{ is differentia le in } J\}.$$
Then $D(J, \mathbb{R})$ is a subring of $C(J, \mathbb{R})$ and $C(J, \mathbb{R})$ is subring of $\mathcal{F}(J, \mathbb{R})$.

6. Let $n \geq 2$. Then the subgroup $n\mathbb{Z}$ is a not a subring of $\mathbb{Z}$, since $1 \notin n\mathbb{Z}$.

7. Let R and S be two rings. Then $R \times \{0\}$ and $\{0\} \times S$ are not subrings of $R \times S$. If $R = S$, then the subset $D(R) := \{(x, x) : x \in \mathbb{R}\}$ a subring of $R \times R$.

8. Let R be a ring. Then
   (a) R is a subring of $R[X]$.
   (b) $R[X]$ is a subring $R[\![X]\!]$.
   (c) $R[X]$ is a subring of the Laurent polynomial ring over R.
   (d) $R[\![X]\!]$ is a subring of the ring of Laurent series over R.

Exercise 3.4.3. 1. Let R be a ring and a be a nonzero element in R. As in groups, we define the centralizer of a as $C(a) := \{x \in R : ax = xa\}$. Show that $C(a)$ is a subring of R. If R is not a commutative ring then $C(a)$ is a proper subring for some a in R.

This is an easy verification.

2. Let R be a ring. Let the *center* of R defined as

$$\{x \in R : xa = ax \text{ for all } a \in R\} = \cap_{a\in R} C(a).$$

Show that the center is a subring of R. This will be a proper subring if R is not commutative.

3. Let R be a ring. Let $n \geq 1$ and

$$R_n := \{f(X) \in R[X] : \deg f(X) = n\} \cup \{0, 1\}.$$

Show that R_n is not a subring of $R[X]$.

Let $f(X) = X^n + X^{n-1}$ and $g(X) = -X^n$. Then their sum $f(X) + g(X) = X^{n-1}$ is not in R_n.

4. Let R be a ring and $R(\mathbb{C})$ be its complexfication. Show that

(a) $R \times \{0\} := \{(r, 0) : r \in R\}$ is a subring of $R(\mathbb{C})$ and

(b) $\{0\} \times R := \{(0, r) : r \in R\}$ is not a subring of $R(\mathbb{C})$.

5. Let R be a ring and $R(\mathbb{C})$ be its complexfication. Find subrings which are not equal to $R \times \{0\} := \{(r, 0) : r \in R\}$.

6. Let $A = \mathbb{R}^2$. Then A is a commutative ring with identity under the componentwise addition and multiplication. What are the subrings of A?

7. Let R be an integral domain. Show that

(a) the product ring $R \times R$ is not an integral domain and

(b) the complexification $R(\mathbb{C})$ of R is an integral domain.

8. Let R be a field. Show that its complexfication $R(\mathbb{C})$ is also a field. Further show that the subring $R \times \{0\} := \{(r, 0) : r \in R\}$ of $R(\mathbb{C})$ is also a field.

9. Let $R := \left\{\sum_{k\geq 0} a_k X^k : a_k \in \mathbb{R} \text{ and } \sum_{k\geq 0} |a_k| < \infty\right\}$. Show that R is a subring of $\mathbb{R}[\![X]\!]$, the ring of formal power series. (We need analysis to show this.)

Let $S := \left\{\sum_{k\geq 0} a_k X^k : a_k \in \mathbb{R} \text{ and } \sum_{k\geq 0} a_k < \infty\right\}$. Is S a subring of $\mathbb{R}[\![X]\!]$?

10. Let A be a subring of $\mathbb{C}$ such that $\mathbb{R} \subseteq A$. Show that either $A = \mathbb{R}$ or $A = \mathbb{C}$.

Suppose $\mathbb{R}$ is a proper subset of the subring A. Then there exists $s + it$ in A such that $t \neq 0$. Since A is a subring of $\mathbb{C}$, we see that $it = s + it - s$ is in A. Since $\mathbb{R}$ is a subset of A, we see that $i = \frac{it}{t}$ is also in A. Therefore $\alpha + i\beta$ is in A for all α and β in $\mathbb{R}$. Thus $A = \mathbb{C}$.

Example 3.4.4. Let $k \geq 2$ and

$$R_k := \left\{ \sum_{n \geq 0} \frac{a_n}{k^n} : a_0 \in \mathbb{Z} \text{ and } |a_n| \leq k-1 \text{ for } n \geq 1 \right\}.$$

Then R_k is a subring of $\mathbb{R}$.

Let $\sum_{n \geq 0} \frac{a_n}{k^n}$ be an element of R. Then $|\frac{a_n}{k^n}| \leq \frac{k-1}{k^n} \leq \frac{1}{k^{n-1}}$ for $n \geq 1$. Therefore

$$|\sum_{n \geq 0} \frac{a_n}{k^n}| \leq |a_0| + \sum_{n \geq 0} \frac{1}{k^n}.$$

We know that this series this converges etc.,

Let R be a ring. If a and b are two elements in R such that a divides b, then we can write $b = ac$ for some c in R. This shows that $b \in \{ar : r \in R\}$. Hence $\{bs : s \in R\} = \{acs : s \in R\} \subseteq \{ar : r \in R\}$.

Conversely let a and b be two elements in R such that $\{bs : s \in R\} \subseteq \{ar : r \in R\}$. Then there exists an element c in R such that $b = ac$ and this means that a divides b in R.

Thus we have shown that, if a and b are two elements in R, then a divides b in R iff $\{bs : s \in R\} \subseteq \{ar : r \in R\}$.

For example, in $\mathbb{Z}$, we already know that an integer k divides another integer l iff $l\mathbb{Z} \subseteq k\mathbb{Z}$.

Now we want to study analogous properties in rings and we start with the following definition.

Definition 3.4.5 (Ideal). Let R be a ring. A subset I in R is called an ideal in R if

1. $(I, +)$ is a subgroup of $(R, +)$ and
2. for every $x \in I$ and $a \in R$, the product ax is also in I.

Remark 3.4.6. If the ring R is not commutative, then the product ax may not be same as the product xa for elements x and a in R. In view of this we say a subset is a left ideal, if it satisfies the condition 2 of the definition above.

We say that the subset is a right ideal if the condition 2 is replaced by xa is in I for $x \in I$ and a in R.

We say that I is a two sided ideal if for x in I and a in R, the products ax and xa are in I.

Examples 3.4.7.

1. Let $n \geq 2$ and $I = n\mathbb{Z}$. Then I is an ideal in $\mathbb{Z}$. We can show that any ideal in $\mathbb{Z}$ is of this form.

   Let I be an ideal in $\mathbb{Z}$. Then $(I, +)$ is a subgroup of $(\mathbb{Z}, +)$. Therefore there exists $k \in \mathbb{N} \cup \{0\}$ such that $I = k\mathbb{Z}$.
   Conversely any subgroup of $(\mathbb{Z}, +)$ is an ideal of $(\mathbb{Z}, +, .)$.

2. Let R be a ring and $a \in R$. Let $I := \{ra : r \in R\}$. Then I is an ideal in R.

Definition 3.4.8 (Principal Ideal). An ideal I in R is called a principal ideal if there exists an element a in R such that

$$I = \{ar : r \in R\}.$$

In this case we also say that the ideal I is a generated by a and we denote it by (a).

Examples 3.4.9. 1. Let S be a subset of a ring R and let

$$I(S) := \left\{ \sum_{a \in F} r_a\, a : F \text{ is a finite subset of } S \text{ and } r_a \in R \right\}.$$

This set $I(S)$ is an ideal and it is called the ideal generated by the set S. Sometimes we also denote it by (S)
If $S = \{s\}$ is a singleton, then $I(S) = \{r\, s : r \in R\}$ is called a principal ideal generated by s and denoted by (s). If $S = \{a, b\}$ is a two element set then $I(S) = \{ra + sb : r, s \in R\}$ is also denoted as (a, b). In general if $S = \{a_1, a_2, \cdots, a_k\}$ is a finite set with k-elements, then we denote $I(S)$ as $(a_1, a_2, \cdots, a_k)$.

2. Let I be an ideal in R and a be an element in R. Then the set $(I, a) := \{x + sa : s \in R\}$ is an ideal in R. If a is not in I, then I is strictly contained in (I, a).

3. Let

$$R := \{(a_n) : a_n \in \mathbb{Q} \text{ and } (a_n) \text{ is a cauchy sequence}\}$$

and

$$I := \left\{ (a_n) \in R : \lim_{n \to \infty} a_n = 0 \right\}.$$

Then I is an ideal in R.

It is easy to show that $(I, +)$ is a subgroup. Let (a_n) be an element in I and (b_n) be an element in R. Since the sequence is cauchy, it is bounded. Hence there exists a constant $M > 0$ such that $|b_n| \leq M$ for all $n \in \mathbb{N}$.
Now $0 \leq |a_n b_n| \leq M|a_n|$ for all n in $\mathbb{N}$. Given $\varepsilon > 0$, there exists N in $\mathbb{N}$ such that $|a_n| < \frac{\varepsilon}{M}$ for all $n \geq N$. As a consequence, it follows that $|a_n b_n| \leq M|a_n| < M\frac{\varepsilon}{M} = \varepsilon$ for all $n \geq N$. This proves that the sequence $(a_n b_n)$ also converges to 0.

We can do the same as above with real sequences too!

4. Let

$$R = \{(a_n) : a_n \in \mathbb{R} \text{ and } (a_n) \text{ is convergent }\}$$

and

$$I := \{(a_n) \in R : a_n = 0 \text{ for almost all } n\}.$$

Then I is an ideal in R.

Exercise 3.4.10. 1. Let a and b be two elements in a ring R. Show that $a|b$ in R iff the principal ideal (b) is contained in the principal ideal (a).

2. Let a and b be two nonzero, nonunits and non-associates in an integral domain R. If a divides b in R show that the principal ideal (b) is strictly contained in the principal ideal (a).

3. Let R be a ring and let

$$I := \{a_0 + a_1X + \cdots + a_nX^n \in R[X] : \sum_{i=1}^{n} a_i = 0\}.$$

Show that I is an ideal in $R[X]$. Is the set

$$J := \{a_0 + a_1X + \cdots + a_nX^n \in R[X] : \sum_{i=1}^{n} a_i = 1\}$$

an ideal in $R[X]$?

Let $I = \{a_0 + a_1X + \cdots + a_nX^n \in R[X] : \sum_{i=1}^{n} a_i = 0\}$. The zero polynomial is in I. Hence I is nonempty.
Let $f(X) = a_0 + a_1X + \cdots + a_nX^n$ and $g(X) = b_0 + b_1X + \cdots + b_mX^m$. Then $f(X) + g(X) = (a_0 + b_0) + (a_1 + b_1)X + \cdots + (a_n + b_n)X^n + \cdots + b_mX^m$ if $n \leq m$ and the sum of the coefficients is

$$\sum_{i=0}^{n}(a_i + b_i) + \sum_{i \geq n+1} b_i = \sum_{i=0}^{n} a_i + \sum_{j=0}^{m} b_j = 0 + 0 = 0.$$

If $f(X) = a_0 + a_1X + \cdots + a_nX^n$, then $-f(X) = -a_0 - a_1X - \cdots - a_nX^n$ is also in I. Thus $(I, +)$ is a subgroup of $R[X]$.
Let $f(X) = a_0 + a_1X + \cdots + a_nX^n$ be an element of I and $g(X) = b_0 + b_1X + \cdots + b_mX^m$ be a polynomial in $R[X]$. Then $g(X)f(X) = \sum_{k=0}^{n+m} \sum_{i+j=k} a_ib_jX^k$.
We will now show that the sum $\sum_{k=0}^{n+m} \sum_{i+j=k} a_ib_j = 0$.
Let us first observe that for any polynomial $p(X) = c_0 + c_1X + c_2X^2 + \cdots + c_rX^r$, the sum of the coefficients $\sum_{k=0}^{r} c_k = p(1)$. Thus in our case $\sum_{k=0}^{n+m} \sum_{i+j=k} a_ib_j = g(1)f(1)$. Since we are given that $f(1) = 0$, it follows that $\sum_{k=0}^{n+m} \sum_{i+j=k} a_ib_j = g(1)f(1) = 0$.
This proves that I is an ideal in $R[X]$.
The set J is not even a subgroup since the zero element is not in J.

4. Let $R = C(J, \mathbb{R})$ the ring of continuous functions on the open interval J. Let $x_0 \in J$ and $I = \{f \in R : f(x_0) = 0\}$. Show that I is an ideal in R.

5. Let J be an interval and F be a finite subset of J. Let

$$I(F) := \{f \in R : f(x) = 0 \text{ for } x \in F\}.$$

Show that $I(F)$ is an ideal. What happens if F is infinite?

We will now study some properties of ideals in a ring R.

Proposition 3.4.11. *Let R be a ring and I and J be two ideals in R. Then*

1. *$I + J := \{a + b : a \in I \text{ and } b \in J\}$ is an ideal in R,*

2. *$I \cap J$ is an ideal in R, and*

3. *$I = R$, if $1 \in I$.*

Proof. It is easy to show that $I + J$ and $I \cap J$ are ideals in R.

Let I be an ideal and $1 \in I$. Let r be an element of R. Then $r = r.1$ is in I by definition of ideal. Thus $R = I$. □

Proposition 3.4.12. *Let R be a ring. Let $I = (a)$ and $J = (b)$ two nonzero principal ideals in R. Then the following holds.*

1. *If $I + J = (d)$ is a principal ideal, then $d = \gcd(a, b)$.*

2. *If $I \cap J = (c)$ is a principal ideal, then $c = \operatorname{lcm}(a, b)$.*

Proof. The ideal $I + J = \{ar + bs : r, s \in R\}$ and $I + J = (d)$ for an element d in R. By definition the sum $ar + bs$ is in $(d) = \{dx : x \in R\}$ for every choice of elements r and s in R. If we choose $r = 1$ and $s = 0$, we see that the element a is in (d) and therefore can be written as $a = dx$ for some x in R. This means that d divides a in R.

Similarly we take $r = 0$ and $s = 1$ and conclude that d divides b as well

Now let an element e divide a and b in R. Then there exist elements u and v in R such that $a = eu$ and $b = ev$. Since $(d) = I + J$ there exist elements r and s in R such that $d = ar + bs$. Hence $d = eur + evs = e(ur + vs)$ and this proves that e divides d in R. Thus d is a gcd of the elements a and b in R.

The second part about lcm is similar and we leave it to the reader. □

Exercise 3.4.13. 1. Let R be a ring and I and J be two ideals in R. Show that $I \cup J$ is an ideal in R iff one of them is contained in the other.

2. A ring R is a field iff it has no proper ideals.

3. Let R be a commutative ring with identity and let $I_1 \subseteq I_2 \subseteq \ldots I_n \subseteq \ldots$ be an increasing sequence of ideals in R. Then $\cup_{n\geq} I_n$ is an ideal in R.

   Let $I := \cup_{n\geq 1} I_n$. Since 0 is in I_n for all n, the set I is non-empty. Let a and b be two elements in I. Then there exist natural numbers k and l such that the element a is in I_k and the element b is in I_l. If $k \leq l$, then $I_k \subseteq I_l$ and hence the element a is also in I_l. Therefore $a + b$ is in I_l. Thus I is a subgroup of R.
   Let a be an element in I and r in R. Since a is in $I = \cup_{n\in\mathbb{N}} I_n$, there exists k such that a is in I_k. Therefore the element ra is in $I_k \subseteq I$. This shows that I is an ideal in R.

4. Let I be an ideal in R and an element u in I is a unit. What can you conclude about the ideal I?

5. Let R be a ring. Show that $I = \{x \in R : x \text{ is nilpotent in } R\}$ is an ideal in R.

6. Let X be a non-empty set and I be a proper ideal in $\mathcal{F}(X, \mathbb{R})$. Show that there exists a function f in $\mathcal{F}(X, \mathbb{R})$ and a point x_0 in X such that $f(x_0) = 0$.

7. Let $A = \mathbb{R}^2$. What are the ideals in A?

Let R be a commutative ring with identity. Let I and J be two ideals in R. Then the set $IJ := \{ab : a \in I \text{ and } b \in J\}$ need not be an ideal in R: the sum $a_1b_1 + a_2b_2$ of two elements a_1b_1 and a_2b_2 need not be of the form ab for a in I and b in J.

For example let $R = \mathbb{Z}[X]$. Let $I := (2, X)$ and $J := (3, X^3)$. If $a_1 = 2$, $b_1 = X^3$, $a_2 = X$ and $b_2 = 3$ then $a_1b_1 + a_2b_2 = 2X^3 + 3X^2$ is not of the form ab for a in I and b in J.

> First we write $2X^3 + 3X^2 = X(3 + 2X^2)$. If $2X^3 + 3X^2$ is of the form ab, then, since $a = X$ is in I, the element $3 + 2X^2$ is in J. Therefore $2X^2$ is in J and if we write $2X^2 = 3f(X) + X^3g(X)$ for some polynomials $f(X) = a_0 + a_1X + \ldots + \alpha_rX^r$ and $g(X)$ in $\mathbb{Z}[X]$. This shows that $2 = 3a_2$, a contradiction.

Let K be an ideal containing elements ab for a in I and b in J. Then this ideal must contain the sum $a_1b_1 + a_2b_2$ of the elements a_1b_1 and a_2b_2 of the set IJ. Therefore, we define the product of two ideals I and J to be the set

$$\left\{\sum_{i=1}^{k} a_ib_i : a_i \text{ in } I, b_i \text{ in } J, \text{ and } k \in \mathbb{N}\right\}.$$

This set can easily be seen to be an ideal and we denote it by IJ. This ideal is contained in K.

Observe that the product ideal IJ is contained in both I and J. Hence IJ is contained in $I \cap J$ also.

Question 3.4.14. Let I and J be two ideals in a ring R. Then, what is IJ if $I \subseteq J$?

For example what is $(m)\,(n)$ in $\mathbb{Z}$ for m and n in $\mathbb{Z}$

Example 3.4.15. 1. Let $I := \left(1 + \sqrt{-5}\right)$ and $J := \left(1 - \sqrt{-5}\right)$ be two ideals in the ring $\mathbb{Z}[\sqrt{-5}]$. Then the product ideal IJ is the principal ideal (6) in $\mathbb{Z}[\sqrt{-5}]$.

> The integer $6 = (1 + \sqrt{-5})(1 - \sqrt{-5})$ in $\mathbb{Z}[\sqrt{-5}]$. Hence the integer 6 is in the product ideal IJ. As a consequence the principal ideal (6) is contained in the product ideal IJ.
> Conversely if an element $(1 + \sqrt{-5})(a + b\sqrt{-5})$ is in the ideal I and an element $(1 - \sqrt{-5})(c + d\sqrt{-5})$ is in the ideal J, then the product
>
> $$\begin{aligned}&(1 + \sqrt{-5})(1 - \sqrt{-5})(a + b\sqrt{-5})(c + d\sqrt{-5})\\ =\;& 6(a + b\sqrt{-5})(c + d\sqrt{-5})\end{aligned}$$
>
> is in the principal ideal (6). This shows that the ideal IJ is contained in the principal ideal (6).

2. Let $R = \mathbb{Z}[i\sqrt{2}]$ and $I = \left(1 + i\sqrt{2}\right)$ and $J = \left(1 - i\sqrt{2}\right)$. Then $IJ = (3)$, the principal ideal generated by the integer 3.

Exercise 3.4.16. Let D be a square free integer. Let $I := \left(a + b\sqrt{D}\right)$ and $J := \left(a - b\sqrt{D}\right)$. What is IJ?

Example 3.4.17. The ideal $I = (2, X)$ is not a principal ideal in the polynomial ring $\mathbb{Z}[X]$.

Suppose I is a principal ideal in $\mathbb{Z}[X]$. Then there exists a polynomial $f(X)$ in $\mathbb{Z}[X]$ such that $I = (f(X))$.

Let $g(X)$ in $\mathbb{Z}[X]$ be such that $2 = f(X)g(X)$. Therefore $\deg(2) = \deg(f(X)g(X)) = \deg(f(X)) + \deg(g(X))$. Hence $\deg f(X)$ is zero which in turn implies that $f(X)$ is a constant polynomial, say c. Therefore $I = (c)$.

Since the polynomial X is in $I = (2, X) = (c)$, there exists a polynomial $h(X) = b_0 + b_1X + \cdots + b_rX^r$ such that $X = ch(X)$. Comparing the coefficients, we see that $cb_1 = 1$. Thus c is a unit and $I = (2, X) = \mathbb{Z}[X]$, the whole ring. Therefore there exist two polynomials $\varphi(X) = a_0 + a_1X + \cdots + a_lX^l$ and $\psi(X)$ such that $1 = 2\varphi(X) + X\psi(X)$. Again by comparing the coefficients we get $2a_0 = 1$, a contradiction. Hence I is a not a principal ideal in the ring $\mathbb{Z}[X]$.

Observe that I is the sum of two principal ideals (2) and (X). However its sum is not a principal ideal.

However if the ring R is a field, then we can show that every ideal in $R[X]$ is a principal ideal.

Proposition 3.4.18. *Let R be a field. Then every ideal in $R[X]$ is a principal ideal. Conversely if R is an integral domain and every ideal in $R[X]$ is principal then R is a field.*

Proof. Let $I \neq (0)$ be an ideal in $R[X]$. We use division algorithm to show that I is a principal ideal.

Let $k = \min\{\deg f(X) : f(X) \in I \setminus \{0\}\}$. If $k = 0$, then there exists a nonzero element $a \in R$ such that a is in I. Since R is a field, a has multiplicative inverse a^{-1} and $1 = a \cdot a^{-1}$ is in I. This proves that $I = R[X]$.

Assume $k \geq 1$ and $f(X)$ in I be such that $\deg f(X) = k$. Let $g(X)$ be an element in I. By division algorithm there exist unique $q(X)$ and $r(X)$ in $R[X]$ such that $g(X) = q(X)f(X) + r(X)$, where $r(X) \equiv 0$ or $0 \leq \deg(r(X)) < k$. Since I is an ideal, the polynomial $q(X)f(X)$ is in I and $r(X) = g(X) - q(X)f(X)$ is also in the ideal I. If $r(X) \neq 0$, then $\deg r(X) < k$, contradicts the minimality of the degree of elements in I. Thus $r(X) \equiv 0$ and the proof is complete.

Conversely let us assume that every ideal in $R[X]$ is a principal ideal. We will show that every nonzero element in R is a unit in R.

Let a be a nonzero element in R and $I = (a, X)$ be the ideal generated by the elements a and X. By our assumption I is a principal ideal in $R[X]$. Hence there exists a nonzero polynomial $f(X)$ in $R[X]$, such that $I = (f(X))$, the principal ideal generated by $f(X)$.

Let $a = f(X)g(X)$ for some polynomial $g(X)$ in $R[X]$. Since R is an integral domain, it follows that $0 = \deg(a) = \deg f(X) + \deg g(X)$. Hence $\deg f(X) = 0$ and $f(X)$ is a constant polynomial, say c. Similarly we write $X = c\psi(X)$ for some polynomial $\psi(X) = a_0 + a_1X + \cdots$ in $R[X]$. By comparing the coefficients we conclude that $1 = ca_1$. This proves that c is a unit and hence $I = R[X]$. Therefore there exist polynomials $\varphi(X)$ and $\theta(X)$ in $R[X]$ such that $1 = a\varphi(X) + X\theta(X)$. If $\varphi(X) = b_0 + b_1X + \cdots$, by comparing the coefficients of the constant term, we get that $1 = ab_0$. This shows that a is a unit in R. Thus we have shown that every nonzero element in R is a unit. Hence R is a field. □

Exercise 3.4.19. Let R be a ring and I be an ideal in $R[X]$. Let $k = \min\{\deg f(X) : f(X) \in I \setminus \{0\}\}$. Assume that $f(X) = a_0 + a_1X + \cdots + a_kX^k$ is a polynomial of degree k in I such that a_k is a unit in R. Show that I is a principal ideal in $R[X]$ generated by $f(X)$.

Let R be a ring and $P := \{I \subseteq R : I \text{ is a proper ideal in } R\}$. On this set there is a natural partial order $\leq$.

Let I and J be two ideals in R. We say $I \leq J$ iff $I \subseteq J$. It is very easy to verify that this is a partial order on the set P.

Let us recall the definition of a maximal element in a partially ordered set. Let $(P, \leq)$ be a partially ordered set. An element a is called a *maximal element* if there exists an element x in P such that $a \leq x$, then $x = a$.

Definition 3.4.20 (Maximal Ideal)**.** Let R be a ring. We say that a proper ideal I is a maximal ideal in R, if I is a maximal element in the set of all proper ideals of R.

The following proposition is an immediate consequence of this definition.

Proposition 3.4.21. *Let R be a ring and I be an ideal in R. Then the following statements are equivalent.*

1. *I is a maximal ideal in R.*

2. *If there exists an ideal J such that $I \subseteq J \subseteq R$, then $I = J$ or $J = R$.*

Proof. The proof is easy. □

Definition 3.4.22 (Prime Ideal)**.** Let R be a ring. An ideal I in R is called a prime ideal if a and b in R such that $ab \in I$, then either a or b is in I.

Example 3.4.23.

1. Let p be a prime number and $I = (p)$ be the principal ideal generated by the integer p. Then I is a maximal ideal in $\mathbb{Z}$.

   Let J be an ideal in $\mathbb{Z}$ such that $I \subseteq J \subseteq \mathbb{Z}$. Since every ideal in $\mathbb{Z}$ is principal, there exists an integer k such that $J = (k)$. The integer p is in I and hence in J. Therefore there exists an integer l such that $p = kl$. On the other hand, since p is prime it follows that $k = \pm p$ or ± 1. If $k = \pm p$, then $I = J$ and if $k = \pm 1$, then $J = \mathbb{Z}$. Thus I is a maximal ideal in $\mathbb{Z}$.

2. Let R be a field. Then (0) is a maximal ideal in R.

3. Let R be an integral domain. Then (0) is a prime ideal in R.

4. Let X be a non-empty set and $R = \mathcal{F}(X, \mathbb{R})$. Let x_0 be a point in X. Then $I = \{f \in \mathcal{F}(X, \mathbb{R}) : f(x_0) = 0\}$ is a maximal ideal in R.

Suppose J is an ideal in R such that $I \subsetneqq J$. If $I \neq J$, then there exists a function g in J such that $g(x_0) \neq 0$.

Let $\varphi : X \to \mathbb{R}$ be the function defined by $\varphi(x) = 1 - \frac{1}{g(x_0)} g(x)$. Then φ is in $\mathcal{F}(X, \mathbb{R})$ and $\varphi(x_0) = 0$. Thus φ is in $I \subseteq J$. Since J is an ideal, it follows that $1 = \varphi - \frac{1}{g(x_0)} g$ is also in J. This proves that $J = \mathcal{F}(X, \mathbb{R})$.

5. Let J be an interval in $\mathbb{R}$ and $R := C(J, \mathbb{R})$ and x_0 be a point in J. Then $Z(x_0) := \{f \in C(J, \mathbb{R}) : f(x_0) = 0\}$ is a maximal ideal in $C(J, \mathbb{R})$.

6. Let $R = C(J, \mathbb{R})$. Let $x \neq y$ be two distinct points in J. Then

$$Z(\{x, y\}) := \{f \in C(J, \mathbb{R}) : f(x) = 0 = f(y)\}$$

is not a prime ideal in $C(J, \mathbb{R})$.

The functions

$$\varphi : J \to \mathbb{R} \text{ defined by } \varphi(t) = t - x$$

and

$$\psi : J \to \mathbb{R} \text{ defined by } \varphi(t) = t - y$$

are not in $Z(\{x, y\})$. However the product $\varphi \cdot \psi$ is in $Z(\{x, y\})$.

7. The ideal $I = (2, X)$ is a maximal ideal in $\mathbb{Z}[X]$.

Let J be an ideal in $\mathbb{Z}[X]$ such that $I \subseteq J$. We will show that $I = J$ or $J = \mathbb{Z}[X]$

Let $f(X) = a_0 + a_1 X + a_2 X^2 + \cdots + a_r X^r$ in J. Since the polynomial X is in I, all the powers X^k of X are also in I and consequently in J. This shows that the constant term $a_0 = f(X) - \left(a_1 X + a_2 X^2 + \cdots + a_r X^r\right)$ is in J. That is, if $f(X)$ is in J, then the constant term a_0 of $f(X)$ is also in J.

Let

$$S := \{a \in R : a \text{ is the constant term of some } f(X) \in J\}.$$

If all the elements in S are even integers, then every element of S is a multiple of 2 and hence contained in I. This proves that, if $f(X) = a_0 + a_1 X + \cdots$ is in J, then a_0 is in I. We have already shown that the sum $a_1 X + a_2 X^2 + \cdots$ is in I. Hence the polynomial $f(X) = a_0 + a_1 X + a_2 X^2 + \cdots$ is also in I. Thus $I = J$.

Assume S contains an odd integer, say $2k + 1$ for some $k \geq 0$. Since the integer 2 is in I, we conclude that $1 = 2k + 1 - 2k$ is in J. Therefore $J = \mathbb{Z}[X]$ and we conclude that I is a maximal ideal in $\mathbb{Z}[X]$.

Exercise 3.4.24. 1. Let $k \geq 2$ be an integer and let $I = (k)$ be the principal ideal generated by k in $\mathbb{Z}$. Is the ideal I maximal in $\mathbb{Z}$?

2. Let I and J be two ideals in $\mathbb{Z}$. When is $I + J$ a maximal ideal in $\mathbb{Z}$?

3. Let p be a prime number and $I := (p, X)$ be an ideal in $\mathbb{Z}[X]$. Show that I is a maximal ideal in $\mathbb{Z}[X]$.

4. Let $k \geq 2$ be an integer. Is the ideal (k, X) a maximal ideal in $\mathbb{Z}[X]$?

5. Let p be a prime number and $t \geq 2$ be an integer. Is the ideal $I := (p^t, X)$ a maximal ideal in $\mathbb{Z}[X]$? Is it a prime ideal in $\mathbb{Z}[X]$?

6. Let R be a commutative ring. Show that the ideal (X)

   (a) is prime in $R[\![X]\!]$ if R is an integral domain.

   (b) is maximal in $R[\![X]\!]$ if R is a field.

7. Let R be a ring and M be an ideal in R. Let $U := \{x \in R : x \notin M\}$ be the set of all units in R. (That is $U = U(R)$). Show that M is the unique maximal ideal in R.

The solution of Example 3.4.23(4) is not ad hoc as the following proposition shows.

Proposition 3.4.25. *Let R be a commutative ring. If an ideal I is maximal, then for every element x in $R \setminus I$, there exists an element a in R such that $1 - ax$ is in I.*

Conversely if I is an ideal such that for every element x in $R \setminus I$, there exists an element a in R such that $1 - ax$ is in I, then I is a maximal ideal in R.

Proof. Let I be a maximal ideal in R and x be an element in $R \setminus I$. Then the set $\{t + ax : t \in I \text{ and } a \in R\}$ is an ideal in R.(Why?) Further the ideal I is strictly contained in $\{t + ax : t \in I \text{ and } a \in R\}$. Since the ideal I is maximal, it follows that the ideal $\{t + ax : t \in I \text{ and } a \in R\} = R$. Hence $1 = t + ax$ for some a in R and therefore $1 - ax = t \in I$. This completes one part of the proof.

Conversely, let J be an ideal in R such that $I \subseteq J \subseteq R$. If $I \neq J$, then there exists an element x in J but not in I. By hypothesis there exists an element a in R such that $1 - ax$ is in I. Since I is contained in J, the element $1 - ax$ is in J also. Therefore $1 = (1 - ax) + ax$ is in J. This proves that $J = R$ and the proof is complete. □

Proposition 3.4.26. *Let R be a commutative ring and I be a maximal ideal. Then I is a prime ideal.*

Proof. Let I be a maximal ideal. Let a and b be two elements in R such that their product ab is in I. We need to show that either a is in I or b is in I.

The ideal I is contained in the ideal $\{s + ar : s \in I \text{ and } r \in R\}$. Since I is a maximal ideal it follows that either $\{s + ar : s \in I \text{ and } r \in R\} = I$ or $\{s + ar : s \in I \text{ and } r \in R\} = R$. If $\{s + ar : s \in I \text{ and } r \in R\} = I$ then a is in I and we are through.

If $\{s + ar : s \in I \text{ and } r \in R\} = R$, there exist $s \in I$ and $x \in R$ such that $1 = s + ax$. We multiply this equation by b to get $b = sb + x(ab)$. This proves that b is in I and the proof is complete. □

Remark 3.4.27. The converse of the result is not true as shown in the following example.

Example 3.4.28. The principal ideal $I = (X)$ is prime in $\mathbb{Z}[X]$. But this is not a maximal ideal in $\mathbb{Z}[X]$.

Let $f(X) = a_0 + a_1X + a_2X^2 + \cdots + a_rX^r$ and $g(X) = b_0 + b_1X + b_2X^2 + \cdots + b_sX^s$ be two elements in $\mathbb{Z}[X]$ such that the product $f(X)g(X)$ is in the ideal $I = (X)$. Therefore there exists a polynomial $\psi(X) = c_0 + c_1X + c_2X^2 + \cdots + c_tX^t$ such that $f(X)g(X) = X\psi(X)$. By comparing the coefficients we get that $a_0b_0 = 0$. This means that either $a_0 = 0$ or $b_0 = 0$ and as a consequence we get either $f(X) = X(a_1 + a_2X + \cdots + a_rX^{r-1})$ is a multiple of X or $g(X) = X(b_1 + b_2X + \cdots + b_sX^{s-1})$ is a multiple of X. This proves that either $f(X)$ is in I or $g(X)$ is in I. Hence the ideal (X) is a prime ideal.

The ideal (X) is strictly contained in the ideal $(2, X)$. Hence it is not a maximal ideal in $\mathbb{Z}[X]$.

Example 3.4.29. Let R be an integral domain that is not a field. Then (0) is a prime ideal but not maximal in R.

The next Theorem 3.4.30 assures us the existence of maximal ideals in a commutative ring with identity.

Theorem 3.4.30. *Let R be a commutative ring with identity. Then R has a maximal ideal.*

Proof. If R is a field then (0) is a maximal ideal in R. Hence we will prove the result when R is not a field.

Let

$$\mathcal{C} := \{I \subseteq R : I \text{ is a proper ideal in } R\}.$$

This set is non-empty, because, for every nonzero and nonunit a in R, the principal ideal (a) is in $\mathcal{C}$.

We have already seen that $\mathcal{C}$ is a partially ordered set under the set inclusion of subsets. Let $\mathcal{T}$ be a totally ordered subset of $\mathcal{C}$. We claim that $\mathcal{T}$ has an upper bound in $\mathcal{C}$.

Let $J = \cup_{I\in\mathcal{T}} I$. Then we can easily prove that J is an ideal in R. Each of the ideal I in $\mathcal{T}$ is contained in J. Hence J is an upper bound of $\mathcal{T}$.

We claim that J is a proper ideal in R. Suppose $J = R$. Since the multiplicative identity 1 is in $R = J = \cup_{I\in\mathcal{T}} I$, it follows that 1 is in I for some I in $\mathcal{T}$, a contradiction.

Hence by Zorn's Lemma the set $\mathcal{C}$ has a maximal element, say K. By very definition, the ideal K is a maximal ideal in R. □

Remark 3.4.31. We have crucially used the fact that R is a ring with multiplicative identity to conclude that the ideal J is proper. There are rings without identity which do not have maximal ideal.

3.4.1 Ideals in $M(n, \mathbb{R})$

This section is devoted to showing that $M(n, \mathbb{R})$ does not have non-trivial two sided ideals.

Theorem 3.4.32. *Let R be a commutative ring and I be a two sided ideal in $M(n, R)$. Then there exists an ideal J in R such that $I = M(n, J)$.*

Before starting with the proof of the theorem some preliminaries below.

For $1 \leq i, j \leq n$, we denote by E_{ij} the elementary matrix with 1 in the (i, j)-th entry and 0 elsewhere. That is

$$(F_{ij})_{k,l} = \begin{cases} 1 & \text{if } (k, l) = (i, j) \\ 0 & \text{otherwise} \end{cases}.$$

We fix a pair (i, j) and write $E_{ij} := (e_{rs})$. Let $A = (a_{kl})$ be an element of $M(n, \mathbb{R})$. Then $E_{ij}A = (e_{rs})(a_{kl})$ is the matrix $(\sum_s e_{us}a_{sl})$ and

$$\sum_s e_{us}a_{sl} = \begin{cases} a_{jl} & \text{if } (u, s) = (i, j) \\ 0 & \text{otherwise} \end{cases}.$$

This shows that the i-th row of the matrix $E_{ij}A$ is the j-th row of the matrix A.

Let us now look at the product AE_{ij}. The (u, v)-th entry of AE_{ij} is

$$\sum_l a_{kl}u_{lv} = \begin{cases} a_{ui} & \text{if } (l, v) = (i, j) \\ 0 & \text{otherwise} \end{cases}.$$

This shows that the j-th column of the matrix AE_{ij} is the i-th column of the matrix A.

Finally let us look at the product $E_{pq}AE_{rs}$. We know that $E_{pq}A$ is the matrix whose p-th row is equal to the q-th row of A. If we write

$$E_{pq}A = \begin{pmatrix} R_1 \\ R_2 \\ \vdots \\ R_n \end{pmatrix}$$

as a row matrix, then

$$R_i = \begin{cases} (a_{q1}, a_{q2}, \ldots, a_{qn}) & \text{if } i = p \\ (0, 0, \ldots, 0) & \text{if } i \neq p \end{cases}.$$

That is if we denote by $B = E_{pq}A$, then

$$b_{ij} = \begin{cases} 0 & \text{if } i \neq q \\ a_{qj} & \text{if } i = q \end{cases}.$$

Let us denote the entries of the matrix E_{rs} by $e_{jk}^{r,s}$. Then

$$\sum_j b_{ij}e_{jk}^{r,s} = \begin{cases} b_{ir} & \text{if } k = s \\ 0 & \text{otherwise} \end{cases}.$$

Thus $E_{pq}AE_{rs}$ is the matrix whose (p, s)-th entry is a_{qr} and all other entries are zero. We summarize this as follows. Let A be a matrix. Then

- $E_{ij}A$ is the matrix whose i-th row is equal to the j-th row of A and the rest of the rows are zero.

- AE_{ij} is the matrix whose j-th column is equal to the i-th column of A and other columns are zero.

- $E_{pq}AE_{rs}$ is the matrix whose (p, s)-th entry is a_{qr} and all other entries are zero.

Now we begin with the proof of the theorem.

Proof. To understand the ideas in the proof of the theorem, we work out the details for $n = 2$.

Let I and ideal in $M(2, R)$ and $A = \begin{pmatrix} a & b \\ c & d \end{pmatrix}$ be an element in I.

We will now show that the matrices

$$\begin{pmatrix} a & 0 \\ 0 & 0 \end{pmatrix}, \begin{pmatrix} 0 & b \\ 0 & 0 \end{pmatrix}, \begin{pmatrix} 0 & 0 \\ c & 0 \end{pmatrix} \text{ and } \begin{pmatrix} 0 & 0 \\ 0 & d \end{pmatrix}$$

are also in I. This is achieved as follows:

- $\begin{pmatrix} 1 & 0 \\ 0 & 0 \end{pmatrix}\begin{pmatrix} a & b \\ c & d \end{pmatrix} = \begin{pmatrix} a & b \\ 0 & 0 \end{pmatrix}$.

- $\begin{pmatrix} a & b \\ 0 & 0 \end{pmatrix}\begin{pmatrix} 1 & 0 \\ 0 & 0 \end{pmatrix} = \begin{pmatrix} a & 0 \\ 0 & 0 \end{pmatrix}$

- $\begin{pmatrix} a & b \\ 0 & 0 \end{pmatrix}\begin{pmatrix} 0 & 0 \\ 0 & 1 \end{pmatrix} = \begin{pmatrix} 0 & b \\ 0 & 0 \end{pmatrix}$

- $\begin{pmatrix} 0 & 0 \\ 0 & 1 \end{pmatrix}\begin{pmatrix} a & b \\ c & d \end{pmatrix} = \begin{pmatrix} 0 & 0 \\ c & d \end{pmatrix}$

- $\begin{pmatrix} 0 & 0 \\ c & d \end{pmatrix}\begin{pmatrix} 0 & 1 \\ 0 & 0 \end{pmatrix} = \begin{pmatrix} 0 & 0 \\ c & 0 \end{pmatrix}$

- $\begin{pmatrix} 0 & 0 \\ c & d \end{pmatrix}\begin{pmatrix} 0 & 0 \\ 0 & 1 \end{pmatrix} = \begin{pmatrix} 0 & 0 \\ c & 0 \end{pmatrix}$

- $\begin{pmatrix} 0 & 0 \\ c & d \end{pmatrix}\begin{pmatrix} 0 & 0 \\ 0 & 1 \end{pmatrix} = \begin{pmatrix} 0 & 0 \\ 0 & d \end{pmatrix}$

If we denote by E_{ij}, for $1 \leq i, j \leq 2$, to be the elementary matrix defined by

$$(E_{ij})_{kl} = \begin{cases} 1 & \text{if } (i,j) = (k,l) \\ 0 & \text{otherwise} \end{cases}$$

then, we have shown that aE_{11}, bE_{12}, cE_{21} and dE_{22} are in I and further

$$\begin{pmatrix} a & b \\ c & d \end{pmatrix} = aE_{11} + bE_{12} + cE_{21} + dE_{22}.$$

Let

$$J := \{a \in R : aE_{ij} \text{ is in } I \text{ for some } (i,j)\}$$

We have shown above that $I \subseteq M(n, J)$ and $M(n, J) \subseteq I$.

We claim that J is an ideal in R. Let a and b be two elements in J. Then there exists pairs (i, j) and (k, l) such that aE_{ij} and bE_{kl} is in I. Then the matrix

$$bE_{ij} = E_{ik}(bE_{kl})E_{lj}$$

is in I. Hence $(a+b)E_{ij} = aE_{ij} + bE_{ij} = aE_{ij} + E_{ik}(bE_{kl})E_{lj}$ is in I.

This proves that $a+b$ is also in J. Thus J is closed under addition. Now to show that J is an ideal is very easy. This proves that $I = M(n, J)$ for an ideal J in R. □

Corollary 3.4.33. *If R is a field, then there are no two sided ideals in $M(n, R)$.*

Proof. Easy exercise. □

Fields can be characterized in terms of ideals as follows.

Proposition 3.4.34. *A ring R is a field iff it has no proper ideals.*

Proof. Let R be a field. Let $I \neq (0)$ be an ideal in R. If $\alpha \neq 0$ is in the ideal I, then, since R is a field, $1 = \alpha \cdot \alpha^{-1}$ is in I. This proves that $I = R$ and hence no proper ideal in R.

Conversely assume that R has no proper ideals. Let a be a nonzero element in R and $I := (a)$ be the principal ideal generated by a. Since a is nonzero, the ideal $I \neq (0)$, the zero ideal. Hence $I = R$ and this means that there exists an element b in R such that $1 = ab$. This proves that a is a unit in R completing the proof that R is a field. □

3.5 Quotient Rings

An important construction in rings is the quotient rings. This construction gives rise to new rings. We will do this construction in this section.

Let R be a ring and I be an ideal in R. Since $(R, +)$ is abelian the subgroup $(I, +)$ is a normal subgroup of $(R, +)$. Hence the quotient R/I becomes an abelian group under addition defined by

$$(x + I) + (y + I) := x + y + I$$

for $x + I$ and $y + I$ in R/I. Now we want to define a multiplication on R/I so as to make it a ring.

Let a, b in R and x, y in the ideal I. Then the product $(a + x)(b + y) = ab + xb + ay + xy$ is in the coset $ab + I$. This means that the set

$$(a + I)(b + I) := \{(a + x)(b + y) : x, y \text{ in } I\} \subseteq ab + I.$$

Since the cosets of the ideal I partition the group $(R, +)$, it follows that $ab + I$ is the unique coset containing the product

$$(a + I)(b + I) := \{(a + x)(b + y) : x, y \text{ in } I\}.$$

Hence the multiplication $\cdot : R/I \times R/I \to R/I$defined by

$$(a+I)(b+I) := ab+I$$

is a binary operation on R/I. We will now show that, with this multiplication, $(R/I, +, \cdot)$ becomes a ring .

- We have seen in group theory that $(R/I, +)$ is an abelian group.
- Let $a+I$, $b+I$ and $c+I$ be three elements of R/I. Then, using the associativity of multiplication in R, we get that

$$\begin{aligned}(a+I)\left((b+I)(c+I)\right) &= (a+I)(bc+I) \\ &= a(bc)+I \\ &= (ab)c+I \\ &= (ab+I)(c+I) \\ &= \left((a+I)(b+I)\right)(c+I).\end{aligned}$$

- $(a+I)(1+I) = a+I$ for all cosets $a+I$ in R/I.
- If $a+I$, $b+I$ and $c+I$ are elements of R/I, we use distributivity of multiplication over addition in R to show that

$$\begin{aligned}(a+I)\left((b+I)+(c+I)\right) &= (a+I)(b+c+I) \\ &= a(b+c)+I \\ &= (ab+ac)+I \\ &= (ab+I)+(ac+I).\end{aligned}$$

Thus $R(/I, +, \cdot)$ is a ring. Just as in quotient groups, we continue to denote the coset $a+I$ of the ideal by $\overline{a}$.

If the ideal I is prime or maximal in R, then we can say more about the quotient ring R/I.

Theorem 3.5.1. *Let R be a ring. Let I be an ideal in R. Then*

1. *I is a prime ideal iff R/I is an integral domain.*
2. *I is a maximal ideal iff R/I is a field.*

Proof. Let I be a prime ideal. We will show that R/I is an integral domain.

Let a, b in R such that the product $\overline{a}\cdot\overline{b} = \overline{ab} = \overline{0}$. The zero element in R/I is the ideal I. Therefore $\overline{ab} = ab+I = I$ implies that the element ab is in the ideal I. Since I is a prime ideal, it follows that either a or b is in I. Hence either $\overline{a} = \overline{0}$ or $\overline{b} = \overline{0}$. This proves that R/I is an integral domain.

Conversely let us assume that R/I is an integral domain. Let a and b two elements of the ring R such that the product ab is in I. Hence $\overline{a}\overline{b} = \overline{ab}$ is zero in the ring R/I. This implies that either $\overline{a} = 0$ or $\overline{b} = 0$. If $\overline{a} = 0$ then a is in I and if $\overline{b} = 0$ then b is in the ideal I. This proves that I is a prime ideal and finishes the proof of the first part of theorem.

Let I be a maximal ideal. We claim that R/I is a field. We will show that any nonzero element $\overline{a}$ is a unit in R/I. Since the ideal I is a maximal ideal it is strictly contained in the ideal $\{s + ar : r \in R\}$. Hence $\{s + ar : r \in R\} = R$ and there exists b in R and s in I such that $s + ab = 1$. This means that $\overline{a}\overline{b} = \overline{1}$. Thus we have shown that R/I is a field.

Conversely assume that R/I is a field. Let J be an ideal in R such that $I \subseteq J$. If $I \neq J$ there exists an element a in J such that a is not in I. Therefore the element $\overline{a}$ is nonzero in R/I. Since R/I is a field there exists an element $\overline{b}$ in R/I such that $\overline{a}\overline{b} = \overline{1}$. Hence the element $1 - ab$ is in I. The ideal $I \subseteq J$ and the element a is in J. Hence $1 = (1 - ab) + ab$ is also in J. Thus $J = R$ and the proof is complete. □

Examples 3.5.2. 1. The principal ideal (X) is a prime ideal in $\mathbb{Z}[X]$.

2. Let p be a prime number. Then the principal ideal (p) is a prime ideal in $\mathbb{Z}[X]$.

3. The ideal (p, X) is a maximal ideal in $\mathbb{Z}[X]$ for every prime number p.

4. Let X be a set and $x \neq y$ be two distinct points in X. Then

$$Z(\{x, y\}) := \{f \in \mathcal{F}(X, \mathbb{R}) : f(x) = f(y) = 0\}$$

is not a prime ideal in $\mathcal{F}(X, \mathbb{R})$.

We will see many examples of quotient rings after the section on homomorphisms.

3.6 Homomorphisms

Definition 3.6.1. Let R and S be two rings. A map $f : R \to S$ is a homomorphism if

1. $f(x + y) = f(x) + f(y)$,

2. $f(xy) = f(x)f(y)$ and

3. $f(1) = 1$

for all x and y in R.

Remark 3.6.2. The first condition shows that f is a group homomorphism between the additive groups $(R, +)$ and $(S, +)$. Hence the additive identity 0 of the ring will be mapped to the additive identity of the ring S. That is $f(0) = 0$. However, since, the nonzero elements of R do not form group under multiplication, we can not conclude that $f(1) = 1$. Hence we need to list it separately, if we want the image of f to be a subring of S. This third condition puts a lot of restriction on the existence of homomorphism between rings.

Examples 3.6.3. 1. Let R be a ring. The identity map $x \mapsto x$ is a homomorphism.

2. Let $\varphi : \mathbb{Z}[X] \to \mathbb{R}$ be defined by $\varphi(X) := \sqrt{2}$ and extended to a ring homomorphism by defining
$$\varphi(a_0 + a_1X + \cdots + a_rX^r) = a_0 + a_1\sqrt{2} + \cdots + a_r(\sqrt{2})^r.$$

3. Let $\varphi : \mathbb{R}[X] \to \mathbb{C}$ be defined by $\varphi(X) = i$ and extended as in the previous example.

4. Let R be a ring and $\chi : \mathbb{Z} \to R$ be defined by $\chi(k) := \overbrace{1 + 1 + \cdots + 1}^{k-\text{times}}$. This homomorphism is called the characteristic homomorphism.

5. Let $R := \left\{ A = \begin{pmatrix} a & -b \\ b & a \end{pmatrix} : a, b \in \mathbb{R} \right\}$ and $\varphi : R \to \mathbb{C}$ be defined by $\varphi(A) = a + i\,b$. Then φ is a homomorphism.

6. Let D be a square free integer and
$$R = \left\{ \begin{pmatrix} a & b\,D \\ b & a \end{pmatrix} : a, b \text{ in } \mathbb{Z} \right\}.$$
Then the map
$$\begin{pmatrix} a & b\,D \\ b & a \end{pmatrix} \mapsto a + b\sqrt{D}$$
is a ring homomorphism between R and $\mathbb{Z}[\sqrt{D}]$.

The map φ is an additive homomorphism is easy.

Let $A = \begin{pmatrix} a & b\,D \\ b & a \end{pmatrix}$ and $B = \begin{pmatrix} c & d\,D \\ d & c \end{pmatrix}$ be two elements in R. Then

$AB = \begin{pmatrix} ac + bd\,D & (ad + bc)D \\ ad + bc & ac + bd\,D \end{pmatrix}$ and
$$\begin{aligned} f(AB) &= ac + bd\,D + \sqrt{D}(ad + bc) \\ &= (a + b\sqrt{D})(c + d\sqrt{D}) \\ &= f(A)f(B). \end{aligned}$$

The result is true if we replace $\mathbb{Z}$ by $\mathbb{Q}$.

7. Let R be a ring and $D : R[X] \to R[X]$ be the map defined by
$$D\left(\sum_{n=0}^{k} a_nX^n\right) = \sum_{n=1}^{k} na_nX^{n-1}$$
is **not** a ring homomorphism. For, $D(1) = 0 \neq 1$.
However this is a group homomorphism.

Another similar example:

Let R be a ring and $D : R[\![X]\!] \to R[\![X]\!]$ be the map defined by
$$D\left(\sum_{n=0}^{\infty} a_nX^n\right) = \sum_{n=1}^{\infty} na_nX^{n-1}$$

is **not** a ring homomorphism.
Notice that $D(f(x)g(X)) = D(f(X))g(X) + f(X)D(g(X))$ for any two power seris $f(X)$ and $g(X)$ in $R[\![X]\!]$.

8. Let $R = C(J, \mathbb{R})$ be the ring of real valued continuous functions on the closed and bounded interval $J = [a, b]$. Then the map $f \mapsto \int_a^b f(x)dx$ is additive group homomorphism but not a **ring** homomorphism.

For example, let $f(x) = x = g(x)$ for $x \in [a, b]$. Then

$$\int_a^b f(x)g(x)dx = \int_a^b x^2 dx = \frac{b^3 - a^3}{3} \text{ and}$$
$$\left(\int_a^b f(x)dx\right)\left(\int_a^b g(x)dx\right) = \left(\int_a^b x dx\right) = \left(\frac{b^2 - a^2}{2}\right)^2.$$

This need not be equal!

Exercise 3.6.4. Let R and S be two rings and $\varphi : R \to S$ be a ring homomorphism. Show that $\varphi(u)$ is a unit in S, if u is a unit in R.

Let $\varphi : R \to S$ be a ring homomorphism. Let u be a unit in R and v be its inverse. Then

$$1 = \varphi(1) = \varphi(uv) = \varphi(u)\varphi(v).$$

Thus $\varphi(u)$ is a unit in S. In particular $\varphi(u) \neq 0$ for a unit u in R.

Remark 3.6.5. The converse of this exercise is not true.

1. Consider the inclusion of $\mathbb{Z}$ in to $\mathbb{Q}$. This is a ring homomorphism and every nonzero integer is a unit in $\mathbb{Q}$ but not in $\mathbb{Z}$!
2. Let p be a prime number and consider the natural map $k \mapsto [k]$ from $\mathbb{Z}$ to $\mathbb{Z}_p$. If k is not a multiple of p, then $[k]$ is a unit in $\mathbb{Z}_p$ but k is not a unit in $\mathbb{Z}$, if $|k| \geq 2$.

Exercise 3.6.6. Show that a homomorphism between two rings need not map zero divisors to zero divisors.

The image of a zero divisor could even be a unit and image of a nonzero divisor could be a zero divisor. Here we give examples illustrating both the possibilities.

1. Let k and l be two co-prime integers and $\varphi : \mathbb{Z}_{kl} \to \mathbb{Z}_k$ be the homomorphism defined by $\varphi([x]) = [x]$ where $[x]$ is the equivalence class of x in $\mathbb{Z}_k$. Since k and l are co-prime, the image of $[l]$ in $\mathbb{Z}_k$ is a unit in $\mathbb{Z}_k$ while $[l]$ is a zero divisor in $\mathbb{Z}_{kl}$.
2. Let $n \geq 2$ be a composite number and consider the natural map $\mathbb{Z} \to \mathbb{Z}_n$. If k is a proper divisor of n, then $[k]$ is a zero divisor in $\mathbb{Z}_n$ but k is not a zero divisor in $\mathbb{Z}$.

Exercise 3.6.7. Show that the function $f(x) = \sin x$ is a zero divisor in $\mathcal{F}(\mathbb{R}, \mathbb{R})$ but not in $C(\mathbb{R}, \mathbb{R})$.
Hint: What can you about $\mathbb{R} \setminus \{x \in \mathbb{R} : \sin x = 0\}$?

Exercise 3.6.8. 1. Find all rings homomorphisms of $\mathbb{Z}$.

Let $\varphi : \mathbb{Z} \to \mathbb{Z}$ be a ring homomorphism. Since $\varphi(1) = 1$, it follows that

$$\varphi(k) = \varphi(\overbrace{1 + 1 + \cdots + 1}^{k\text{-times}}) = \overbrace{f(1) + f(1) + \cdots + f(1)}^{k\text{-times}} = k$$

for all $k \in \mathbb{N}$. Since $\varphi(-l) = -\varphi(l)$ for every integer l, we see that $\varphi(k) = k$ for all $k \in \mathbb{Z}$. Thus the identity map is the only ring homomorphism of $\mathbb{Z}$.

2. Show that the only ring homomorphism of $\mathbb{Q}$ is identity.

3. Let n and m be two natural numbers. Find all ring homomorphisms between $\mathbb{Z}_m$ and $\mathbb{Z}_n$.

Let $\varphi : \mathbb{Z}_m \to \mathbb{Z}_n$ be a ring homomorphism. Then $\text{Im}(\varphi)$ is a subring of $\mathbb{Z}_n$. Hence by Lagrange's theorem $|\text{Im}(\varphi)|$ must divide the order of m and n.

If m and n are relatively prime this is not possible. (Why?). Thus if m and n are relatively prime then there are no ring homomorphism between $\mathbb{Z}_m$ and $\mathbb{Z}_n$.

Let us now assume that $d = \gcd(m, n) \neq 1$.

Let $\varphi : \mathbb{Z}_m \to \mathbb{Z}_n$ be a ring homomorphism. Then $\varphi([1]) = [1]$, the identity element in $\mathbb{Z}_m$. Let $[k]$ be an element in $\mathbb{Z}_m$. Then $[k] = \overbrace{[1] + [1] + \cdots + [1]}^{k\text{-times}}$ and $\varphi([k]) = \overbrace{[1] + [1] + \cdots + [1]}^{k\text{-times}} = k[1]$ in $\mathbb{Z}_n$. If $k = m$, then $\varphi([m]) = \varphi([0]) = [0]$ in $\mathbb{Z}_n$ and this can also be written as $\varphi([m]) = \overbrace{[1] + [1] + \cdots + [1]}^{m\text{-times}} = m[1]$. Thus $m[1]$ must be zero in $\mathbb{Z}_n$. This is possible iff n divides m. This proves that there is a ring homomorphism between $\mathbb{Z}_m$ and $\mathbb{Z}_n$ iff $m \geq n$ and n divides m. Further this homomorphism is unique.

4. Does there exist a ring homomorphism

   (a) $\varphi : \mathbb{Q} \to \mathbb{Z}$?

   (b) $\varphi : \mathbb{R} \to \mathbb{Q}$?

   (c) $\varphi : \mathbb{C} \to \mathbb{R}$?

5. Let R be a ring and $R[X]$ be the polynomial ring over R. Show that the inclusion map $i : R \to R[X]$, $a \mapsto a$, the constant polynomial is a homomorphism.

6. Let R be a ring and I be an ideal in R. Show that the map $\varphi : R \to R/I$ defined by $\varphi(a) = a + I$ is a ring homomorphism.

7. Let R be a ring and J be an ideal in the polynomial ring $R[X]$. Show that the natural map $\varphi \circ i : R \to R[X]/J$ is a ring homomorphism. Further show that the homomorphism is on to iff X is in the image of $\varphi \circ i$.

It is easy to show that $\varphi \circ i$ is a ring homomorphism.
If $\varphi \circ i : R \to R[X]/J$ is on-to, then $X + J = \varphi(k)$ for some element k in R.
For the converse assume that $X+J$ is in the image and $\varphi(k) = X+J$ for some element k in R. Let $f(X) = a_0 + a_1X + a_2X^2 + \ldots + a_rX^r$ be a polynomial in $R[X]$. Then

$$\begin{aligned}\varphi\left(\sum_{i=0}^{r} a_i k^i\right) &= \sum_{i=0}^{r}\varphi(a_i)\varphi(k^i)\\ &= \sum_{i=0}^{r}(a_i + J)(X+J)^i\\ &= \sum_{i=0}^{r}(a_i + J)(X^i + J)\\ &= \left(\sum_{i=0}^{r} a_i X^i\right) + J.\end{aligned}$$

Thus $\varphi \circ i : R \to R[X]/J$ is on to.

8. Show that there does exist any ring homomorphism from the ring $\mathbb{Z}[i]$ in to the ring $\mathbb{Z}/2\mathbb{Z}$.

9. Let D be a square free integer Does there exist a ring homomorphism between the rings $\mathbb{Z}[\sqrt{D}]$ and $\mathbb{Z}[\sqrt{2}]$, if $|D| \neq 2$? Do this exercise by replacing $\mathbb{Z}$ with $\mathbb{Q}$, the field of rational numbers.

   Hint: Let $\varphi : \mathbb{Z}[\sqrt{D}] \to \mathbb{Z}[\sqrt{2}]$ be a ring homomorphism. Then $\varphi(D) = D$. If $\varphi(\sqrt{D}) = a + b\sqrt{2}$, then $D = \varphi(D) = [\varphi(\sqrt{D})]^2 = (a + b\sqrt{2})^2 = a^2 + 2b^2 + 2ab\sqrt{2}$.

10. Let D be a square free intger. Does there exist a ring homomorphism between the rings $\mathbb{Z}[\sqrt{D}]$ and $\mathbb{Z}[i]$, if $|D| \neq 1$?

11. Let D be a square free integer. Does there exist a ring homomorphism from $\mathbb{Z}[i]$ in to $\mathbb{Z}[\sqrt{-D}]$?

12. Let $n \geq 3$ and ω be a primitive n-th root of unity. Show that there does not exist any ring homomorphism between $\mathbb{Z}[\omega]$ and $\mathbb{Z}[\sqrt{2}]$.

13. Let $n, m \geq 3$ and $n \neq m$ be two integers. Let ω be a primitive n-th root of unity and ζ be a primitive m-th root of unity. Does there exist a ring homomorphism between the two rings $\mathbb{Z}[\omega]$ and $\mathbb{Z}[\zeta]$?

14. Does there exist a ring homomorphism $\varphi : \mathbb{Q}[i] \to \mathbb{Q}[\sqrt{2}]$?

    No!

15. Let D_1 and D_2 be two square free integers. Does there exist a ring homomorphism between the rings

    (a) $\mathbb{Z}[\sqrt{D_1}]$ and $\mathbb{Z}[\sqrt{D_2}]$?
    (b) $\mathbb{Q}[\sqrt{D_1}]$ and $\mathbb{Q}[\sqrt{D_2}]$?

16. Let R and S be two rings and $\varphi : R \to S$ be a ring homomorphism. Show that there exists a ring homomorphism $f : R(\mathbb{C}) \to S(\mathbb{C})$ such that $\varphi(r) = f(r, 0)$.

17. Find all ring homomorphisms $\varphi : \mathbb{R} \to \mathbb{R}$.

    Let $\varphi : \mathbb{R} \to \mathbb{R}$ is a ring homomorphism. Then $\varphi(r) = r$ for all rational numbers r. (Why?)

    - $\varphi(x) > 0$ if x is a positive real number.
      Let x be a positive real number. Then there exists a real number y such that $y^2 = x$. Therefore $\varphi(x) = \varphi(y^2) = (\varphi(y))^2 \geq 0$. Since y is nonzero it is a unit in $\mathbb{R}$ and hence $\varphi(y)$ is also a unit in $\mathbb{R}$. Thus $\varphi(x)$ is positive.
      If $x < 0$ then $t = -x > 0$. Therefore $\varphi(t) > 0$. That is $\varphi(-x) = -\varphi(x) > 0$. This shows that $\varphi(x) < 0$ if x is negative.
    - The map φ is increasing:
      Let $x < y$. Then $y - x > 0$ and hence $\varphi(y) - \varphi(x) = \varphi(y - x) > 0$.
    - φ is identity.
      Suppose there exists x in $\mathbb{R}$ such that $x \neq \varphi(x)$. If $x < \varphi(x)$ choose a rational number r such that $x < r < \varphi(x)$. Since φ is increasing, we see that $\varphi(x) < \varphi(r) = r < \varphi(x)$, a contradiction. Similarly $\varphi(x) > x$ is not possible either. Thus $\varphi(x) = x$ for all x.

18. Find all ring homomorphisms of $\mathbb{C}$ such that $\varphi(r)$ is in $\mathbb{R}$ if r is in $\mathbb{R}$.

    If $\varphi(x)$ is in $\mathbb{R}$ for x in $\mathbb{R}$, then we have a very simple descriptions of all the homomorphisms of $\mathbb{C}$.

    Let us observe that $\mathbb{R}$ can be considered as a subring of $\mathbb{C}$ through the homomorphism $x \mapsto x + i \cdot 0$.

    Let $\varphi : \mathbb{C} \to \mathbb{C}$ be a ring homomorphism. Since we assume $\varphi(x)$ to be a real number for x in $\mathbb{R}$, it follows that $\varphi : \mathbb{R} \to \mathbb{R}$ is a ring homomorphism. Hence $\varphi(x) = x$ for all x in $\mathbb{R}$.

    Let $z = s + it$ be a complex number. Then $\varphi(s + i\,t) = \varphi(s) + \varphi(i)\varphi(t) = s + \varphi(i)\,t$. Thus it is enough to find the image of i. Now

    $$-1 = \varphi(i^2) = (\varphi(i))^2.$$

    This shows that $\varphi(i) = \pm\, i$ and the only homomorphism of $\mathbb{C}$ which maps real numbers to real numbers are the maps

    $$z \mapsto z \text{ and } z \mapsto \overline{z}.$$

    There are other homomorphisms if we don't insist on reals going to reals. We are not doing it here.

Proposition 3.6.9. *Let R and S be two rings and $\varphi : R \to S$ be a ring homomorphism. Then*

1. $\varphi(0) = 0$,

2. *the set* $\ker(\varphi) := \{r \in R : \varphi(r) = 0\}$ *is an ideal in R called the kernel of the homomorphism f and*

3. *the set* $Im(\varphi) := \{s \in S : s = \varphi(r)$ *for some* r *in* $R\}$ *is a subring of* S *called the image of the homomorphism* φ.

Proof. If $\varphi : R \to S$ is a ring homomorphism, then $\ker(\varphi)$ is a subgroup of $(R, +)$.

Let $a \in R$ and $r \in \ker(\varphi)$. Then $\varphi(ar) = \varphi(a)\varphi(r) = \varphi(a) \cdot 0 = 0$. Thus ar is in $\ker(\varphi)$. This proves that $\ker(\varphi)$ is an ideal in R.

We leave it to the reader to verify that $\text{Im}(\varphi)$ is a subring of S. □

Proposition 3.6.10. *Let* R *and* S *be two rings and* $\varphi : R \to S$ *be a homomorphism. Then*

1. *the set* $\varphi^{-1}(J) := \{r \in R : \varphi(r) \in J\}$ *is an ideal in* R *for every ideal* J *in* S.

2. *the ideal* $\varphi^{-1}(J)$ *is a prime ideal in* R, *if* J *is a prime ideal in* S.

3. *the ideal* $\varphi^{-1}(J)$ *need not be a maximal ideal if* J *is a maximal ideal.*

Proof. Let J be an ideal in S. Then $\varphi^{-1}(J)$ is non-empty, because 0 is in $\varphi^{-1}(J)$.

Let a and b be two elements in $\varphi^{-1}(J)$. Then $\varphi(a)$ and $\varphi(b)$ are in the ideal J. Hence

$$\varphi(-a) = -\varphi(a) \text{ and } \varphi(a + b) = \varphi(a) + \varphi(b)$$

are also in J, proving that $-a$, $-b$ and $a + b$ are in $\varphi^{-1}(J)$. Now it is very easy to show that $\varphi^{-1}(J)$ is a group under addition.

Let r be an element in R and a in $\varphi^{-1}(J)$. Then

$$\varphi(ar) = \varphi(a)\varphi(r)$$

is in J. This proves that $\varphi^{-1}(J)$ is an ideal in R.

Now let J be a prime ideal in S. Let a and b be two elements in R such that the product ab is in $\varphi^{-1}(J)$. Using the fact that J is an ideal in S, we conclude that $\varphi(ab) = \varphi(a)\varphi(b)$ is in J. Since J is a prime ideal either $\varphi(a)$ or $\varphi(b)$ is in J, which in turn proves that either a or b is in $\varphi^{-1}(J)$. Thus $\varphi^{-1}(J)$ is a prime ideal in R.

We have the natural homomorphism from $\mathbb{Z}[X]$ in to $\mathbb{Q}[X]$. The ideal (X) is maximal in $\mathbb{Q}[X]$(Why?) but its inverse image is not maximal in $\mathbb{Z}[X]$ as we have seen earlier. □

Exercise 3.6.11. Find the kernel of the following ring homomorphisms.

1. $\varphi : \mathbb{Z}[X] \to \mathbb{R}$ defined by $\varphi(X) = \sqrt{2}$.

2. $\varphi : \mathbb{R}[X] \to \mathbb{C}$ mapping X to i.

3. $\varphi : \mathbb{R}[X] \to \mathbb{C}$ defined by $\varphi(X) = \omega$ where $\omega \neq 1$ is a cube root of unity.

4. $\varphi : \mathbb{R}[X] \to \mathbb{C}$ defined by $\varphi(X) = \zeta_n$, where ζ_n is a primitive n-th root of unity for $n \geq 3$.

Exercise 3.6.12. 1. Let $\mathbb{F}$ be a field and R be a ring. Show that any ring homomorphism $\varphi : \mathbb{F} \to R$ is one-one.

2. Let $\varphi : R \to S$ be a homomorphism between the rings R and S with kernel $\ker(\varphi)$. Let $f : R(\mathbb{C}) \to S(\mathbb{C})$ be the extension of φ. (See Exercise 3.6.8(16)). What is the kernel of f?

Exercise 3.6.13. Let $n \geq 2$. What are the homomorphisms from $M(n, \mathbb{R})$ to $\mathbb{R}$?

Let $\varphi : M(n, \mathbb{R}) \to \mathbb{R}$ be a ring homomorphism. Then $\ker(\varphi)$ is a two sided ideal in $M(n, \mathbb{R})$. We know that there are no proper two sided ideals in $M(n, \mathbb{R})$. Hence $\varphi : M(n, \mathbb{R}) \to \mathbb{R}$ is one-one. Since $\mathbb{R}$ is commutative, the ring $M(n, \mathbb{R})$ as a subring of $\mathbb{R}$ must be commutative, a contradiction. Hence there does not exist any ring homomorphism $\varphi : M(n, \mathbb{R}) \to \mathbb{R}$ for $n \geq 2$.

3.6.1 Characteristic of a ring

Let G be a group and x be an element of G. We know that there exists a unique group homomorphism from $f : \mathbb{Z} \to G$ such that $f(1) = x$.

Let R be a commutative ring with identity. Since $(R, +)$ is an abelian group there exists a unique group homomorphism $\chi : \mathbb{Z} \to R$ such that $\chi(1) := 1_R$, the multiplicative identity of the ring R. Let k be a natural number. Then $\overbrace{1 + 1 + \cdots + 1}^{k-\text{times}}$ and $\overbrace{(-1) + (-1) + \cdots + (-1)}^{k-\text{times}}$ are also in R.

We denote $\overbrace{1 + 1 + \cdots + 1}^{k-\text{times}}$ by $k \cdot 1$ and $\overbrace{(-1) + (-1) + \cdots + (-1)}^{k-\text{times}}$ by $k \cdot (-1)$. Notice that $k \cdot 1 + k \cdot (-1) = 0$.

Therefore $\chi(k) = \begin{cases} k \cdot 1 & \text{if } k \text{ is positive} \\ (-k) \cdot (-1) & \text{if } k \text{ is negative} \end{cases}$.

We denote this by $k \cdot 1$ for all $k \in \mathbb{Z}$. We will now show that this map is a ring homomorphism.

Proposition 3.6.14. *The map $\chi : \mathbb{Z} \to R$ is a ring homomorphism.*

This map is called the characteristic homomorphism *of the ring R.*

Proof. Let k and l be two integers. Then, since $\chi : \mathbb{Z} \to R$ is a group homomorphism,

$$\begin{aligned} \chi(kl) &= \chi(\overbrace{l + l + \cdots + l}^{k-\text{ times}}) \\ &= \overbrace{\chi(l) + \chi(l) + \cdots + \chi(l)}^{k-\text{ times}} \\ &= k(\chi(l)) \\ &= (k.1)\chi(l) \\ &= \chi(k)\chi(l). \end{aligned}$$

This proves that $\chi : \mathbb{Z} \to R$ is a ring homomorphism. □

Let $\ker(\chi)$ be the kernel of the homomorphism. It is an ideal in $\mathbb{Z}$. Hence there exists a non-negative integer n such that $\ker(\chi) = n\mathbb{Z}$. By definition

$n \cdot 1 = 0$ and this is the smallest non-negative integer n with this property. The integer n is called the *characteristic* of the ring R and it is also denoted by $\mathrm{ch}(R)$.

Examples 3.6.15. 1. Let $R = \mathbb{Z}$. Then the map χ is identity and its kernel is trivial. Thus the characteristic of the ring of integers is zero.

2. The characteristic of the rings $\mathbb{Q}$, $\mathbb{R}$, $\mathbb{C}$ and their polynomial rings are all zero.

3. Let k be a natural number and $\chi : \mathbb{Z} \to \mathbb{Z}_k$ be the characteristic homomorphism. Then the characteristic of the ring $\mathbb{Z}_k$ is k.

Exercise 3.6.16. 1. Let n and m be two natural numbers bigger than 1. What is the characteristic of $\mathbb{Z}_n \times \mathbb{Z}_m$.

2. Let R and S be two rings. Is there a relation between $\mathrm{ch}(R \times S)$, $\mathrm{ch}(R)$ and $\mathrm{ch}(S)$?

3. Let R be a ring. What is the characteristic of $R(\mathbb{C})$, the complexification of R?

4. Let R and S be two rings of different characteristics. Does there exist a ring homomorphism between R and S?

Proposition 3.6.17. *Let R be an integral domain. Then the characteristic of R is either zero or a prime number p.*

Proof. Let R be an integral domain and $\chi : \mathbb{Z} \to R$ be the characteristic homomorphism.

If the characteristic k is nonzero and composite, we write $k = k_1 k_2$ where k_1 and k_2 are proper divisors of k. Since χ is a ring homomorphism, it follows that

$$0 = k \cdot 1 = \chi(k) = \chi(k_1)\chi(k_2) = (k_1 \cdot 1)(k_2 \cdot 1).$$

Since R is an integral domain, either $k_1 \cdot 1 = 0$ or $k_2 \cdot 1 = 0$. This is a contradiction to the minimality of k. Thus k is either zero or a prime number. □

A simple application of characteristic homomorphism is the following proposition.

Proposition 3.6.18. *Let R be a ring of characteristic $k \neq 0$. Then $(a+b)^k = a^k + b^k$ for all elements a and b in R.*

Proof. Let a and b be two elements of the ring R. Then it follows from binomial theorem that

$$(a+b)^k = a^k + \sum_{r=1}^{k-1} \binom{k}{r} a^{k-r} b^r + b^k.$$

Since k divides $\binom{k}{r}$ for $1 < r < k$, it follows that $\binom{k}{r} a^{k-r} b^r = 0$. Hence

$$(a+b)^k = a^k + b^k$$

for all elements a and b in R. □

Corollary 3.6.19. *Let R be a ring of characteristic $k \neq 0$. Then the map $a \mapsto a^k$ is a ring homomorphism of R.*

3.6.2 Fundamental Theorem of Homomorphism

In this section we prove the following theorem.

Theorem 3.6.20 (Fundamental Theorem of Homomorphism)**.** *Let R and S be two rings and $\varphi : R \to S$ be an onto ring homomorphism. Then there exists a unique isomorphism $\overline{\varphi} : R/\ker(\varphi) \to S$ such that $\varphi(x) = \overline{\varphi} \circ \pi(x)$ where $\pi : R \to R/\ker(\varphi)$ is the quotient homomorphism.*

Proof. Let $\varphi : R \to S$ be an onto ring homomorphism between the rings R and S. We denote the ideal $\ker(\varphi)$ by K. Therefore K is a normal subgroup of R and by first fundamental theorem of homomorphism of groups (See Theorem 2.7.2) that there exists a unique isomorphism $\overline{\varphi} : R/K \to S$ of groups. This map is defined by $\overline{\varphi}(a + K) = \varphi(a)$ for $a + K \in R/K$.

Now we need to check this map is a ring homomorphism.

Let $a + K$ and $b + K$ be two elements in R/K. Then

$$\begin{aligned}\overline{\varphi}((a+K)(b+K)) &= \overline{\varphi}(ab+K)\\ &= \varphi(ab)\\ &= \varphi(a)\varphi(b)\\ &= \overline{\varphi}(a+K)\overline{\varphi}(b+K).\end{aligned}$$

This completes the proof. □

The Fundamental Theorem of Homomorphism 3.6.20 for rings has many applications and some of them are done below.

Examples 3.6.21.

1. Let $\varphi : \mathbb{R}[X] \to \mathbb{C}$ be the homomorphism mapping $X \mapsto i$. This homomorphism is on-to and the kernel is the principal ideal generated by the polynomial $X^2 + 1$. Hence by Theorem 3.6.20, the quotient ring $\mathbb{R}[X]/\left(X^2 + 1\right)$ is isomorphic to the field of complex numbers.

2. Let $\varphi : \mathbb{Q}[X] \to \mathbb{R}$ be the homomorphism mapping $X \mapsto \sqrt{2}$. As in the previous example the kernel of φ is the principal ideal $I = \left(X^2 - 2\right)$. Hence by Fundamental Theorem of Homomorphism (Theorem 3.6.20), the quotient ring $\mathbb{Q}[X]/I$ is isomorphic to the ring $\mathbb{Q}[\sqrt{2}]$.

3. Let $R := \{(a_n) : a_n \in \mathbb{Q}$ and (a_n) is a Cauchy sequence$\}$ and let $I := \{(a_n) \in R : \lim_n a_n = 0\}$. Then R/I is isomorphic to $\mathbb{R}$, the field of real numbers.

   We know that every cauchy sequence of real numbers is convergent in $\mathbb{R}$. Hence $\lim_n a_n$ exists for every element (a_n) in R. This suggests a map $\varphi : R \to \mathbb{R}$ defined by $\varphi((a_n)) = \lim_n a_n$. This map is well defined.

   Using algebra of convergent sequences, we can show that φ is a ring homomorphism and the kernel is I.

We use density of rationals in $\mathbb{R}$ to show that this map is onto: Since $\mathbb{Q}$ is dense in $\mathbb{R}$, given a real number a, there is a sequence (a_n) of rational converging to a. Hence R/I is isomorphic to $\mathbb{R}$.

4. Let R be a ring and I be an ideal. Let $I[X] = \{\sum_{i=0}^{n} a_i X^i : a_i \in I\}$. Then $I[X]$ is an ideal in $R[X]$ and the polynomial ring $(R/I)[X]$ defined over the quotient ring R/I is isomorphic to the quotient ring $R[X]/I[X]$.

It is easy to show that $I[X]$ is an ideal in $R[X]$. There is a natural map $\varphi : R[X] \to (R/I)[X]$ defined by

$$\varphi(a_0 + a_1 X + \cdots a_r X^r) = \overline{a}_0 + \overline{a}_1 X + \cdots + \overline{a}_r X^r$$

where $\overline{a_i} = a_i + I$.

It is easy to prove that φ is a group homomorphism.

Let $f(X) = \sum_{i=0}^{n} a_i X^i$ and $g(X) = \sum_{j=0}^{m} b_j X^j$ be two polynomials in $R[X]$. Then

$$\begin{aligned} f(X)g(X) &= \sum_{k=0}^{n+m} \sum_{i+j=0}^{k} a_i b_j X^{i+j} \text{ and} \\ \varphi(f(X)g(X)) &= \sum_{k=0}^{n+m} \sum_{i+j=0}^{k} \overline{a_i b_j} X^{i+j} \\ &= \sum_{k=0}^{n+m} \sum_{i+j=0}^{k} \overline{a}_i \overline{b}_j X^{i+j} \\ &= (\overline{a}_0 + \overline{a}_1 X + \cdots + \overline{a}_n X^n) \\ &\quad (\overline{b}_0 + \overline{b}_1 X + \cdots + \overline{b}_m X^m) \\ &= \varphi(f(X))\varphi(g(X)). \end{aligned}$$

Thus $\varphi : R[X] \to (R/I)[X]$ is a ring homomorphism.

Suppose $f(X) = a_0 + a_1 X + \cdots + a_n X^n$ be a polynomial in $R[X]$ with all the coefficients a_i's are in the ideal I. Then $\overline{\varphi}(f(X)) = \overline{a}_0 + \overline{a}_1 X + \cdots + \overline{a}_n X^n = 0$. This is because the coefficients $\overline{a}_i$'s are zero in R/I and this proves that the ideal $I[X]$ is contained in the kernel $\ker \overline{\varphi}$.

Conversely let $f(X) = a_0 + a_1 X + \cdots + a_n X^n$ be in the kernel. This means that $\overline{\varphi}(f(X)) = \overline{a}_0 + \overline{a}_1 X + \cdots + \overline{a}_n X^n = 0$ in the ring $(R/I)[X]$. Hence all the elements a_i are in I and the polynomial $f(X)$ is in the ideal $I[X]$. It is very easy to see that the homomorphism $R[X] \to (R/I)[X]$ is on-to. Thus by Fundamental Theorem of homomorphism (Theorem 3.6.20), the ring $R[X]/I[X]$ is isomorphic to the ring $(R/I)[X]$. This completes the proof.

We can use fundamental theorem of homomorphism to check if and ideal is a maximal ideal or a prime ideal in a ring R.

Example 3.6.22. 1. The principal ideal $I = (1+i)$ is a maximal ideal in $\mathbb{Z}[i]$.

The ideal I is a maximal ideal in $\mathbb{Z}[i]$ iff $\mathbb{Z}[i]/I$ is a field. The characteristic homomorphism $\chi : \mathbb{Z} \to \mathbb{Z}[i]/I$ is given by $\chi(k) = k + I$.

The integer $2 = (1 - i)(1 + i)$ is in the ideal I. That is the kernel of χ contains the ideal (2). Since (2) is a maximal ideal, it follows that the kernel is the ideal J. By Fundamental Theorem of Homomorphism, the ring $\mathbb{Z}_2 = \mathbb{Z}/(2)$ is isomorphic to the ring $\mathbb{Z}[i]/I$. Since $\mathbb{Z}_2 = \mathbb{Z}/(2)$ is a field we conclude that I is a maximal ideal in $\mathbb{Z}[i]$.

Exercise 3.6.23. 1. Identify the quotient ring $\mathbb{Q}[X]/(X^2 + 1)$ with a subring of $\mathbb{C}$.

2. Let $R = \mathbb{Z}[i]$ and $J = (1 + 2i)$. Show that J is a maximal ideal in R and R/J is isomorphic to $\mathbb{Z}_5$.

3. Let $R = \mathbb{Z}[i]$ and $J = (4 + 2i)$. Identify the quotient ring.

4. Let $R = \mathbb{Z}[i]$ and $J = (a + ib)$. If the integers a and b are mutually prime, show that R/J is isomorphic to $\mathbb{Z}_{a^2+b^2}$.

   Let $\chi : \mathbb{Z} \to R$ be the characteristic homomorphism.
   We will first determine the kernel of the homomorphism even if the integers a and b are not relatively prime. Let $d = gcd(a, b)$ and $a = da_1$, $b = db_1$ for a_1, b_1 in $\mathbb{Z}$.

   Let $k \in \mathbb{N}$ be the generator of the kernel of χ. Then k is in the ideal J and therefore there exists $u + iv$ such that $k = d(a_1 + ib_1)(u + iv)$. If we solve this equation, we get that $k = d\lambda(a_1^2 + b_1^2)$ for some $\lambda \geq 1$. On the other hand $(a_1^2 + b_1^2) = (a_1 + ib_1)(a_1 - ib_1)$ is in the ideal. Since k is the smallest integer in the ideal, it follows that $\lambda = 1$ and $k = d(a_1^2 + b_1^2)$. Thus $\ker(\chi) = (d(a_1^2 + b_1^2))$ and the characteristic of R is $d(a_1^2 + b_1^2)$.

   We will now show that, if a and b are relatively prime, the homomorphism $\chi : \mathbb{Z} \to R$ is on to. It suffices to show that the element $i + J$ is in the image of χ. Let u and v be integers such that $au + bv = 1$. Then $(a + ib)(u + iv) = (au - bv) + i(au + bv) = (au - bv) + i$. Since $(a + ib)(u + iv)$ is in J, it follows that $(au - bv) + J = i + J$. That is $\chi(au - bv) = (au - bv) + J = i + J$ and the map χ is on to.

5. Show that the ideal $J = (2X - 3, X - 10)$ is maximal in $\mathbb{Z}[X]$ and the quotient ring R/J is isomorphic to $\mathbb{Z}_{17}$.

Exercise 3.6.24. 1. Let R be a commutative ring with identity and $\varphi : R[X] \to R$ be the ring homomorphism $f(X) \mapsto f(1)$. Show that the kernel of this homomorphism is a

   (a) prime ideal iff the ring R is an integral domain.
   (b) maximal ideal iff the ring R is a field.

   First observe that the homomorphism is onto. (Why?)

2. Let R be a ring and P be a prime ideal. Show that the ideal $P[X]$ is a prime ideal in $R[X]$.

3. Let R be a ring M be a proper maximal ideal. Show that the ideal $M[X]$ is not maximal ideal in $R[X]$.

Definition 3.6.25. Let R and S be two rings. A ring homomorphism $\varphi : R \to S$ is called an isomorphism if φ is one-one and onto.

If $R = S$ an isomorphism is called an automorphism of the ring R.

Examples 3.6.26.

1. Let p be a prime number and $R = \mathbb{Z}_p$. Then the map $a \mapsto a^p$ is an automorphism of R.
2. Let $R = \mathbb{Z}[\sqrt{D}]$ and $\varphi : R \to R$ be the map $\varphi(a + b\sqrt{D}) = a - b\sqrt{D}$. Then φ is an automorphism.

Exercise 3.6.27. Let R be a ring. Let $\varphi : R[X] \to R[X]$ be the map defined by $\varphi(f(X)) = f(X + 1)$. Is φ an isomorphism?

Exercise 3.6.28.

1. Let $R = \mathbb{Z}[\sqrt{D}]$ and $\varphi : R \to R$ be an automorphism. Show that $\varphi = Id$ or $\varphi(a + b\sqrt{D}) = a - b\sqrt{D}$ for all elements $a + b\sqrt{D}$ in R. (This should remind us of what we did in the case of finding homomorphisms from $\mathbb{C} \to \mathbb{C}$.)
2. Let $\omega \neq 1$ be a cube root of unity and let

$$R = \left\{a + b\omega + c\omega^2 : a, b, c \in \mathbb{Q}\right\}.$$

Find all the automorphisms of R.

Let $\varphi : R \to R$ be an automorphism. Since $\varphi(1) = 1$, it follows that $\varphi(k) = k$ for all integers k. Consequently $\varphi(r) = r$ for every rational number r. Therefore we need only to find the image of ω. Since $\omega^3 = 1$, it follows that $1 = \varphi(1) = \varphi(\omega^3) = \varphi(\omega)^3$. This shows that $\varphi(\omega) \neq 1$ is also a root of unity. Hence $\varphi(\omega) = \omega$ or $\varphi(\omega) = \omega^2$. If $\varphi(\omega) = \omega$, then φ is identity on R and if $\varphi(\omega) = \omega^2$, then $\varphi(a + b\omega + c\omega^2) = a + b\omega^2 + c\omega$ for all elements $a + b\omega + c\omega^2$ in R.

This shows that there are only two automorphisms of R, namely the identity map

$$a + b\omega + c\omega^2 \mapsto a + b\omega + c\omega^2$$

and

$$a + b\omega + c\omega^2 \mapsto a + b\omega^2 + c\omega$$

for $a + b\omega + c\omega^2$ in R.

3. Find all automorphisms of $\mathbb{Z}[\omega]$, if
   (a) ω is a p-th root of unity where p is a prime number.
   (b) ω is a n-th root of unity for $n \geq 3$.
   (c) ω is a primitive n-th root of unity for $n \geq 3$.
4. Find all automorphisms of $\mathbb{Q}[\omega]$, if
   (a) ω is a p-th root of unity where p is a prime number.
   (b) ω is a n-th root of unity for $n \geq 3$.
   (c) ω is a primitive n-th root of unity for $n \geq 3$.

Exercise 3.6.29. Let D be a square free integer. Find all automorphisms of $\mathbb{Q}[\sqrt{D}]$.

Exercise 3.6.30. Let R and S be two rings. If $\varphi : R \to S$ is a ring isomorphism, show that $\varphi : U(R) \to U(S)$ is an isomorphism between the multiplicative group of units $U(R)$ of R and $U(S)$ of S respectively.

Exercise 3.6.31.

1. Let m and n be two mutually prime natural numbers. Using Exercise 3.6.30 show that the group $U(\mathbb{Z}_{mn})$ of units in $\mathbb{Z}_{mn}$ is isomorphic to the group $U(\mathbb{Z}_m) \times U(\mathbb{Z}_n)$, the group of units in the ring $\mathbb{Z}_m \times \mathbb{Z}_n$.

2. If $n = p_1^{k_1} p_2^{k_2} \cdots p_r^{k_r}$, use the exercise above to conclude that $U(\mathbb{Z}_n)$ is isomorphic to the group $\prod_{j=1}^{r} U(\mathbb{Z}_{p_j^{k_j}})$.

3. If φ denotes the Euler's phi-function and $n = p_1^{k_1} p_2^{k_2} \cdots p_r^{k_r}$ show that $\varphi(n) = \prod_{j=1}^{r} \varphi(p_1^{k_1})\varphi(p_2^{k_2}) \cdots \varphi(p_r^{k_r}) = n \prod_{p|n} \left(1 - \frac{1}{p}\right)$.

3.6.3 Substitution Principle

An important principle in homomorphism is what we call as substitution principle. This gives us a method to construct new rings.

Theorem 3.6.32. *Let $\varphi : R \to S$ be a homomorphism and $s_0 \in S$. Then there exists a unique homomorphism $\overline{\varphi} : R[X] \to S$ such that $\overline{\varphi}(X) = s_0$ and $\overline{\varphi}(r) = \varphi(r)$ for all r in R.*

Proof. Given a ring homomorphism $\varphi : R \to S$ and s_0 in S, we define $\overline{\varphi} : R[X] \to S$ by

$$\overline{\varphi}(a_0 + a_1 X + \cdots + a_n X^n) := \varphi(a_0) + \varphi(a_1)s_0 + \varphi(a_2)s_0^2 + \cdots + \varphi(a_n)s_0^n.$$

It is an easy exercise to show that $\overline{\varphi}$ is a additive group homomorphism between $R[X]$ and S.

Let $f(X) = \sum_{i=0}^{n} a_i X^i$ and $g(X) = \sum_{j=0}^{r} b_j X^j$. Then

$$\begin{aligned}
\overline{\varphi}(f(X)g(X)) &= \overline{\varphi}((\sum_{i=0}^{n} a_i X^i)(\sum_{j=0}^{r} b_j X^j)) \\
&= \overline{\varphi}(\sum_{i+j=0}^{n+r} a_i b_j X^{i+j}) \\
&= \sum_{i+j=0}^{n+r} \overline{\varphi}(a_i b_j X^{i+j}) \\
&= \sum_{i+j=0}^{n+r} \varphi(a_i b_j) s_0^{i+j} \\
&= \sum_{i+j=0}^{n+r} \varphi(a_i)\varphi(b_j) s_0^{i+j} \\
&= \left(\sum_{i=0}^{n} a_i s_0^i\right)\left(\sum_{j=0}^{r} b_j s_0^j\right) \\
&= \overline{\varphi}(f(X))\overline{\varphi}(g(X)).
\end{aligned}$$

This proves that $\overline{\varphi} : R[X] \to S$ is a ring homomorphism. □

Here are some examples to illustrate the substitution principle.

Examples 3.6.33. Let R and S be two rings and S and let $\varphi : R \to S$ be a homomorphism. The ring S is a subring of $S[X]$ and let $i : S \hookrightarrow S[X]$ be the inclusion homomorphism. Then the map $i \circ \varphi : R \to S[X]$ is a ring homomorphism and by the substitution principle there exists unique homomorphism $\overline{\varphi} : R[X] \to S[X]$ such that $\overline{\varphi}(X) = X$ and $\overline{\varphi}(r) = \varphi(r)$ for all r in R.

We have another example: Let R be a ring and $R[X, Y]$ be the polynomial ring in two variables:
$R[X, Y] := \left\{\sum_{0 \le i+j \le k} a_{ij} X^i Y^j : a_{ij} \in R \text{ and } i, j, k \in \mathbb{N} \cup \{0\}\right\}$. Then $R[X, Y]$ is a ring. Using substitution principle we prove the following

Corollary 3.6.34. *The rings $R[X, Y]$ and $R[X][Y]$ are isomorphic.*

Proof. The ring $R[X]$ is a subring of $R[X, Y]$.(Why?) Therefore by Theorem 3.6.32 there is a unique homomorphism $\varphi : R[X][Y] \to R[X, Y]$.

Observe that R is subring of $R[X][Y]$. Hence by substitution principle there exists a unique homomorphism $\psi : R[X, Y] \to R[X][Y]$. It is a routine exercise to check that $\varphi \circ \psi$ and $\psi \circ \varphi$ are identity maps. Hence the two rings $R[X][Y]$ and $R[X, Y]$ are isomorphic. □

Using substitution principle we can construct new rings and we will also identify many quotient rings with some familiar rings.

Examples 3.6.35. 1. Let $i : \mathbb{Z} \to \mathbb{R}$ be the inclusion map. Then there exists a unique ring homomorphism $\varphi : \mathbb{Z}[X] \to \mathbb{R}$ such that $\varphi(X) = \sqrt{2}$. The image is the subring $\mathbb{Z}[\sqrt{2}]$ and it is isomorphic to the quotient ring $\mathbb{Z}[X]/\left(X^2 - 2\right)$.

By substitution principle there is a unique ring homomorphism $\varphi : \mathbb{Z}[X] \to \mathbb{R}$ such that $\varphi(X) = \sqrt{2}$.

Let $a + b\sqrt{2}$ be an element in $\mathbb{Z}[\sqrt{2}]$. Then $\varphi : \mathbb{Z}[X] \to \mathbb{Z}[\sqrt{2}]$ is onto. We will now show that the kernel is the principal ideal $\left(X^2 - 2\right)$.

Observe that the kernel does not contain any linear polynomial. Therefore the kernel contains only polynomials of degree bigger than or equal to two. The polynomial $X^2 - 2$ is in the kernel of φ. Now let $f(X)$ be in the kernel. Then by division algorithm(Theorem 3.3.7), there exist unique polynomials $q(X)$ and $r(X)$ such that $f(X) = (X^2 - 2)q(X) + r(X)$ where $r(X) \equiv 0$ or $\deg(r(X)) < 2$. Since $f(\sqrt{2}) = 0$, it follows that $r(\sqrt{2}) = 0$. Since a linear polynomial is not in the kernel the polynomial $r(X)$ must be zero polynomial. Thus $f(X) = (X^2 - 2)q(X)$ and the kernel is the principal ideal $\left(X^2 - 2\right)$.

2. Let $i : \mathbb{Z} \to \mathbb{Q}$ be the inclusion homomorphism and $a \neq \pm 1$ be an integer. Then by substitution principle(Theorem 3.6.32) there exists a unique homomorphism $i_a : \mathbb{Z}[X] \to \mathbb{Q}$ such that $i_a(X) = \frac{1}{a}$. The image is a subring isomorphic to the ring $\mathbb{Z}[X]/\left(aX - 1\right)$. The image is denoted as $\mathbb{Z}[\frac{1}{a}]$ and this is the smallest subring of $\mathbb{Q}$ in which the integer a is a unit.

For example, if we take $a = 1/2$, we get the smallest ring R in $\mathbb{Q}$ in which 2 is a unit.

By substitution principle there exists a unique homomorphism

$$\varphi : \mathbb{Z}[X] \to \mathbb{Q} \qquad \text{such that} \qquad \varphi(X) = \frac{1}{a}.$$

We will show that the kernel is the ideal $(aX - 1)$ in $\mathbb{Z}[X]$.

Let $f(X) = a_0 + a_1X + a_2X^2 + \cdots + a_nX^n$ be a polynomial in the kernel of φ. Then $a_0 + a_1\frac{1}{a} + a_2\frac{1}{a^2} + \cdots + a_n\frac{1}{a^n} = 0$ and we get $a_0a^n + a_1a^{n-1} + a_2a^{n-2} + \cdots + a_n = 0$. Therefore $a_n = -(a_0a^n + a_1a^{n-1} + \cdots + a_{n-1}a)$. We substitute this in $f(X) = a_0 + a_1X + a_2X^2 + \cdots + a_nX^n$ to get

$$\begin{aligned}
\sum_{i=0}^{n} a_nX^n &= a_0 + a_1X + a_2X^2 + \cdots \\
&\quad -(a_0a^n + a_1a^{n-1} + \cdots + a_{n-1}a^{n-1})X^n \\
&= a_0(1 - (aX)^n) + a_1X(1 - (aX)^{n-1}) + \cdots \\
&\quad + a_{n-1}(1 - aX) \\
&= a_0(1 - aX)f_{n-1}(X) + a_1X(1 - aX)f_{n-2}(X) \\
&\quad + \cdots + a_{n-1}(1 - aX) \\
&= (1 - aX)\left(\sum_{j=0}^{n-1} a_jf_{n-1-j}(X)\right).
\end{aligned}$$

In the computation above, for $k \geq 2$, we have written $(1-(aX)^k)$ as $(1-(aX)^k) := (1-aX)f_{k-1}(X)$ where $f_{k-1}(X)$ is a polynomial of degree $k-1$. Thus the polynomial $f(X)$ is a multiple of the polynomial $(aX-1)$ and hence it is in the principal ideal $(aX-1)$. On the other hand the polynomial $aX-1$ is in the kernel of the homomorphism φ. Hence the kernel of φ is the principal ideal $(aX-1)$ and the image is isomorphic to the ring $\mathbb{Z}[X]/(aX-1)$.

3. Let

$$\begin{aligned} J &:= \{(X^2+2)f(X)+7g(X) : f(X), g(X) \in \mathbb{Z}[X]\} \text{ and} \\ I &:= \{(X^2-5)\varphi(X)+(7X+14)\psi(X) : \varphi(X), \psi(X) \in \mathbb{Z}[X]\}. \end{aligned}$$

Show that

(a) the ideals I and J are equal, and

(b) the ring $\mathbb{Z}[X]/I$ is isomorphic to the ring $\mathbb{Z}_7[\sqrt{-2}]$.

Let us observe that $7 = (X-2)(7X+14) - 7(X^2-5)$ and

$$\begin{aligned} X^2+2 &= X^2-5+7 \\ &= (X^2-5)+(X-2)(7X+14)-7(X^2-5) \\ &= (X-2)(7X+14)-6(X^2-5). \end{aligned}$$

This shows that the integer 7 and the polynomial X^2+2 are in the ideal I and hence the ideal $J \subseteq I$.

On the other hand $X^2-5 = X^2+2-7$ and $7X+14 = 7(X+2)$. As a consequence the ideal $I \subseteq J$. Thus $I = J$ and the first part is proved.

Let $\chi : \mathbb{Z} \to \mathbb{Z}_7[\sqrt{-2}]$ be the characteristic homomorphism. Then by substitution principle(Theorem 3.6.32) there exists a homomorphism $\varphi : \mathbb{Z}[X] \to \mathbb{Z}_7[\sqrt{-2}]$ such that $\varphi(X) = \sqrt{-2}$ and $\varphi(k) = \chi(k)$ for all integers k.

Then the map φ is onto and X^2+2 is in the kernel. If $f(X)$ is in the kernel, then using division algorithm we can write $f(X) = (X^2+2)q(X)+r(X)$ where $r(X) = aX+b$ is a polynomial of degree at most 1. Since $f(\sqrt{-2}) = 0$ in $\mathbb{Z}_7[\sqrt{-2}]$, it follows that $a\sqrt{-2}+b = 0$. This is true iff $a \equiv 0 \equiv b$ in $\mathbb{Z}_7$. Hence kernel is I and the result follows.

Exercise 3.6.36. 1. Let $R = \mathbb{Z}[X]$ and $I = (2X-3, X-10)$. Show that R/I isomorphic to $\mathbb{Z}_{17}$.

Let us first prove that I is a proper ideal in $\mathbb{Z}[X]$. Suppose $I = \mathbb{Z}[X]$. Then there exist polynomials $f(X)$ and $g(X)$ in $\mathbb{Z}[X]$ such that $1 = (2X-3)f(X)+(X-10)g(X)$. We can show that this is not possible by comparing the degrees of the polynomials.

Let $\chi : \mathbb{Z} \to R/I$ be the characteristic homomorphism. We will find the kernel of the characteristic homomorphism.

The integer $17 = (2X-3) - 2(X-10)$. Hence 17 is in the kernel of the characteristic homomorphism χ. Since 17 is a prime number, the kernel contains the maximal ideal (17). This means that either the kernel is (17) or the full ring $\mathbb{Z}$. Since χ is a ring homomorphism the kernel is (17).

2. Let $i : \mathbb{R} \to \mathbb{C}$ be the inclusion homomorphism. Let α be a complex number and $\mathbb{R}[\alpha]$ be the subring got by the substitution principle. What is the subring $\mathbb{R}[\alpha]$?

3.7 Chinese Remainder Theorem

In Chapter 1, we proved that *if n_1 and n_2 are two relatively prime integers and if a_1 and a_2 are two nonzero integers, then there exists an integer x in $\mathbb{Z}$ such that $x \equiv a_1 \pmod{n_1}$ and $x \equiv a_2 \pmod{n_2}$* .

In this section we will prove an analog of this result in a commutative ring R.

Definition 3.7.1. Let R be a commutative ring.

1. Two ideals I and J are called co-maximal if $I + J = R$.
2. A finite sequence of ideals $I_1, I_2, \ldots, I_k$ are called pairwise co-maximal ideals in R, if I_r and I_s are co-maximal for $1 \leq r \neq s \leq k$.

Examples 3.7.2. 1. Let $p_1 \neq p_2$ be two prime numbers. Then the ideals (p_1) and (p_2) are co-maximal in $\mathbb{Z}$.

> Since p_1 and p_2 are two distinct prime numbers, they are mutually co-prime too. Hence there exists integers k_1 and k_2 such that $p_1k_1 + p_2k_2 = 1$. This proves that 1 is in the ideal $(p_1) + (p_2)$. Thus the two ideals are co-maximal in $\mathbb{Z}$.

2. Let k and l be two mutually co-prime integers. Then the ideals (k) and (l) are co-maximal in $\mathbb{Z}$.
3. If $k_1, k_2, \ldots, k_r$ are pairwise mutually co-prime integers, then the ideals $I_j := (k_j)$ for $1 \leq j \leq k$ are co-maximal in $\mathbb{Z}$.
4. Let I and J be two maximal ideals in R such that $I \neq J$. Then the ideals I and J are co-maximal in R.

> The ideals I and J are strictly contained in the ideal $I + J$. Hence $I + J = R$.

Exercise 3.7.3. 1. Let I be a maximal ideal and and J be a proper ideal in R such that $I \neq J$. Show that the ideals I and J are co-maximal in R.

2. Show that the ideals $(2, X)$ and $\left(3, X^2\right)$ are co-maximal in $\mathbb{Z}[X]$.
3. Show that the ideals $\left(\sqrt{2}\right)$ and $\left(3 + \sqrt{2}\right)$ are co-maximal in $\mathbb{Z}[\sqrt{2}]$.
4. Let D and k be two mutually co-prime integers. Show that the ideals $\left(\sqrt{D}\right)$ and $\left(k + \sqrt{D}\right)$ are co-maximal in $\mathbb{Z}[\sqrt{D}]$.
5. Show that the ideals $(1 + i)$ and $(1 - i)$ are co-maximal in $\mathbb{Z}[i]$.
6. Show that ideals (3) and $(1 + i)$ are co-maximal in $\mathbb{Z}[i]$.

7. Let $k_1, k_2, \ldots, k_r$ be pairwise co-prime integers. Then the two ideals (k_1) and $(k_2k_3 \cdots k_r)$ are co-maximal in $\mathbb{Z}$.

Exercise 3.7.4. Let I, J and K be three pairwise co-maximal ideals in a ring R. Then

1. $IJ = I \cap J$, $JK = J \cap K$ and $IK = I \cap K$.

2. the ideals I and JK are co-maximal in R.

The product ideal IJ is contained in I and J. Hence $IJ \subseteq I \cap J$ is obvious.

Since I and J are co-maximal, there exists x in I and y in J such that $1 = x + y$. If z is an element of $I \cap J$, then $z = z(x + y) = zx + zy$. The elements zx and zy are in IJ and as a consequence, we see that $z = zx + zy$ is also in IJ. Thus $I \cap J = IJ$.

Since I, J and K are co-maximal in R, there exists $x, y \in I$, $z \in J$ and $w \in K$ such that $x + z = 1$ and $y + w = 1$. Hence $1 = (x + z)(y + w) = xy + yz + xw + wz$. The sum $xy + yz + xw$ is in I and wz is in JK. This shows that I and JK are co-maximal ideals in R.

Theorem 3.7.5 (Chinese Remainder Theorem): *Let I and J be two co-maximal ideals in a commutative ring R. Then the natural ring homomorphism*

$$\varphi : R \to R/I \times R/J \quad \textit{defined by} \quad \varphi(x) := (x + I, x + J)$$

is on-to with kernel $I \cap J$.

Proof. It is easy to prove that the natural map

$$\varphi : R \to R/I \times R/J \quad \text{defined by} \quad \varphi(x) = (x + I, x + J)$$

is a ring homomorphism.

If $\varphi(x) = (x + I, x + J) = (0, 0)$, then x is in $I \cap J$. Conversely if x is an element in $I \cap J$, then $\varphi(x) = (0, 0)$. Thus $\ker \varphi = I \cap J$.

We will now show that if I and J are co-maximal in R, then φ is onto.

Since I and J are co-maximal in R, there exist two elements a in I and b in J such that $1 = a + b$.

Let $(r + I, s + J)$ be an element of $R/I \times R/J$. Consider $x = br + as$. Then

$$\begin{aligned} \varphi(x) &= (x + I, x + J) \\ &= (br + as + I, br + as + J) \\ &= ((1 - a)r + as + I, br + (1 - b)s + J) \\ &= (r - a(r - s) + I, s + b(r - s) + J) \\ &= (r + I, s + J) \end{aligned}$$

This completes the proof. □

Corollary 3.7.6. *Let I_j be a sequence of pairwise co-maximal ideals in R for $1 \leq j \leq k$. Then the ring $R/\cap_{j=1}^{k} I_j$ is isomorphic to the product $R/I_1 \times R/I_2 \times \cdots \times R/I_k$.*

Proof. We just need to observe that the two ideals I_1 and the product $I_2I_3 \cdots I_k$ are co-maximal in R and apply Theorem 3.7.5. □

3.8 Field of fractions

In this section we will show that, if R is an integral domain there exists a field $\mathbb{F}$ such that

1. R is a subring of $\mathbb{F}$ and

2. if R is a subring of a field $\mathbb{K}$, then $\mathbb{F}$ is a subring of the field $\mathbb{K}$.

The construction of of such a field mimics the construction of the field of rational numbers $\mathbb{Q}$ from the ring of integers $\mathbb{Z}$.

Let us recall that a fraction a/b in $\mathbb{Q}$ is same as the fraction ka/kb for any nonzero integer k. For example $1/2 = 2/4 = 3/6 = \cdots$ etc.,

In general two fractions a/b and c/d are equal iff $ad = bc$. We use this equation to define an equivalence relation.

Let R be an integral domain and $R^* := \{a \in R : a \neq 0\}$. Let $S := R \times R^*$. We say that two elements (a, b) and (c, d) are related iff $ad = bc$. (Written as $(a, b) \sim (c, d)$). Now it is easy to show that this relation is an equivalence relation on S. For example, if (a, b) is related to (c, d) and (c, d) is related to (e, f), then

$$ad = bc \quad \text{and} \quad cf = de.$$

If we now multiply the equation $ad = bc$ by e, then we get $ade = bce$. Using the equation $de = cf$, we write $acf = ade = bce$. Since $c \neq 0$ and R is an integral domain, it follows that $af = be$. Hence the element (a, b) is related to (e, f) and the relation is transitive.

Let $\mathbb{F} := S/\sim$. We denote an equivalence class containing an element (a, b) of S by $[(a, b)]$ and also by the *fraction* a/b. Let us observe that if (a, b) is an element of S and if $r \neq 0$ in R, then $(ra, rb) \sim (a, b)$. Hence $ra/rb = a/b$ for $r \neq 0$ in R.

If a/b and c/d are two elements of $\mathbb{F}$, we define

$$(a/b) + (c/d) := ad + bc/bd \quad \text{and} \quad (a/b).(c/d) := ac/bd.$$

Now we need to show that if $a/b = a_1/b_1$ and $c/d = c_1/d_1$, then $\frac{ad+bc}{bd} = \frac{a_1d_1+b_1c_1}{b_1d_1}$ for the addition to be well defined.

Since $a/b = a_1/b_1$ and $c/d = c_1/d_1$, we get $ab_1 = a_1b$ and $cd_1 = c_1d$. Therefore, by repeatedly making use of these two equations we get

$$\begin{aligned}
\frac{ad+bc}{bd} &= \frac{adb_1 + bcb_1}{bb_1d} \\
&= \frac{a_1bd + bb_1c}{bb_1d} \\
&= \frac{a_1d + b_1c}{b_1d} \\
&= \frac{a_1dd_1 + b_1cd_1}{b_1dd_1} \\
&= \frac{a_1d_1 + b_1c_1}{b_1d_1} \\
&= a_1/b_1 + c_1/d_1.
\end{aligned}$$

Similarly one can show that the multiplication is also well defined.

We can now show very easily that $(\mathbb{F}, +, \cdot)$ is a field.

1. If a/b, c/d and e/f are in $\mathbb{F}$ then

$$\begin{aligned} a/b + (c/d + e/f) &= a/b + \frac{cf+de}{df} \\ &= \frac{adf + bcf + bde}{bdf} \\ &= \frac{(ad+bc)f + bde}{bdf} \\ &= \frac{ad+bc}{bd} + e/f \\ &= (a/b + c/d) + e/f. \end{aligned}$$

2. $\frac{0}{1}$ is the additive identity.
3. If a/b is in $\mathbb{F}$, then $-a/b$ is the additive inverse of a/b.

This proves that $(\mathbb{F}, +)$ is an abelian group.

1. Associativity of the multiplication is easy.
2. The element a/a is the multiplicative identity.
3. If a/b is nonzero then $a \neq 0$. Hence b/a is in $\mathbb{F}$ and it is the inverse of a/b.
4. Distribution addition over multiplication is very easy.

Thus $(\mathbb{F}, +, \cdot)$ is a field.

Let $\varphi : R \to \mathbb{F}$ be the map $a \mapsto \frac{a}{1}$. Then

$$\varphi(a+b) = \frac{a+b}{1} = \frac{a}{1} + \frac{b}{1}$$

and

$$\varphi(ab) = \frac{ab}{1} = \frac{a}{1}\frac{b}{1}$$

for all a, b in R. If $\varphi(a) = \frac{a}{1} = \frac{0}{1}$, then $a = 0$. Hence φ is one-one and R is a subring of the field $\mathbb{F}$. We write the elements $\frac{a}{1}$ of $\varphi(R)$ as a itself.

Suppose $\mathbb{K}$ is a field and $\theta : R \to \mathbb{K}$ is a ring homomorphism. Then the map $\theta : R \to \mathbb{K}$ extends to a ring homomorphism $\widetilde{\theta} : \mathbb{F} \to \mathbb{K}$ defined by

$$\widetilde{\theta}(a/b) := \theta(a)(\theta(b))^{-1}$$

where $\theta(b)^{-1}$ is the multiplicative inverse of the element $\theta(b)$.

If a/b and c/d are in $\mathbb{F}$, then

$$\begin{aligned} \widetilde{\theta}(a/b + c/d) &= \widetilde{\theta}(\frac{ad+bc}{bd}) \\ &= \theta(ad+bc)\theta(bd)^{-1} \\ &= \theta(ad)\theta(bd)^{-1} + \theta(bc)\theta(bd)^{-1} \\ &= \widetilde{\theta}(ad/bd) + \widetilde{\theta}(bc/bd) \\ &= \widetilde{\theta}(a/b) + \widetilde{\theta}(c/d) \end{aligned}$$

and

$$\begin{aligned}\widetilde{\theta}(ac/bd) &= \theta(ac)\theta(bd)^{-1}\\ &= \theta(a)\theta(c)\theta(b)^{-1}\theta(d)^{-1}\\ &= \widetilde{\theta}(a/b)\widetilde{\theta}(c/d).\end{aligned}$$

This proves that $\widetilde{\theta}$ is ring homomorphism. Since $\mathbb{F}$ and $\mathbb{K}$ are fields this homomorphism is one one and $\mathbb{F}$ is a subfield of the field $\mathbb{K}$.

We summarize the above discussion in the following theorem.

Theorem 3.8.1. *Let R be an integral domain. Then there exists a unique field $\mathbb{F}$ such that*

1. *R is a subring of $\mathbb{F}$ and*
2. *if there exists a field $\mathbb{K}$ and a ring homomorphism $\theta : R \to \mathbb{K}$ then θ extends to a ring homomorphism $\widetilde{\theta} : \mathbb{F} \to \mathbb{K}$. (This property is known as the universal mapping property.)*

Example 3.8.2. 1. If $R = \mathbb{Z}$ then $\mathbb{F} = \mathbb{Q}$, the field of rational numbers.

2. If R is a field then $\mathbb{F}$ is isomorphic to R.

3. Let $\mathbb{F}$ be a field and $R = \mathbb{F}[X]$, the polynomial ring over the field $\mathbb{F}$. Then R is an integral domain and the field of fractions is

$$\left\{\frac{f(X)}{g(X)} : f(X), g(X) \in \mathbb{F}[X] \text{ and } g(X) \neq 0\right\}.$$

This field is called the *rational function field* over the field $\mathbb{F}$ and it is denoted by $\mathbb{F}(X)$.

If $\mathbb{F} = \mathbb{R}$ or $\mathbb{C}$, then $\mathbb{F}(X)$ is the field of rational polynomials.

Exercise 3.8.3. 1. Let R be an integral domain and $\mathbb{F}$ be the quotient field of R. If $S = R[X]$ then the quotient field of S is the rational function field $\mathbb{F}(X)$ of the field $\mathbb{F}$.

For example if $R = \mathbb{Z}$ and $S = \mathbb{Z}[X]$ then $\mathbb{Q}(X)$ is isomorphic to the quotient field of $\mathbb{Z}[X]$.

2. If $\mathbb{F}$ is a field and $R = \mathbb{F}[\![X]\!]$, then the field of fraction of R denoted by $\mathbb{F}((X))$ is

$$\left\{\frac{\sum_{k\geq 0} a_k X^k}{\sum_{k\geq 0} b_k X^k} : \sum_{k\geq 0} a_k X^k, \sum_{k\geq 0} b_k X^k \in \mathbb{F}[\![X]\!] \text{ and } b_0 \neq 0\right\}$$

and it is isomorphic to the field of Laurent series over the field $\mathbb{F}$.

Example 3.8.4. Let $R = \mathbb{Z}[\sqrt{2}]$. Then $\mathbb{Z}$ is a subring of $\mathbb{R}$. Then by the universal mapping property (Theorem 3.8.1) it follows that $\mathbb{Q}[\sqrt{2}]$ is the field of fractions of R.

3.8.1 Ring of Fractions

If we go through the construction of the field of fractions of an integral domain, we notice that we have constructed a ring S containing R as a subring such that all the nonzero elements in R are *inverted* in S. That is all the nonzero elements in R are units in S.

All the nonzero elements of an integral domain are not zero divisors. Therefore it is natural to ask the following question.

Question 3.8.5. Let R be a commutative ring with identity and

$$U := \{a \in R : a \text{ is not a zero divisor in } R\}.$$

Does there exists a ring S containing R such that every element of U is a unit in S?

Before we answer this question, we show that if a is a zero divisor in a ring R, then there does not exist any ring S containing R such that a is a unit in S.

Let R be a commutative ring with identity and $a \neq 0$ be a zero divisor in R. Let $b \neq 0$ in R be such that $ab = 0$. Suppose S is a ring containing R such that a is a unit in R. Let $c \in S$ be the inverse of a. Then

$$b = b.1 = b(ac) = (ba)c = 0$$

a contradiction.

Let us now come back to our question and answer it in the affirmative.

Let R be a commutative ring with unity and

$$U := \{a \in R : a \text{ is not a zero divisor in } R\}.$$

We mimic the construction in the field of fractions to construct a ring S such that every element in U is a unit in S.

Let $X := R \times U$. We say that $(a, b) \sim (c, d)$ iff $ad = bc$. Then this is an equivalence relation on X.

Let $S := X/\sim$. If (a, b) is an element of X, we denote the equivalence class of (a, b) by $[(a, b)]$ and also by a/b. If r is in U, then $[(ra, rb)] = [(a, b)]$ for every element (a, b) in X.

For elements a/b and c/d in S we define

$$a/b + c/d := \frac{ad + bc}{bd} \quad \text{and} \quad (a/b)(c/d) := ad/bc.$$

We need to show that this addition and multiplication are well defined and this is done exactly as in the construction of field of fractions. Moreover we can show that S is a ring . The multiplicative identity of the ring S is $[(1, 1)] = [(a, a)]$ for every a in U.

Let a be an element of U. Then $[(1, a)]$ is the inverse of $[(a, 1)]$.

Let $\varphi : R \to S$ be the map $a \mapsto \frac{a}{1}$. Then φ is a ring homomorphism.

Let b be an element in R such that $\varphi(b) = 0$. This means that $\frac{b}{1} = \frac{0}{a}$ for every nonzero element a in U. That is $ba = 0$. Since a is not a zero divisor in

R, we see that $b = 0$. Thus $\varphi : R \to S$ is a one one ring homomorphism and we can consider R as a subring of S.

In this ring S every element of $U \subseteq R$ is a unit.

We will now show that S is the smallest subring containing R that has this property. Let S_1 be a commutative ring and $\theta : R \to S_1$ be a one one ring homomorphism such that $\theta(b)$ is a unit in S_1 for every element b in U. Then θ extends to a ring homomorphism $\widetilde{\theta} : S \to S_1$ defined by $\widetilde{\theta}(a/b) = \theta(a)\theta(b)^{-1}$. We can easily check that $\widetilde{\theta}$ is a one-one ring homomorphism.

Remark 3.8.6. Suppose we start with a subset A of U. Then we can construct a ring S such that R is a subring of S and every element of A is a unit in S.

Let R be a commutative ring with identity. Suppose a in R is not a zero divisor. Let $A := \{a^n : n \geq 0\}$. Let $X := R \times A$. On this set we define a relation $(r, a^k) \sim (s, a^l)$ iff $ra^l = sa^k$. We can check that this is an equivalence relation on the set X and we denote the set of equivalence classes by $S := X/\sim$. Then S becomes a commutative ring with identity under the addition and multiplication defined as above. In this ring a is a unit and it is the smallest ring with this property. We also denote this ring by $R[\frac{1}{a}]$.

For example, if $R = \mathbb{Z}$ and $a = 1/2$, then the small ring S containing $\mathbb{Z}$ in which 2 is a unit is the ring

$$\left\{\frac{m}{2^k} : m \in \mathbb{Z} \text{ and } k = 0, 1, 2, \cdots\right\}.$$

This ring is denoted by $\mathbb{Z}\left[\frac{1}{2}\right]$.

Exercise 3.8.7. Let R be a commutative ring with identity. If and element a in R is not a zero divisor, show that the ring $R[X]/(aX - 1)$ is isomorphic to the ring $R[\frac{1}{a}]$.

By substitution principle there exists a unique homomorphism

$$\varphi : R[X] \to R\left[\frac{1}{a}\right] \qquad \text{such that} \qquad \varphi(X) = \frac{1}{a}.$$

If we now show that this map is on-to and the kernel is the ideal $(aX - 1)$ then we are through.

Let r/a^k be an element in the ring $R[\frac{1}{a}]$. Then $\varphi(rX^k) = r/a^k$. Thus the map φ is onto.

Let $f(X) = a_0 + a_1X + a_2X^2 + \cdots + a_nX^n$ be a polynomial in the kernel of φ. Then $a_0 + a_1\frac{1}{a} + a_2\frac{1}{a^2} + \cdots + a_n\frac{1}{a^n} = 0$. Multiplying this by a^n, we get $a_0a^n + a_1a^{n-1} + a_2a^{n-2} + \cdots + a_n = 0$. Therefore $a_n = -(a_0a^n + a_1a^{n-1} + a_2a^{n-2} + \cdots + a_{n-1}a^{n-1})$ and we substitute this in the polynomial $f(X) = a_0 + a_1X + a_2X^2 + \cdots + a_nX^n$ to get

$$\begin{aligned}
\sum_{k=i}^{n} a_kX^k &= a_0 + a_1X + a_2X^2 + \cdots \\
&\quad -(a_0a^n + a_1a^{n-1} + a_2a^{n-2} + \cdots + a_{n-1}a^{n-1})X^n \\
&= a_0(1-(aX)^n) + a_1X(1-(aX)^{n-1}) + \cdots \\
&\quad + a_{n-1}(1-aX) \\
&= a_0(1-aX)f_{n-1}(X) + a_1X(1-aX)f_{n-2}(X) + \cdots \\
&\quad + a_{n-1}(1-aX) \\
&= (1-aX)\left(a_0f_{n-1}(X) + a_1Xf_{n-2}(X) + \cdots + a_{n-1}\right)
\end{aligned}$$

where we have written the polynomial $(1-(aX)^k)$ for $k \geq 2$ as $(1-(aX)^k) := (1-aX)f_{k-1}(X)$ and $f_{k-1}(X)$ is a polynomial of degree $k-1$. Thus the polynomial $f(X)$ is a multiple of the polynomial $(aX-1)$ and hence it is in the principal ideal $(aX-1)$. On the other hand the polynomial $aX-1$ is in the kernel of the homomorphism φ. Hence the kernel of φ is the principal ideal $(aX-1)$ and the ring $R[X]/(aX-1)$ is isomorphic to the ring $R[\frac{1}{a}]$.

3.9 Factorization

In the first chapter on number theory we have proved *The Fundamental Theorem of Arithmetic* which states that *every nonzero integer $k \geq 2$ can be written as a product $k = \pm p_1^{r_1}p_2^{r_2}\cdots p_t^{r_t}$ where the integers p_i are prime numbers, $t \geq 1$ and $r_i \geq 1$. Further this expression is unique except for the order in which the prime factors $p_j^{r_j}$ occur.*

In this section we will study rings which have this property.

Let us recall some of the important results used in the proof of Fundamental theorem of arithmetic.

- (Division Algorithm). Let a be a nonzero integer. If b is an integer, then there exist integers q and r such that $b = aq + r$, where $0 \leq r < |a|$. Further the integers q and r are unique.
- (Bezout Identity). If k and l are two integers and d is the gcd of k and l, then there exist integers x and y such that $kx + ly = d$.
- Bezout identity is used to show that a natural number is prime iff it is irreducible.

However not all rings have these properties listed above. We have seen earlier that

1. if $R[X]$ is a polynomial ring over a field then the ring $R[X]$ satisfies all these three properties and
2. if $R = \mathbb{Z}[X]$, the gcd of the integer 2 and the polynomial X is 1. However we can't write $1 = 2f(X) + Xg(X)$ with $f(X)$ and $g(X)$ in $\mathbb{Z}[X]$.

In this section our aim is to study rings in which fundamental theorem of arithmetic is true.

We will first start with the class of rings in which *Division algorithm* is true.

3.9.1 Euclidean Domains

Let R be an integral domain. A function $\sigma : R \setminus \{0\} \to \mathbb{Z}^+$, the set of non-negative integers is called a size function.

Definition 3.9.1. An integral domain R is called a Euclidean domain, if there exists a size function $\sigma : R \setminus \{0\} \to \mathbb{Z}^+$ such that

1. $\sigma(a) \leq \sigma(ab)$ for all elements a and b in $R \setminus \{0\}$ and
2. if a in R is nonzero, nonunit and b is in R, there exist q and r in R such that $b = aq + r$ where either $r = 0$ or $\sigma(r) < \sigma(a)$. This condition is called the *division algorithm.*

In many text books euclidean domains are defined as integral domains with a size function σ satisfying condition 2 in Definition 3.9.1. However we have chosen to add one more condition. This does not create much of a problem. This is because, given a size function σ satisfying division algorithm, we can construct another size function $\widetilde{\sigma}$ out of σ, which satisfies the condition 1 and 2 of Definition 3.9.1.

Further, two of the important results of Euclidean domain, namely Proposition 3.9.8 and Theorem 3.9.10 are true in a Principal ideal domain also. Therefore, in many books first it is shown that every Euclidean domain is a Principal ideal domain and then use this result to prove Theorem 3.9.10. Moreover one does not need condition 1 of Definition 3.9.1 to show that every Euclidean domain is a Principal ideal domain.

We will prove Proposition 3.9.8 and Theorem 3.9.10 and other results directly.

Proposition 3.9.2. *Let R be an integral domain and $\sigma : R \setminus \{0\} \to \mathbb{Z}+$ be a size function satisfying the condition 2 in Definition 3.9.1. Then there exists a size function $\widetilde{\sigma} : R \setminus \{0\} \to \mathbb{Z}_+$ satisfying conditions 1 and 2 in Definition 3.9.1.*

Proof. Let R be an integral domain with a size function $\sigma : R \setminus \{0\} \to \mathbb{Z}+$ satisfying the condition 2 in Definition 3.9.1.

We define a new function $\widetilde{\sigma} : R \setminus \{0\} \to \mathbb{Z}+$ by

$$\widetilde{\sigma}(a) := \min\{\sigma(ab) : b \in R \setminus \{0\}\}.$$

Let a and b be two elements in $R \setminus \{0\}$. By definition $\widetilde{\sigma}(a) \leq \sigma(ax)$ for every element x in $R \setminus \{0\}$. Then $\widetilde{\sigma}(a) \leq \widetilde{\sigma}(aby)$ for every element y in $R \setminus \{0\}$. Hence $\widetilde{\sigma}(a) \leq \min\{\widetilde{\sigma}(aby) : y \in R \setminus \{0\}\} = \widetilde{\sigma}(ab)$.

We will now prove the division algorithm with respect to the size function $\widetilde{\sigma}$.

Let $a \neq 0$ be a nonunit and b be an element in R. Let $c \in R$ be such that $\widetilde{\sigma}(a) = \sigma(ac)$.

For the pair (b, ac) we apply division algorithm in (R, σ). By definition there exist elements q_1 and r_1 in R such that $b = acq_1 + r_1$ and $\sigma(r_1) < \sigma(ac) = \widetilde{\sigma}(a)$. To prove division algorithm in $(R, \widetilde{\sigma})$, we rewrite $b = acq_1 + r_1$ as $b = a(cq_1) + r_1$. If we now show that $\widetilde{\sigma}(r_1) \leq \widetilde{\sigma}(a)$, we are through. By definition $\widetilde{\sigma}(r_1) = \min\{\sigma(r_1 x) : x \in R \setminus \{0\}\} \leq \sigma(r_1)$. Therefore $\widetilde{\sigma}(r_1) \leq \sigma(r_1) < \widetilde{\sigma}(a)$. Thus we have written $b = aq + r$ with $\widetilde{\sigma}(r) < \widetilde{\sigma}(a)$ where $q = cq_1$. □

In view of Proposition 3.9.2, we always assume that the size function satisfies the conditions 1 and 2 of Definition 3.9.1.

As we will see below division algorithm allows us to conclude that in an euclidean domain

1. every ideal is a principal ideal and
2. euclidean algorithm for finding the gcd of two elements is valid.

The condition 1 of Definition 3.9.1 allows to show directly that every nonzero and nonunit in a euclidean domain R is a product of finitely many irreducible elements in R.

Let us now look at some examples of Euclidean domains.

Examples 3.9.3. 1. Let $R = \mathbb{Z}$ and $\sigma(k) = |k|$, the absolute value of the integer k. Then σ is a size function. Division algorithm says that it is Euclidean domain. Further let us observe that in the definition of a Euclidean domain, we have not talked about the uniqueness of the quotient q and remainder r.

For example, if $a = 2$ and $b = -5$, then we can write $-5 = (-2)2+(-1) = (-3)2+1$. Thus the quotient and remainder need not be unique.

2. Let $\mathbb{F}$ be a field and $R = \mathbb{F}[X]$. Then $\sigma(f(X)) = \deg f(X)$ is a size function which satisfies division algorithm. In this case the quotient and remainder are unique as we have shown in division algorithm for polynomials. (See Theorem 3.3.7).

3. The ring $R = \mathbb{Z}[i]$ is a Euclidean domain with the size function $\sigma(a+ib) = a^2 + b^2$ for $a + ib \neq 0$ in R.

   Let $c+id$ be an element of R. If $a+ib$ is a nonzero complex number, by multiplying and dividing $c + id$ by $a + ib$ in $\mathbb{C}$, we can write $c + id = (a + ib)\frac{c+id}{a+ib}$. If we write $x + iy = \frac{c+id}{a+ib}$ then $x + iy = \frac{1}{a^2+b^2}(c + id)(a - ib) = \frac{1}{a^2+b^2}((ac + bd) + i(ad - bc))$.
   Let k and l be two integers such that such that $k \leq x < k + 1$ and $l \leq y < l + 1$. Therefore either $|k - x| \leq \frac{1}{2}$ or $|k + 1 - x| \leq \frac{1}{2}$ and $|l - y| \leq \frac{1}{2}$ or $|l + 1 - y| \leq \frac{1}{2}$. Thus there exist integers u and v such that $|u - x| \leq \frac{1}{2}$ and $|v - y| \leq \frac{1}{2}$.
   If we now let $s + it = (c + id) - (a + ib)(u + iv)$, then

   $$\begin{aligned} s + it &= (a + ib)(x + iy) - (a + ib)(u + iv) \\ &= (a + ib)\left((x - u) + i(y - v)\right) \end{aligned}$$

   and

   $$\begin{aligned} |s + it| &= |(a + ib)((x - u) + i(y - v))| \\ &= |a + ib|\sqrt{(x - u)^2 + (y - v)^2} \\ &\leq |a + ib|\sqrt{\frac{1}{2}} \\ &< |a + ib|. \end{aligned}$$

   Thus we have written $c + id = (a + ib)(u + iv) + (s + it)$ in $\mathbb{Z}[i]$ such that $0 \leq |s + it| < |a + ib|$.

4. Let $R = \mathbb{Z}[\sqrt{2}]$ and $\sigma : R \setminus \{0\} \to \mathbb{Z}^+$ be defined by $\sigma(a + b\sqrt{2}) = N(a + b\sqrt{2}) = |a^2 - 2b^2|$ the norm function defined earlier. Then R is a Euclidean domain with σ as size function.

   Let $a + b\sqrt{2}$ and $c + d\sqrt{2}$ be two nonzero elements of R. Then $|a^2 - 2b^2| \geq 1$ and $|c^2 - 2d^2| \geq 1$. Therefore

$$\sigma(a + b\sqrt{2}) \leq \sigma((a + b\sqrt{2})(c + d\sqrt{2}))$$

   for nonzero elements $a + b\sqrt{2}$ and $c + d\sqrt{2}$ in R.
   Let $a + b\sqrt{2}$, $c + d\sqrt{2}$ be two elements in R with $a + b\sqrt{2}$ nonzero. We want to show that there exist two elements $u + v\sqrt{2}$ and $\alpha + \beta\sqrt{2}$ in R such that

$$c + d\sqrt{2} = (u + v\sqrt{2})(a + b\sqrt{2}) + (\alpha + \beta\sqrt{2})$$

   and $\sigma(\alpha + \beta\sqrt{2}) < \sigma(a + b\sqrt{2})$.
   We write

$$\begin{aligned} c + d\sqrt{2} &= \frac{c + d\sqrt{2}}{a + b\sqrt{2}}(a + b\sqrt{2}) \\ &= \frac{(c + d\sqrt{2})(a - b\sqrt{2})}{a^2 - 2b^2}(a + b\sqrt{2}) \\ &= \frac{(ac - 2bd) + (ad - bc)\sqrt{2}}{a^2 - 2b^2}(a + b\sqrt{2}) \end{aligned}$$

   in $\mathbb{Q}[\sqrt{2}]$.
   Just as in the previous example, we can choose two integers u and v such that $|\frac{ac - 2bd}{a^2 - 2b^2} - u| \leq 1/2$ and $|\frac{ad - bc}{a^2 - 2b^2} - v| \leq 1/2$.
   Let $\alpha + \beta\sqrt{2} = (c + d\sqrt{2}) - (u + v\sqrt{2})(a + b\sqrt{2})$. Then

$$c + d\sqrt{2} = (u + v\sqrt{2})(a + b\sqrt{2}) + (\alpha + \beta\sqrt{2}).$$

   If $\alpha + \beta\sqrt{2} \neq 0$, then

$$\begin{aligned} &\alpha + \beta\sqrt{2} \\ &= \frac{c + d\sqrt{2}}{a + b\sqrt{2}}(a + b\sqrt{2}) - (u + v\sqrt{2})(a + b\sqrt{2}) \\ &= \left[\frac{(c + d\sqrt{2})(a - b\sqrt{2})}{a^2 - 2b^2} - (u + v\sqrt{2})\right](a + b\sqrt{2}) \\ &= \left[\frac{(ac - 2bd) + (ad - bc)\sqrt{2}}{a^2 - 2b^2} - (u + v\sqrt{2})\right](a + b\sqrt{2}) \\ &= \left[\left(\frac{ac - 2bd}{a^2 - 2b^2} - u\right) + \left(\frac{ad - bc}{a^2 - 2b^2} - v\right)\sqrt{2}\right](a + b\sqrt{2}) \end{aligned}$$

and

$$\begin{aligned}
&\sigma(\alpha+\beta\sqrt{2})\\
&= \sigma\left(\left(\frac{ac-2bd}{a^2-2b^2}-u\right)+\left(\frac{ad-bc}{a^2-2b^2}-v\right)\sqrt{2})(a+b\sqrt{2})\right)\\
&\le \sigma\left(\left(\frac{ac-2bd}{a^2-2b^2}-u\right)+\left(\frac{ad-bc}{a^2-2b^2}-v\right)\sqrt{2}\right)\sigma(a+b\sqrt{2})\\
&= \left|\left(\frac{ac-2bd}{a^2-2b^2}-u\right)^2-2\left(\frac{ad-bc}{a^2-2b^2}-v\right)^2\right|\sigma(a+b\sqrt{2})\\
&\le \left[\left(\frac{ac-2bd}{a^2-2b^2}-u\right)^2+2\left(\frac{ad-bc}{a^2-2b^2}-v\right)^2\right]\sigma(a+b\sqrt{2})\\
&\le (1/4+2(1/4))\sigma(a+b\sqrt{2})\\
&< \sigma(a+b\sqrt{2}).
\end{aligned}$$

Exercise 3.9.4. 1. Let D be a square free integer and $R = \mathbb{Z}[\sqrt{D}]$. Then R is a Euclidean domain, if $|D| \le 3$.

Since the norm function is multiplicative the first condition for the ring to be a euclidean domain is satisfied.

For the second part, we can mimic what is done in the previous two problems. We leave the details.

2. Is the ring $\mathbb{Z}[\sqrt{5}]$ a Euclidean domain?

3. Show that the rings $\mathbb{Z}[\sqrt{-2}]$ and $\mathbb{Z}[\sqrt{6}]$ are Euclidean domains.

We will now record some of the consequences of condition 1 of the Definition of a euclidean domain.

Proposition 3.9.5. *Let R be a euclidean domain. Then the following holds.*

1. *If an element b is not a unit, then $\sigma(a) < \sigma(ab)$ for all nonzero elements a in R.*

2. *If an element b is a unit in R, then $\sigma(a) = \sigma(ab)$ for all nonzero elements a in R.*

3. *An element a is a unit in R iff $\sigma(a) = \sigma(1)$.*

Proof. Let b in R be a nonunit. We claim that $\sigma(a) < \sigma(ab)$ for all a in R.

Suppose there exist an element $a \neq 0$ in R such that $\sigma(a) = \sigma(ab)$. We apply division algorithm to the pair (a, ab) and write $a = (ab)q + r$ where $r = 0$ or $\sigma(r) < \sigma(ab) = \sigma(a)$. We re-write $a = (ab)q + r$ as $r = a(1 - bq)$. Since b is a nonunit $1 - bq \neq 0$ and hence $a(1 - bq) \neq 0$. Now $\sigma(a) \le \sigma(a(1 - bq)) = \sigma(r) < \sigma(a)$. This is a contradiction. Hence $\sigma(a) < \sigma(ab)$ if b is nonzero and nonunit.

Let b be an element in R. Then $\sigma(b) \le \sigma(ab)$ for all $a \in R \setminus \{0\}$. If a is a unit, then $\sigma(b) \le \sigma(ab) \le \sigma(a^{-1}(ab)) = \sigma(b)$. Thus we have proved that if a is a unit in R, then $\sigma(b) = \sigma(ab)$ for all b in $R \setminus \{0\}$.

Let $a \in R$ be a unit. Then by condition 1 of Definition 3.9.1 of Euclidean domain it follows that

1. $\sigma(a) \leq \sigma(aa^{-1}) = \sigma(1)$ and

2. $\sigma(1) \leq \sigma(1.a) = \sigma(a)$.

Combining these two inequalities, we get that $\sigma(a) = \sigma(1)$.

Conversely assume that $\sigma(a) = \sigma(1)$. If a is not a unit then it folllows from the first part of the proposition that $\sigma(1) > \sigma(1.a) = \sigma(a)$. □

As a consequence of this proposition it follows that $\sigma(a) \geq \sigma(1)$ for every nonzero element a in R. We may assume that $\sigma(1) = 1$. For, if $\sigma(1) \neq 1$, we may define a new size function $\sigma(a) - \sigma(1) + 1$. This size function satisfies the two conditions in Definition 3.9.1 of an Euclidean domain.

Exercise 3.9.6. 1. Let R be a Euclidean domain and a be a nonzero element such that $\sigma(a) = \min\{\sigma(b) : b \in R \setminus \{0\}\}$. Show that a is a unit and conclude that $U(R) := \{a \in R \setminus \{0\} : \sigma(a) \text{ is minimum}\}$ is the set of units in R. Further show that $U(R)$ is a multiplicative group.

Let a in R be such that $\sigma(a) = \min\{\sigma(b) : b \in R \setminus \{0\}\}$. Let b be an element of R. By division algorithm there exist q and r in R such that $b = aq + r$ with $r = 0$ or $\sigma(r) < \sigma(a)$. Since $\sigma(a)$ is the minimum of all $\sigma(b)$ for b in $R \setminus \{0\}$, it follows that $r = 0$. Therefore $b = aq$, proving that a divides b. Thus we have shown that the element a divides all nonzero elements of the ring R. In particular $1 = aq$. Hence a is a unit.

2. Let R be a Euclidean domain and a in R be such that $\sigma(a) = 2$. Show that a is irreducible in R.

Let $a = a_1a_2$ for a_1 and a_2 in R. Then $1 \leq \sigma(a_i) \leq \sigma(a_1a_2)$. If a_1 is a nonunit, it follows from Proposition 3.9.5 that $\sigma(a_2) < \sigma(a_1a_2) = \sigma(a)$. This means that $\sigma(a_2) = \sigma(1)$ and we conclude from Proposition 3.9.5 that a_2 is a unit in R.

Similarly, if a_2 is not unit, we conclude that a_1 is a unit in R. This proves our claim.

3. Let R be a Euclidean domain and a in R be nonzero and nonunit. If a is reducible, let b be a proper divisor of a such that $\sigma(b) := \min\{\sigma(c) : c \text{ divides } a\}$. Then b is irreducible in R.

If b is reducible let $b = b_1b_2$ where b_1 and b_2 are proper divisors of b. Since b_i's divide b, they also divide a. Further

$$\sigma(b_i) < \sigma(b) \leq \sigma(a).$$

This is a contradiction to our assumption $\sigma(b)$ is minimal among all divisors of a. Hence b is irreducible in R.

We are now ready to prove the first part of fundamental theorem of arithmetic for elements in a Euclidean domain R. More precisely we prove the following theorem.

Theorem 3.9.7. *Let R be a Euclidean domain and a be a nonzero and nonunit element in R. Then a is a product of finitely many irreducible elements of R.*

Proof. Let R be a euclidean domain and a be nonzero and nonunit element in R. We will prove the result by induction on the size of a.

If $\sigma(a) = 2$, then we conclude from Exercise 3.9.6(2) that a is irreducible in R. Let us now assume that all nonzero nonunits a in R such that $\sigma(a) < n$ are product of finitely many irreducible elements in R for some $n \geq 3$.

Let us now assume that $\sigma(a) = n$. If a is irreducible then there is nothing to prove. Otherwise let a_1 be a proper divisor of a such that $\sigma(a_1) := \min\{\sigma(b) : b \text{ is a proper divisor of } a\}$. We have shown in Exercise 3.9.6(3) that such an a_1 is irreducible in R. Let b_1 in R be such that $a = a_1 b_1$. Since b_1 is nonzero and nonunit, it follows that $\sigma(b_1) < \sigma(a_1 b_1) = \sigma(a) = n$. By induction hypothesis, b_1 is a product of finitely many irreducible elements in R and this produces a factorization of a in to finitely many irreducible elements in R. □

Next we would like to know if this factorization of a in to irreducible elements is unique. To answer this let's quickly go through the proof of uniqueness in fundamental theorem of arithmetic. Uniqueness of factorization of integers followed from the crucial result that if an integer p is irreducible then it is prime. That is if p is irreducible and if p divides the product ab of two integers a and b, then p divides a or b. This in turn depends on Bezout identity, namely if $d = \gcd(a, b)$ then there exists integers x and y such that $d = ax + by$. We shall now show that gcd of any two elements of a Euclidean domain exist and they satisfy Bezout type equation.

First we prove the following Proposition.

Proposition 3.9.8. *Let R be a Euclidean domain. Then every ideal in R is a principal ideal.*

Proof. Let I be a nonzero proper ideal of R and let a in R be such that

$$\sigma(a) := \min\{\sigma(b) : b \in I \setminus \{0\}\}.$$

If b is an element in the ideal I, then by division algorithm there exist q and r in R such that $b = aq + r$ where $r = 0$ or $\sigma(r) < \sigma(a)$.

We write $b = aq + r$ as $r = b - aq$. Since a and b are in the ideal I, the element r is also in I. If $r \neq 0$, then $\sigma(r) < \sigma(a)$ contradicts the minimality of $\sigma(a)$. Therefore $r = 0$ and this proves that every element in I is a multiple of a. Hence $I = (a)$, the principal ideal generated by a. □

Remark 3.9.9. Let us observe that the generator we have found need not be unique. It is unique only up to units: Suppose a and b are two generators of I. Then there exists u and v in R such that $a = bu$ and $b = av$. Hence $a = bu = auv$. Since R is an integral domain, it follows that $uv = 1$ and this proves that a and b are associates.

We will now prove Bezout type identity in a Euclidean domain.

Theorem 3.9.10 (Bezout Identity). *Let R be a Euclidean domain. Let a and b two nonzero elements in R. Then*

1. *gcd of a and b exists in R and*

2. *there exists x and y in R such that* $\gcd(a, b) = ax + by$. *We denote* $\gcd(a, b)$ *by* d.

Proof. Let a and b two nonzero elements in R. Since R is a Euclidean domain, the ideal $I = \{ar+bs : r, s \in R\}$ is a principal ideal in R. (See Proposition 3.9.8). Let d in R be such that $I = (d)$. Now we appeal to Proposition 3.4.12 to conclude that d is a gcd of the elements a and b and there exist two elements x and y in R such that $d = ax + by$. □

Proposition 3.9.11. *Let R be a Euclidean domain. If an element a in R is irreducible in R then a is prime in R.*

Proof. Let a be an irreducible element of R. Let b and c be two elements of R such that a divides the product bc. If a does not divide b, then the elements a and b can not have a common divisor. Hence gcd of a and b is 1 and there exist x and y in R such that $ax + by = 1$. Now we multiply this equation by c to get $acx + bcy = c$. Since a divides bc, we see that a divides $acu + bcv = c$ and the proof is complete. □

We are now ready to prove the uniqueness of irreducible factorization of a nonzero element in a Euclidean domain.

Theorem 3.9.12. *Let R be a Euclidean domain. If*

$$a = a_1 a_2 \cdots a_r$$

and

$$a = b_1 b_2 \cdots b_s$$

is an irreducible factorization of a nonzero, nonunit a in R, then $r = s$ and each a_j is an associate of b_{i_j} for $1 \leq j \leq r$.

Proof. Let $a = a_1 a_2 \cdots a_r$ and $a = b_1 b_2 \cdots b_s$ be two factorizations of a nonzero, nonunit element a in to irreducible factors.

Each of the irreducible factors a_j divides $a = b_1 b_2 \cdots b_s$. Hence a_j divides b_{i_j} for some $i_j \leq s$. Let $b_{i_j} = u_{i_j} a_j$. If $r < s$, then

$$a = a_1 a_2 \cdots a_r = b_1 b_2 \cdots b_s = (a_1 a_2 \cdots a_r)(u_{i_1} u_{i_2} \cdots u_{i_r}) \prod_{t \neq i_j} b_t.$$

By canceling the a_i's, we get $(u_{i_1} u_{i_2} \cdots u_{i_r}) \prod_{t \neq i_j} b_t = 1$. Since each b_t for $t \neq i_j$ is irreducible, the product $\prod_{t \neq i_j} b_t$ is not a unit in R. This is a contradiction. Hence $r = s$ and each of the irreducible element a_j is an associate of b_{i_j} for some i_j. □

We will now prove Euclidean algorithm in an Euclidean Domain.

Let a and b be two nonzero and nonunits in a euclidean domain R. We apply division algorithm repeatedly and get

$$\begin{aligned} b &= aq_1 + r_1 \\ a &= r_1q_2 + r_2 \\ r_1 &= r_2q_3 + r_3 \\ &\vdots \\ r_{n-3} &= r_{n-2}q_{n-1} + r_{n-1} \\ r_{n-2} &= r_{n-1}q_n + r_n \\ r_{n-1} &= r_nq_{n+1} + r_{n+1} \end{aligned}$$

where q_i's and r_i's are in R such that $\sigma(r_i) < \sigma(r_{i-1})$. Since $\sigma(r_i)$'s are natural numbers this will stop after finitely many steps with $r_{n+1} = 0$. The elements q_i's and r_i's need not be unique.

We will now prove that r_n is the gcd of a and b in the following

Theorem 3.9.13. *Let R be a Euclidean domain with a size function σ. Let a and b be two nonzero and nonunits in R. If r_n is the remainder in the euclidean algorithm then*

1. *r_n is a gcd of a and b and*

2. *there exist x and y in R such that $r_n = ax + by$.*

We prove two lemmas needed for the proof this theorem.

Lemma 3.9.14. *Let a and b be two nonzero and nonunits in a Euclidean domain R with a size function σ. If $b = aq + r$ with $r \neq 0$, then any* $\gcd(a, b)$ *is* $\gcd(a, r)$ *and conversely.*

Proof. Let $d := \gcd(a, b)$ and $e := \gcd(a, r)$. By definition d divides a and b. Hence d divides $r = b - aq$. Hence d divides e by definition of e.

By definition e divides a and r and hence e divides $b = aq + r$. Therefore e divides d. This proves that d and e are associates in R. □

Lemma 3.9.15. *Let a be a nonzero and nonunit in a Euclidean domain R. Then*

$$\gcd(a, 0) = a.$$

Proof. We know that $I = (a, 0) = (a)$. This proves the result. □

We will now prove Theorem 3.9.13.

Proof. The proof is just book keeping!

Let $d = \gcd(a,b)$. Then by Lemma 3.9.14

$$\begin{aligned} d &= \gcd(a,b) \\ &= u\gcd(a,r) \\ &= u_1\gcd(r,r_1) \\ &\vdots \\ &= u_{n-1}\gcd(r_{n-1},r_n) \\ &= \gcd(r_n,0) \\ &= r_n. \end{aligned}$$

This finishes the proof of first part.

We will now prove the second part. We will prove this by showing that the ideal $(a,b) := \{ra + sb : r, s \in R\}$ is equal to the principal ideal (r_n).

By definition of gcd the element r_n divides a and b. This proves that $a, b \in (r_n)$. As a consequence the ideal $(a,b) \subseteq (r_n)$. If we now show that the element r_n is in the ideal (a,b), then we are through.

The equation $r_{n-1} = r_n q_{n+1}$ implies that r_n divides r_{n-1}. Now the equation $r_{n-2} = r_{n-1}q_n + r_n$ show that r_n divides r_{n-2} also. This proves that $r_n \in (r_{n-2}, r_{n-1})$. Now working backwards in euclidean algorithm, we conclude from the equation $r_{k-2} = r_{k-1}q_k + r_k$, that r_k divides r_{k-1} and r_{k-2}. This leads to the following chain of inclusions

$$(r_n) \subseteq (r_{n-2}, r_{n-1}) \subseteq \cdots \subseteq (r_1, r) \subseteq (a, r) \subseteq (a, b)$$

ideals. This proves that $(a,b) = (r_n)$ and there exist x and y in R such that $r_n = ax + by$.

We remark that this is not the only way to prove this result. We could have started with backward substitution, namely write r_n in terms of the previous terms and go up to d in the first step. □

We have seen in Example 3.9.3(1), that the quotient and remainder in the division algorithm need not be unique.

As another example consider $R = \mathbb{Z}[i]$. Let $a = 1 + 8i$ and $b = 2 - 4i$. We can write

$$1 + 8i = 2 + 6i + (-1 + 2i) = (2 - 4i)(-1 + i) + (-1 + 2i)$$

and

$$1 + 8i = 10i + (1 - 2i) = (2 - 4i)(-2 + i) + (1 - 2i).$$

However if we insist on the uniqueness of quotient and remainder in the definition of Euclidean domain then the domain must necessarily be a field or a polynomial ring over a field. This is given as exercise.

Exercise 3.9.16. Let R be a Euclidean domain with unique quotient and remainder in the condition 2 of Definition 3.9.1(2). Then R is either a field or a polynomial ring over a field.

We list the steps involved in proving this.

1. $\sigma(a+b) \leq \max\{\sigma(a), \sigma(b)\}$ for all nonzero elements a, b in R.

   Suppose not. Then there exist two elements a and b in R such that $\sigma(a+b) > \sigma(a), \sigma(b)$. Now we can write $a = 0(a+b) + a$ and $a = 1(a+b) - b$. Since -1 is a unit, $\sigma(-b) = \sigma((-1)b) = \sigma(b) < \sigma(a+b)$. This violates our assumption that the quotient and remainder are unique.

2. The set $\mathbb{F} := \{a \in R \setminus \{0\} : \sigma(a) = \sigma(1)\} \cup \{0\}$ is a field.

   Observe that $\mathbb{F}$ is the union of set of units and the zero element.

   Let a and b two nonzero elements in $\mathbb{F}$. If $a + b \neq 0$, then $\sigma(1) \leq \sigma(a+b) \leq \max\{\sigma(a), \sigma(b)\} = \sigma(1)$. This proves that $\sigma(a+b) = \sigma(1)$. Hence $\sigma(a+b)$ is a unit. This proves that $\mathbb{F}$ is a field. If $\mathbb{F} = R$, then we are through.

3. Assume $\mathbb{F} \neq R$. We will now find an element X in $R \setminus \mathbb{F}$ such that $R = \mathbb{F}[X]$.

   Let $X \in R \setminus \mathbb{F}$ be such that $\sigma(X) = \min\{\sigma(a) : a \in R \setminus \mathbb{F}\}$. We will show that R is the polynomial ring over $\mathbb{F}[X]$. Observe that X is a nonunit. Hence $\sigma(X) < \sigma(X^2) < \ldots < \sigma(X^k) < \ldots$ is a strictly increasing sequence of natural numbers.

   Let a be an element in $R \setminus \mathbb{F}$. Then there exists a natural number k such that $\sigma(X^k) \leq \sigma(a) < \sigma(X^{k+1})$. We apply division algorithm to the pair (a, X^k) and write $a = a_k X^k + r_k$ for unique elements a_k and r_k in R such that $\sigma(r_k) < \sigma(X^k)$. If a_k is not a unit then $\sigma(a_k X^k) < \sigma(X^k)$. Hence $\sigma(a) = \sigma(a_k X^k + r_k) \leq \max\{\sigma(a_k X^k), \sigma(r_k)\} < \sigma(X^k)$, a contradiction. This proves that given an element a in R, there exist a natural number k, a unit a_k in $\mathbb{F}$ and an element r_k in R such that $a = a_k X^k + r_k$ and $\sigma(r_k) < \sigma(X^k)$. For this element r_k, we follow the same procedure and get a natural number l, a unit a_l in $\mathbb{F}$ and r_l in R such that $r_k = a_l X^l + r_l$ and $\sigma(r_l) < \sigma(X^l)$. Since $\sigma(r_l) < \sigma(X^l) < \sigma(X^k)$, this process will stop at finitely many steps and we can write $a = a_0 + a_1 X + a_2 X^2 + \cdots + a_k X^k$ where a_i's are in $\mathbb{F}$; here we allow the coefficients a_i to be either zero or unit to write in this form. This shows that any element a in $R \setminus \mathbb{F}$ can be written as polynomial in X with coefficients from the field $\mathbb{F}$.

4. Uniqueness of this expression follows from the uniqueness of quotients and remainders in each step!

3.9.2 Principal Ideal Domains

We have shown above that Fundamental theorem of arithmetic is valid in a Euclid domain. The crucial result we have used to prove this is the fact that in a Euclidean domain every ideal is a principal ideal.

However there are many rings which are not euclidean domains but every ideal is a principal ideal. For example the ring $R = \mathbb{Z}[\frac{1+\sqrt{-19}}{2}]$ is not a euclidean domain but every ideal in R is a principal ideal. For a proof see [3] and [5].

In this section we study integral domains in which every ideal is a principal ideal and prove fundamental theorem of arithmetic in such rings.

Definition 3.9.17. An integral domain R is called a principal ideal domain, PID in short, if every ideal in R is a principal ideal.

Examples 3.9.18. 1. The ring of integers $\mathbb{Z}$ is a PID.

2. If R is a euclidean domain, then R is a PID. Thus the rings $\mathbb{F}[X]$, $\mathbb{Z}[i]$ and $\mathbb{Z}[\sqrt{2}]$ are all PIDs.

3. The ring $\mathbb{Z}[X]$ is not a PID.

   Let $I = (2, X)$ be the ideal generated the integer 2 and the polynomial X. We have already shown in Example 3.4.17 that I is not a principal ideal in $\mathbb{Z}[X]$.

We show that in a PID any two nonzero elements have gcd and the Bezout's identity is valid in R.

Theorem 3.9.19 (Bezout Identity). *Let R be a PID. Let a and b be two nonzero elements of R. Then*

1. *gcd of a and b exists and*

2. *if d is a gcd of a and b in R, there exist x and y in R such that $d = ax+by$.*

Proof. Let R be a PID. Let a and b be two nonzero elements of R. Let $I = (a)$, $J = (b)$ and $K = I + J$. Then K is also an ideal. Since R is a PID, there exists an element d in R such that $(a) + (b) = K = (d)$. Let x and y be in R such that $d = ax + by$.

The elements a and b are in the ideal $K = (d)$. Hence there exist r and s in R such that $a = dr$ and $b = ds$. This proves that d divides a and b.

Let e be an element in R dividing both a and b. Then e divides $ar + bs$ for all r and s in R. In particular e divides $d = ax + by$. Thus d is a gcd of a and b.

Note that if u is a unit in R, then du is also a gcd of a and b. □

We will now prove an important property about irreducible elements in terms of the ideal they generate.

Proposition 3.9.20. *Let R be a PID. A nonzero element a in R is irreducible iff the principal ideal $I = (a)$ is a maximal ideal in R.*

Proof. Let a be an element of R and $I = (a)$ be the principal ideal generated by a.

Assume that a is irreducible in R and let $J = (b)$ be an ideal in R such that $I \subseteq J$. Then there exists an element c in R such that $a = bc$. Since a is irreducible either b is a unit or c is a unit in R. If b is a unit then $J = R$. If c is a unit then $b = ac^{-1}$ and hence $J \subseteq I$. Thus $I = J$. This proves that I is a maximal ideal in R.

Conversely assume that the ideal $I = (a)$ is a maximal ideal in R. If a is not irreducible in R then there exist nonzero and nonunits b and c in R such that $a = bc$. As a consequence it follows that the ideal I is strictly contained in the ideal $J = (b)$, a contradiction. □

We will now prove the main result of this section, namely the fundamental theorem of arithmetic in a PID.

Theorem 3.9.21. *Let R be a PID and a in R be a nonzero and nonunit. Then there exist finitely many irreducible elements $a_1, a_2 \ldots, a_k$ in R such that*

$$a = a_1 a_2 \cdots a_k.$$

Further if

$$a = a_1 a_2 \cdots a_k = b_1 b_2 \cdots b_l.$$

where a_i and b_j are irreducible in R, then $k = l$ and each a_i is an associate of b_{j_i} for some integer j_i.

This theorem will be proved through a series of lemmas.

Lemma 3.9.22. *Let R be a PID. If $I_1 \subseteq I_2 \subseteq \ldots I_n \subseteq \ldots$ is an increasing sequence of ideals, then there exists $k \in \mathbb{N}$ such that $I_k = I_{k+r}$ for all $r \geq 1$.*

Proof. Let $I_1 = (a_1) \subseteq I_2 = (a_2) \subseteq \ldots I_n = (a_n) \subseteq \ldots$ be an increasing sequence of ideals in R. Let $I = \cup_{n \in \mathbb{N}} I_n$. We can easily show that I is an ideal in R. Since R is a PID, there exists an element a in R such that $I = (a)$. The element a is in I. Hence there exists $k \in \mathbb{N}$ such that a is in I_k. As a consequence the ideal $I = (a)$ is contained in I_k. On the other hand I_k is already contained in I. Thus $I_k = I_{k+r}$ for all $r \geq 1$. □

Lemma 3.9.23. *Let R be a PID and a in R be nonzero and nonunit. Then a has an irreducible divisor in R.*

Proof. If a is irreducible we are through. If a is not irreducible then we can write $a = a_1 b_1$ where a_1 and b_1 are nonzero and nonunits in R. Further they are not associates of a in R. If either a_1 or b_1 is irreducible in R then we are through.

Suppose both are reducible elements in R. Then $(a) \subseteq (a_1)$ and the inclusion is strict: If $(a) = (a_1)$, there exists c_1 in R such that $a_1 = ac_1$. Therefore $a = a_1 b_1 = ac_1 b_1$. As a consequence it follows that b_1 is a unit, a contradiction.

Since a_1 is reducible, there exist two nonzero nonunits a_2 and b_2 such that $a_1 = a_2 b_2$. If none of the elements a_2 and b_2 are irreducible, then $(a_1) \subseteq (a_2)$ and $a_2 = a_3 b_3$ where a_3 and b_3 are nonzero and nonunits. Then $(a_2) \subseteq (a_3)$. If a_3 is reducible, then we continue the factorization.

Inductively, if none of the a_j's are irreducible, we get a strictly increasing sequence of ideals

$$(a) \subseteq (a_1) \subseteq (a_2) \ldots$$

This contradicts Lemma 3.9.22. Therefore the element a_k is irreducible in R for some $k \geq 1$.(Why?) □

Lemma 3.9.24. *Let R be a PID and a in R be irreducible in R. Then a is prime in R.*

Proof. Let R be a PID and a in R be irreducible in R. Suppose b and c are two elements in R such that a divides the product bc. Suppose a does not divide b. Let $I := \{ax + by : x, y \in R\}$ be the ideal generated by these two elements a and b in R. Since the ideal (a) is a maximal ideal (See Proposition 3.9.20) and it is strictly contained in I it follows that $I = R$. Therefore there exist x and y

in R such that $ax + by = 1$. If we now multiply the equation $ax + by = 1$ by the element c, we get $c = acx + bcy$. The element a divides ac and bc. Hence a divides $acx + bcy = c$ and the proof is complete. □

We will now prove Theorem 3.9.21.

Proof. Let R be a PID and a in R be nonzero and nonunit. We know from Lemma 3.9.23 that a has an irreducible divisor a_1. Hence there exists b_1 in R such that $a = a_1 b_1$. If b_1 is a unit, then a is irreducible and we are through. If b_1 is irreducible then $a = a_1 b_1$ is a product of two irreducible elements of R and the proof is complete.

Otherwise the element b_1 has an irreducible divisor a_2 and we write $b_1 = a_2 b_2$. Observe that $(a) \subseteq (a_1) \subseteq (a_2)$ and the inclusion is strict. If b_2 is reducible, then this process continues to produce a strictly increasing sequence of ideals

$$(a) \subseteq (a_1) \subseteq (a_2) \subseteq \dots$$

This chain of ideals must stop at some stage. Hence there exists $k \in \mathbb{N}$ such that $a = a_1 a_2 \cdots a_k$ where each of the a_i's are irreducible in R. This proves the existence of factorization in to finitely many irreducible elements in R.

Uniqueness part same as in the case of Euclidean domain and hence we omit the proof. □

3.9.3 Unique Factorization Domains

Definition 3.9.25. An integral domain R is called a unique factorization domain if the following holds.

1. Let a in R be nonzero and nonunit. Then there exists finitely many irreducible elements $p_1, p_2, \dots, p_r$ in R and positive integers k_i such that $a = p_1^{k_1} p_2^{k_2} \cdots p_r^{k_r}$.

2. If there exists irreducible elements $q_1, q_2, \cdots, q_s$ in R and positive integers t_j such that
$$a = p_1^{k_1} p_2^{k_2} \cdots p_r^{k_r} = q_1^{t_1} q_2^{t_2} \cdots q_s^{t_s}$$
then $r = s$ and each p_i is an associate of q_{j_i} in R, for $1 \leq i \leq r$.

We have shown above that Euclidean domains and Principal Ideal domains are unique factorization domains.

The ring $\mathbb{Z}[X]$ is a not a PID. However given a polynomial $f(X)$ in $\mathbb{Z}[X]$ we would like to factorize it in to irreducible elements and also would like this factorization to be unique.

In this section we study this problem and show that $R[X]$ is a unique factorization domain iff R is so.

Before doing so let us prove some simple and important properties of an unique factorization domain.

Proposition 3.9.26. *Let R be a unique factorization domain.*

1. *If an element a in R is irreducible in R then a is prime in R.*

2. *Any two nonzero nonunit elements a and b in R have a gcd d in R.*

Proof. Let a be an irreducible element in R. Let b and c be two nonzero nonunits such that the element a divides the product bc. Then there exists an element w in R such that $bc = aw$.

Let $b = b_1^{k_1} b_2^{k_2} \cdots b_t^{k_t}$ and $c = c_1^{l_1} c_2^{l_2} \cdots c_s^{l_s}$ be the irreducible factorization of the elements b and c. Then

$$aw - b_1^{k_1} b_2^{k_2} \cdots b_t^{k_t} c_1^{l_1} c_2^{l_2} \cdots c_s^{l_s}.$$

Since b_i's and the c_j's are irreducible, by the uniqueness of the factorization, it follows that either $a = b_i$ for some i or $a = c_j$ for some j. Hence a divides either b or c. This proves that an irreducible element in a unique factorization domain is prime.

Let a and b be two nonzero elements. We will show that the elements a and b have a gcd.

Let $a = a_1^{k_1} a_2^{k_2} \cdots a_t^{k_t}$ and $b = b_1^{l_1} b_2^{l_2} \cdots b_s^{l_s}$ be the irreducible factorization of the elements b and c in R. Let $P(a)$ and $P(b)$ be the set of all prime divisors of a and b respectively. Let $P = P(a) \cup P(b)$. Then we can write $a = p_1^{r_1} p_2^{r_2} \cdots p_n^{r_n}$ and $b = p_1^{s_1} p_2^{s_2} \cdots p_n^{s_n}$ where p_i's are in P and r_i, s_i are non-negative integers. It is possible that $r_i = 0$ or $s_j = 0$ for some i and j.

Let $d = p_1^{t_1} p_2^{t_2} \cdots p_n^{t_n}$ where $t_i = \min\{r_i, s_i\}$. Then d divides a and b. Let e be an element such that e divides a and b.

Suppose p is a prime divisor of e. Then p divides a and b too. Therefore the set of prime divisors of e is a subset of P, the set of all prime divisors of a and b. Hence we can write $e = p_1^{v_1} p_2^{v_2} \ldots p_n^{v_n}$ where v_is are non-negative integers. Since e divides a and b, there exists a_1 and b_1 in R such that

$$p_1^{r_1} p_2^{r_2} \cdots p_n^{r_n} = a = ea_1 = p_1^{v_1} p_2^{v_2} \ldots p_n^{v_n} a_1$$

and

$$p_1^{s_1} p_2^{s_2} \cdots p_n^{s_n} = b = eb_1 = p_1^{v_1} p_2^{v_2} \ldots p_n^{v_n} b_1.$$

Hence $v_i \leq r_i, s_i$ for $1 \leq i \leq n$ and $v_i \leq \min\{r_i, s_i\} = t_i$. Now we can write

$$d = p_1^{v_1} p_2^{v_2} \ldots p_n^{v_n} p_1^{t_1 - v_1} p_2^{t_2 - v_2} \ldots p_n^{t_n - v_n} = e(p_1^{t_1 - v_1} p_2^{t_2 - v_2} \ldots p_n^{t_n - v_n}).$$

This proves that e divides d and completes the proof by showing that d is a gcd of a and b. □

Definition 3.9.27. Let R be a UFD. We say that two nonzero elements a and b in R are relatively prime in R, if $\gcd(a, b)$ is a unit in R.

Exercise 3.9.28. Let a and b two relatively prime elements in a UFD R such that $ab = c^n$ for some element c in R. Show that there exist two nonzero elements a_1, b_1 in R and units u and v in R such that $a = ua_1^n$ and $b = vb_1^n$.

The Proposition 3.9.26 above is sometimes very useful to show that some rings are not UFDs. We can produce two nonzero nonunits which have proper divisors but no gcd. One such example is given below.

Example 3.9.29. Let $R = \mathbb{Z}[\sqrt{-5}]$. Let $a = 6$ and $b = 2(1 + \sqrt{-5})$. The elements a and b do not have a gcd in R.

First we write $6 = 2.3 = (1+\sqrt{-5})(1-\sqrt{-5})$. This shows that the elements 2 and $1 + \sqrt{-5}$ divide both a and b.

Suppose d is a gcd of a and b. Then the integer $N(d)$ divides the integers $N(6) = 36$ and $N(2(1 + \sqrt{-5}) = N(2)N(1 = \sqrt{-5}) = 4.6 = 24$. The possible proper divisors of the integers 24 and 36 are 1, 2, 3, 4, 6, 12.

Since 2 and $1 + \sqrt{-5}$ must divide d, it follows that the integers $N(2) = 4$ and $N(1 + \sqrt{-5}) = 6$ must divide $N(d)$. This forces $N(d)$ to be equal to 12. If $d = \alpha + \beta\sqrt{-5}$, then $N(\alpha + \beta\sqrt{-5}) = \alpha^2 + 5\beta^2$ must be equal to 12.

Now $5\beta^2 \leq \alpha^2 + 5\beta^2 = 12$. This forces that $\beta = \pm 1$ or 0. If $\beta = \pm 1$ then $\alpha^2 = 12 - 5 = 7$ and if $\beta = 0$ then $\alpha^2 = 12$. Both are impossible. Hence a and b do not have a gcd.

Exercise 3.9.30. 1. Let $p \geq 3$ be a prime. Then $\mathbb{Z}[\sqrt{-p}]$ is not a UFD.

The proof follows along the same lines as in the previous Example 3.9.29.

2. Show that the ring $R := \mathbb{Z}[i\sqrt{5}]$ is not a UFD by writing a nonzero nonunit as product of irreducible elements in two different ways.

   $6 = 2.3 = (1 + i\sqrt{5})(1 - i\sqrt{5})$. If we now show that 2, 3 $1 \pm i\sqrt{5}$ are irreducible in R then we are through.

   Let us assume that $2 = (a+i\sqrt{5}b)(\alpha+i\sqrt{5}\beta)$ in R. Then $4 = N(2) = (a^2 + 5b^2)(\alpha^2 + 5\beta^2)$. This forces $b = 0 = \beta$ and either $a = \pm 1$ or $\alpha = \pm 1$.(why?) Hence 2 is irreducible in R.

   Similarly we can prove that the other three elements also irreducible in R.

   If 2 and $1 \pm i\sqrt{5}$ are associates, then their norms must be same.(Why?) However $N(2) = 4 \neq N(1 \pm i\sqrt{5}) = 6$.

3. Show that the rings $\mathbb{Z}[\sqrt{d}]$ are not UFDs if $d = -6, -10$.

4. Show that $\mathbb{Z}[\sqrt{5}]$ is not a UFD.

   First we write $4 = (\sqrt{5} + 1)(\sqrt{5} - 1) = 2 \times 2$. Now we show that 2 is irreducible and it does not divide either of the factors $(\sqrt{5} + 1)$ or $(\sqrt{5} - 1)$ of 4.

   Suppose 2 is reducible in R and $2 = (a + b\sqrt{5})(c + d\sqrt{5})$ where $(a + b\sqrt{5})$ and $(c + d\sqrt{5})$ are nonunits in R.

   Now $4 = N(2) = |a^2 - 5b^2||c^2 - 5d^2|$. Therefore $|a^2 - 5b^2| = 2 = |c^2 - 5d^2|$.

   Assume that $a^2 - 5b^2 = \pm 2$. This is possible iff a and b are both odd or both even.

   Suppose $a = 2a_1 + 1$ and $b = 2b_1 + 1$. Then

$$\begin{aligned} \pm 2 &= (2a_1 + 1)^2 - 5(2b_1 + 1)^2 \\ &= 4\left[a_1(a_1 + 1) - 5b_1(b_1 + 1)\right] - 4. \end{aligned}$$

   This shows that $4 \pm 2 = 4\left[a_1(a_1 + 1) - 5b_1(b_1 + 1)\right]$. Therefore $2 \pm 1 = 2\left[a_1(a_1 + 1) - 5b_1(b_1 + 1)\right]$, a contradiction.

If $a = 2a_1$ and $b = 2b_1$, then $\pm 2 = 4\left[a_1^2 - 5b_1^2\right]$. This is again a contradiction. Therefore either $a + b\sqrt{5}$ or $c + d\sqrt{5}$ is a unit in R.

We will now show that 2 does not divide $\sqrt{5} \pm 1$. Suppose it does. Then $1 \pm \sqrt{5} = 2(a + b\sqrt{5})$. This means that $2a = \pm 1$, a contradiction. Therefore $\mathbb{Z}[\sqrt{5}]$ is not a UFD.

5. Let $R = \mathbb{Z}[\sqrt{5}]$. Show that the elements $3 \pm \sqrt{5}$ and $1 \pm \sqrt{5}$ are associates in R.

Now we will prove the main theorem of this section.

Theorem 3.9.31. *Let R be a UFD. Then $R[X]$ is a UFD.*

This will be done in a series of steps. We start with a nonzero nonconstant polynomial $f(X)$ in $R[X]$ and factorize $f(X)$ in to irreducible elements in $\mathbb{F}[X]$, the polynomial ring over the field of fractions $\mathbb{F}$ of R. Then we show that each of these irreducible factors can be written as a product of finitely many irreducible polynomial in $R[X]$ and a nonzero element in the field $\mathbb{F}$. Next we show that the product of these non zero elements of $\mathbb{F}$ is an element in R which can then be written as product of irreducible elements of R.

Definition 3.9.32. A polynomial $f(X) = a_0 + a_1X + a_2X^2 + \cdots + a_kX^k$ in a polynomial ring $R[X]$ is called primitive if the gcd of the coefficients $a_0, a_1, a_2, \ldots, a_k$ is a unit in R.

Exercise 3.9.33. Let R be a UFD and a in R be an irreducible element in R. Then a is irreducible in $R[X]$.

Let $f(X) = a_0 + a_1X + a_2X^2 + \cdots + a_nX^n$ be a nonconstant polynomial of degree d. If $d = \gcd(a_0, a_1, \ldots, a_n)$, we write $a_i = db_i$. Then $f(X) = d(b_0 + b_1X + b_2X^2 + \cdots + b_nX^n)$. Since d is in R we can write as a product of finitely many irreducible elements in R. Therefore it suffices to write the primitive polynomial $b_0 + b_1X + b_2X^2 + \cdots + b_nX^n$ as a product of finitely many irreducible elements in $R[X]$.

Let $f(X) = a_0 + a_1X + a_2X^2 + \cdots + a_nX^n \in R[X]$ be a primitive polynomial. Since $\mathbb{F}[X]$ is a PID, the polynomial $f(X)$ can be written as write $f(X)$ as $f(X) = f_1(X)f_2(X)\cdots f_r(X)$ where $f_1(X)$, $f_2(X), \ldots, f_r(X)$ in $\mathbb{F}[X]$ are irreducible in $\mathbb{F}[X]$. Let $n_i = \deg(f_i(X))$ for $1 \leq i \leq r$.

Let $f_i(X) = \sum_{j=1}^{n_i} \frac{a_{ji}}{b_{ji}} X^j$ where the coefficients $\frac{a_{ji}}{b_{ji}}$ are in $\mathbb{F}$. By clearing the denominators b_{ji} in the polynomial $f_i(X)$, we can write $f_i(X) = b_i \sum_{ji} c_{ji}X^j$ for suitable coefficients c_{ji} in R and b_i in $\mathbb{F}$. If d_i is the gcd of the coefficients c_{ji}, we can write$\sum_{j=1}^{n_i} c_{ji}X^j = d_ig_i(X)$ where $g_i(X)$ is a polynomial in $R[X]$ such that the gcd of the coefficients is 1. Hence we can write $f_i(X) = b_id_ig_i(X) = c_ig_i(X)$ where $c_i = b_id_i$ is in $\mathbb{F}$ and $g_i(X)$ is primitive in $R[X]$.

In the discussion above, we have actually proved the following lemma.

Lemma 3.9.34. *Let $f(X)$ be a nonconstant polynomial in $\mathbb{F}[X]$. Then $f(X) = cg(X)$ where c is in $\mathbb{F}$ and $g(X)$ is a primitive polynomial in $R[X]$.*

Let $f(X) = a_0 + a_1X + a_2X^2 = \cdots + a_nX^n \in R[X]$ be a primitive polynomial and let $f(X) = f_1(X)f_2(X)\cdots f_r(X)$ where $f_i(X)$ are irreducible in

$\mathbb{F}[X]$. By Lemma 3.9.34 each $f_i(X) = c_i g_i(X)$ where $g_i(X)$ is a primitive polynomial in $R[X]$. Therefore the polynomial $f(X) = f_1(X)f_2(X)\cdots f_r(X) = c_1c_2\cdots c_r g_1(X)g_2(X)\cdots g_r(X)$. We will show that the product $c_1c_2\cdots c_r$ is a unit in R. Since the gcd of coefficients of $f(X)$ is 1, to show that $c_1c_2\cdots c_r$ is a unit in R, it suffices to show that the gcd of the coefficients of the product $g_1(X)g_2(X)\cdots g_r(X)$ is a unit in R. This is done in the following lemma.

Lemma 3.9.35 (Gauss Lemma). *Let $f_1(X)$ and $f_2(X)$ be two primitive polynomials in $R[X]$. Then the product $f_1(X)f_2(X)$ is also primitive in $R[X]$.*

Proof. Let $f_1(X) = \sum_{i=0}^{k} a_i X^i$ and $f_2(X) = \sum_{j=0}^{l} b_j X^j$ be two primitive polynomials in $R[X]$. We want to show that if a prime element p in R divides the product $f_1(X)f_2(X)$, then that prime p must divide either $f_1(X)$ or $f_2(X)$.

If possible let p be a prime divisor of the coefficients of the product polynomial $f_1(X)f_2(X) = \sum_{0\leq i+j\leq k+l} a_i b_j X^{i+j}$. In particular p divides the constant term a_0b_0. Since p is prime, it follows that p divides either a_0 or b_0.

Since $f_1(X)$ is primitive, the prime p does not divide a_i for some i. Let $r-1 = \max\{i \leq k : p \text{ divides } a_i\}$. Then $r-1 < k$. Similarly if $s-1 = \max\{j \leq l : p \text{ divides } b_j\}$, then $s-1 < l$. Hence p does not divide the product $a_r b_s$.

We write the product $\sum_{i+j=r+s} a_i b_j$ as

$$\sum_{i+j=r+s} a_i b_j = \left(\sum_{i\leq r-1, j\geq s+1} a_i b_j\right) + a_r b_s + \left(\sum_{i\geq r+1, j\leq s-1} a_i b_j\right).$$

The prime p divides $\sum_{i+j=r+s} a_i b_j$. Further p divides a_i for $i \leq r-1$ and b_j for $j \leq s-1$. Therefore p divides

$$a_r b_s = \sum_{i+j=r+s} a_i b_j - \left(\sum_{i\leq r-1, j\geq s+1} a_i b_j + \sum_{i\geq r+1, j\leq s-1} a_i b_j\right)$$

a contradiction. This completes the proof.

Another Proof: Let p be a prime element in R that divides the product $f_1(X)f_2(X)$.

Since p is prime in R, the principal ideal (p) is a prime ideal and hence the quotient ring $R/(p)$ is an integral domain.

The image of $f_1(X)f_2(X)$ in$(R/(p))[X]$ is $\overline{f_1(X)f_2(X)}$. By assumption this image is zero in the integral domain $(R/(p))[X]$. Therefore either $\overline{f_1(X)} = 0$ or $\overline{f_2(X)} = 0$. This means that the prime p divides all the coefficients of either of the polynomials $f_1(X)$ or $f_2(X)$. This is a contradiction to the fact that the polynomials $f_1(X)$ and $f_2(X)$ are primitive.

□

Now we return to the proof of Theorem 3.9.31.

From Gauss Lemma 3.9.35, we conclude that the gcd of the coefficients of the product $g_1(X)g_2(X)\cdots g_r(X)$ is a unit in R.

Let $c_i = u_i/v_i$ for $1 \leq i \leq r$. We substitute $c_i = u_i/v_i$ in the polynomial $f(X) = c_1c_2 \cdots c_r g_1(X)g_2(X) \cdots g_r(X)$ to get

$$\begin{aligned} f(X) &= \prod_{i=1}^{r} (u_i/v_i)\, g_1(X)g_2(X) \cdots g_r(X) \text{ and} \\ v_1v_2 \cdots v_r f(X) &= u_1u_2 \cdots u_r g_1(X)g_2(X) \cdots g_r(X) \end{aligned}$$

Since the gcd of the coefficients of the polynomials $g_1(X)g_2(X) \cdots g_r(X)$ and $f(X)$ are units, it follows that $u_1u_2 \cdots u_r$ and $v_1v_2 \cdots v_r$ are associates in R. Hence $c_1c_2 \cdots c_r$ is a unit in R.

We will now show that each of these primitive polynomials $g_i(X)$ is irreducible in $R[X]$ in the following lemma.

Lemma 3.9.36. *Let $g(X)$ be a primitive polynomial in $R[X]$. Then $g(X)$ is irreducible in $R[X]$ iff $g(X)$ is irreducible in $\mathbb{F}[X]$.*

Proof. Let $g(X)$ be a primitive polynomial in $R[X]$.

Let us assume that $g(X)$ is irreducible in $R[X]$. Suppose $g(X)$ is reducible in $\mathbb{F}[X]$ and $g(X) = g_1(X)g_2(X)$ where $g_i(X)$ are nonunits in $\mathbb{F}[X]$. In particular the polynomials $g_i(X)$ are nonconstant.

Using Lemma 3.9.34, we write $g_i(X) = c_ih_i(X)$ where c_i is in $\mathbb{F}$ and $h_i(X)$ is a primitive polynomial in $R[X]$. Then $g(X) = c_1c_2h_1(X)h_2(X)$. Since the polynomials $g(X)$ and $h_1(X)h_2(X)$ are primitive, it follows that c_1c_2 is a unit in R. But this contradicts our assumption that $g(X)$ is irreducible in $R[X]$. Hence $g(X)$ is irreducible in $\mathbb{F}[X]$.

Conversely assume that the polynomial $g(X)$ is irreducible in $\mathbb{F}[X]$.

Suppose $g(X) = g_1(X)g_2(X)$ where the polynomials $g_i(X)$ are in $R[X]$. Since $g(X)$ is irreducible in $\mathbb{F}[X]$, it follows that either $g_1(X)$ or $g_2(X)$ is a unit in $\mathbb{F}[X]$. But the units in $\mathbb{F}[X]$ are constant polynomials. However the polynomials $g_i(X)$ are in $R[X]$. Therefore either $g_1(X)$ or $g_2(X)$ is a unit in R and the proof is complete. □

Remark 3.9.37. The lemma is not true if the polynomial is not primitive. Let $f(X) = 2X \in \mathbb{Z}[X]$. Then the polynomial $f(X)$ is the product of the irreducible elements 2 and X in $\mathbb{Z}[X]$. Hence it is reducible in $\mathbb{Z}[X]$. On the other hand, since 2 is a unit in $\mathbb{Q}$, the polynomial $f(X)$ is irreducible in $\mathbb{Q}[X]$.

We are now ready for the proof of the main theorem(Theorem 3.9.31) of this section.

Proof. Let $f(X)$ be a non constant primitive polynomial in $R[X]$ and let $f(X) = f_1(X)f_2(X) \cdots f_r(X)$ be the product of irreducible polynomials in $\mathbb{F}[X]$.

We use Lemma 3.9.34 and write $f_i(X) = c_ig_i(X)$ where $g_i(X)$ are primitive in $R[X]$ and c_i are in $\mathbb{F}$. Further the polynomials $g_i(X)$ are irreducible in $R[X]$.(Why?) Hence $f(X) = c_1c_2 \cdots c_r g_1(X)g_2(X) \cdots g_r(X)$. Since $f(X)$ and the product $g_1(X)g_2(X) \cdots g_r(X)$ are primitive polynomials in $R[X]$, the constant $c_1c_2 \cdots c_r$ is a unit in R. Thus we have shown that the factorization

$$f(X) = f_1(X)f_2(X) \cdots f_r(X) = (c_1c_2 \cdots c_r)g_1(X)g_2(X) \cdots g_r(X)$$

is an irreducible factorization in $R[X]$.

We will now prove the uniqueness of the irreducible factorization of a nonunit in $R[X]$. Let $f(X)$ be a primitive polynomial in $R[X]$ and

$$f(X) = f_1(X)f_2(X)\cdots f_r(X) = g_1(X)g_2(X)\cdots g_s(X)$$

where $f_i(X)$ and $g_j(X)$ are primitive irreducible polynomials in $R[X]$.

It follows from Lemma 3.9.36 that the polynomials $f_i(X)$ and $g_j(X)$ are irreducible in $\mathbb{F}[X]$ too. Since $\mathbb{F}[X]$ is a UFD, by the uniqueness of irreducible factorization in $\mathbb{F}[X]$, we see that $r = s$ and after relabeling if necessary, $f_i(X) = c_i g_i(X)$ for $1 \leq i \leq r$ where c_i is a unit in $\mathbb{F}$.

If $c_i = d_i/e_i$ then $e_i f_i(X) = d_i g_i(X)$. Now we use the fact that $f_i(X)$ and $g_i(X)$ are primitive to conclude that d_i and e_i are associates in R. Hence c_i is a unit in R and $f_i(X)$ and $g_i(X)$ are associates in $R[X]$. This completes the proof. □

Corollary 3.9.38. *Let R be a UFD. Then $R[X_1, X_2, \ldots, X_n]$ is a UFD.*

Proof. The proof follows from the fact that the polynomial ring $R[X, Y]$ in two variables X and Y is isomorphic to the polynomial ring $R[X][Y]$ and induction on n. □

3.9.4 Irreducible polynomials in $R[X]$

We have proved that $R[X]$ is a UFD if R is a UFD. Now we would like to have an efficient method to factorize a polynomial and check if the factors are irreducible or not. In this section we will prove some results in this direction.

Lemma 3.9.39. *Let R be a commutative ring and I be a proper ideal in R. Let $f(X)$ be a nonconstant monic polynomial such that $\overline{f(x)}$ is irreducible in $(R/I)[X]$. Then $f(X)$ is irreducible in $R[X]$.*

Proof. Let $f(X) = X^n + a_{n-1}X^{n-1} + \cdots + a_1X + a_0$ be a monic polynomial in $R[X]$ and $\overline{f(X)} = X^n + \overline{a_{n-1}}X^{n-1} + \cdots + \overline{a_1}X + \overline{a_0}$ be the image in $(R/I)[X]$. Since $f(X)$ is a nonconstant monic polynomial in $R[X]$, the image $\overline{f(X)}$ is a nonconstant polynomial in $(R/I)[X]$.

Suppose

$$f(X) = f_1(X)f_2(X) = (a_rX^r + g_1(X))(b_sX^s + g_2(X))$$

is a factorization in $R[X]$ where $\deg g_1(X) < r$, $\deg g_2(X) < s$ and $r + s = n$. Since the polynomial $f(X)$ is monic, the coefficients a_r and b_s are units in R and hence the image $\overline{f_1(X)}$ and $\overline{f_2(X)}$ are nonzero and nonunits in $(R/I)[X]$. This is a contradiction to our assumption that the polynomial $\overline{f(X)}$ is irreducible in $(R/I)[X]$. □

A slight variation of the above result is the following lemma which is more useful.

Lemma 3.9.40. *Let R be a UFD and let $f(X) = \sum_{i=1}^{n} a_iX^i$ be a polynomial in $R[X]$. Assume that there exists a prime p in R such that*

1. *p does not divide a_n and*

2. *$\overline{f(X)}$ is irreducible in $(R/(p))[X]$*

then $f(X)$ is irreducible in $R[X]$.

Remark 3.9.41. The condition in this lemma is less stringent than in the previous lemma: The prime element p does not divide a_n means that the polynomial $\overline{f(X)}$ is a monic polynomial in $(R/(p))[X]$. However in the previous lemma, we wanted $f(X)$ to be monic in $R[X]$.

Proof. Let $\overline{f(X)} = \bar{a}_n X^n + \bar{a}_{n-1} X^{n-1} + \cdots + \bar{a}_0$ be the image of f in $(R/(p))[X]$. Since p does not divide a_n, the polynomial $\overline{f(X)}$ is nonzero and nonconstant in $(R/(p))[X]$.

Suppose the polynomial $f(X)$ is reducible in $R[X]$ and

$$f(X) = g(X)h(X) = (g_1(X) + b_r X^r)(h_1(X) + c_s X^s).$$

We re write this as $f(X) = \psi(X) + b_r c_s X^{r+s}$. Since p does not divide $a_n = b_r c_s$, it follows that p does not divide b_r and c_s. Therefore

$$\overline{f(X)} = \overline{g(X)h(X)}$$

is a nonzero in $(R/(p))[X]$, a contradiction to our assumption that $\overline{f(X)}$ is irreducible in $(R/(p))[X]$. □

This seems to be a useful method. But there are polynomials in $\mathbb{Z}[X]$ which are reducible in $\mathbb{Z}_p[X]$ for every prime p. For example the polynomial $f(X) = X^4 - 10X^2 + 1$ is one such.

Example 3.9.42. The polynomial $f(X) = X^4 - 10X^2 + 1$ is irreducible in $\mathbb{Z}[X]$ and reducible over $\mathbb{Z}_p$ for every prime number p.

Suppose $f(X)$ is reducible and $f(X) = (X^2 + aX + b)(X^2 + cX + d)$ over $\mathbb{Z}[X]$. By comparing the coefficients, we get

$$a + c = 0, \quad b + d + ac = -10, \quad ad + bc = 0 \quad \text{and} \quad bd = 1.$$

Using the two equations $a + c = 0$ and $ad + bc = 0$, we conclude that $a(d - b) = 0$. Therefore either $a = 0$ or $d = b$. Since $db = 1$, it follows that $b = d = 1$ or $b = d = -1$.

If $a = 0$, then the equation $b + d + ac = -10$ becomes $b + d = -10$, a contradiction.

If $a \neq 0$ and $b = d = 1$, then $b + d + ac = -10$ becomes $a^2 = 12$, a contradiction. Similarly, if $b = d = -1$, then we get $a^2 = 8$, a contradiction. Thus the polynomial $f(X)$ is irreducible over $\mathbb{Z}[X]$.

We will now show that $f(X)$ is reducible in $\mathbb{Z}_p[X]$ for every prime number p.

Let p be a prime number.

Suppose 2 is a square in $\mathbb{Z}_p$ and let $\alpha \in \mathbb{Z}_p$ be such that $\alpha^2 = [2]$ in $\mathbb{Z}_p$. Therefore $f(X) = \left(X^2 + 2\alpha X - 1\right)\left(X^2 - 2\alpha X - 1\right)$ in $\mathbb{Z}_p[X]$.

If 3 is a square in $\mathbb{Z}_p$ and $\beta \in \mathbb{Z}_p$ is such that $\beta^2 = [3]$, then we can write $f(X) = (X^2 + 2\beta X + 1)(X^2 - 2\beta X + 1)$ in $\mathbb{Z}_p[X]$.

If neither 2 or 3 is a square in $\mathbb{Z}_p$, then 6 is a square in $\mathbb{Z}_p$. (For a proof see [9] or [4]). Assume $\gamma \in \mathbb{Z}_p$ is such that $\gamma^2 = [6]$ in $\mathbb{Z}_p$. Then $f(X) = (X^2 - (5 + 2\gamma))(X^2 - (5 - 2\gamma))$ in $\mathbb{Z}_p[X]$.

In spite of the example above, we can use Lemma 3.9.40 many times. For example, if $R = \mathbb{Z}$ and p is a prime number, the ring $\mathbb{Z}_p$ is finite. Hence there are only finitely many polynomials of given degree in $\mathbb{Z}_p[X]$. We may list all of them and verify which one of them are irreducible.

In $\mathbb{Z}_2[X]$, there is only one irreducible polynomial of degree 2, namely the polynomial $X^2 + X + 1$.

Let $f(X) = X^2 + aX + b$ be an irreducible polynomial of degree 2 in $\mathbb{Z}_2[X]$. If $a = b = 0$, then $f(X) = X^2$ is reducible. If $a = 0$ and $b = 1$, then $f(X) = X^2 + 1 = (X + 1)^2$. If $a = 1$ and $b = 0$, then $f(X) = X(X + 1)$ is reducible. Thus $X^2 + X + 1$ is the only irreducible polynomial of degree 2 in $\mathbb{Z}_2[X]$.

We may also try to list other irreducible polynomials of small degree.

Let $f(X) = X^4 - 10X^3 + 24X^2 - 17X + 19$ in $\mathbb{Z}[X]$. Then $f(X)$ is irreducible in $\mathbb{Q}[X]$ can be proved using what we have done above.

The image of $f(X)$ in $\mathbb{Z}_2[X]$ is $X^4 + X + 1$ and this is irreducible in $\mathbb{Z}_2[X]$. Hence by Lemma 3.9.40, the polynomial $f(X)$ is irreducible in $\mathbb{Z}[X]$. Since $f(X)$ is primitive, it follows from Lemma 3.9.36 that $f(X)$ is irreducible in $\mathbb{Q}[X]$.

Let us do one more example. Let $f(X) = X^3 + 9X + 6$. The image of $f(X)$ in $\mathbb{Z}_3[X]$ is X^3. Note that $\overline{f(X)} = X^3$ is reducible in $\mathbb{Z}_3[X]$ and Lemma 3.9.40 is not applicable. Still we shall show that $f(X)$ is irreducible in $\mathbb{Z}[X]$. Therefore the Lemma 3.9.40 gives a sufficient condition but not a necessary condition to check the irreducibility of a polynomial.

Suppose $f(X)$ is reducible in $\mathbb{Z}[X]$ and let

$$f(X) = (aX + b)(\alpha X^2 + \beta X + \gamma) = (a\alpha)X^3 + (a\beta + b\alpha)X^2 + (a\gamma + b\beta)X + b\gamma.$$

Now $\overline{aX + b}$ divides X^3 in $\mathbb{Z}_3[X]$ and

$$\begin{aligned} a\alpha &\equiv 1 \pmod 3 \\ a\beta + b\alpha &\equiv 0 \pmod 3 \\ a\gamma + b\beta &\equiv 0 \pmod 3 \text{ and} \\ b\gamma &\equiv 0 \pmod 3. \end{aligned}$$

We multiply the equation $a\gamma + b\beta \equiv 0 \pmod 3$ by b and conclude that $b^2\beta \equiv 0 \pmod 3$. Now multiply the equation $a\beta + b\alpha \equiv 0 \pmod 3$ by b^2 and conclude that $b^3\alpha \equiv 0 \pmod 3$. Since $a\alpha \equiv 1 \pmod 3$, it follows that b^3 and hence $b \equiv 0 \pmod 3$. Therefore the equation $a\gamma + b\beta \equiv 0 \pmod 3$ be comes $a\gamma \equiv 0 \pmod 3$. Again, we use the equation $a\alpha \equiv 1 \pmod 3$ to conclude that $\gamma \equiv 0 \pmod 3$. This shows that $b\gamma$ is a multiple of 9, a contradiction. Thus

$f(X)$ is irreducible in $\mathbb{Z}[X]$. Since $f(X)$ is monic, by Lemma 3.9.36, it is irreducible in $\mathbb{Q}[X]$ too.

The example above illustrates that we need to have some efficient methods to test irreducibility of a polynomial in $R[X]$. First we prove a useful lemma.

Lemma 3.9.43. *Let R be an integral domain and $f(X)$ be a nonconstant polynomial in $R[X]$ such that the polynomial $f(X)$ divides the monomial X^n in $R[X]$. Then f is a monomial in $R[X]$.*

Proof. Let $f(X)$ be a nonconstant polynomial in $R[X]$ dividing X^n. We will show that $f(X)$ is a monomial by induction on n.

Suppose $f(X)$ divides the monomial X. Let $g(X)$ in $R[X]$ be such that $X = f(X)g(X)$. Since $f(X)$ is nonconstant, by comparing the degrees, we conclude that the degree of $f(X)$ is 1 and $g(X)$ is a constant polynomial, say $c \neq 0$. Therefore $X = cf(X)$ and let $f(X) = aX + b$. We evaluate the polynomial at 0 and arrive at the equation $0 = cf(0) = cb$. Since R is an integral domain, it follows that $f(X) = aX$ and it is a monomial.

Now we assume that if $f(X)$ is a polynomial that divides X^{n-1} for some $n \geq 2$, then $f(X)$ is a monomial.

Let $f(X)$ be a polynomial dividing the monomial X^n and let $g(X)$ be a polynomial in $R[X]$ such that $X^n = f(X)g(X)$. The polynomial X^n has exactly n-zeros, namely $X = 0$. Therefore $0 = f(0)g(0)$. Since R is an integral domain, we get that either $f(0) = 0$ or $g(0) = 0$. If $g(0) = 0$, we can write $g(X) = Xg_1(X)$. By substituting this in the equation $X^n = f(X)g(X)$, we get $X^n = Xf(X)g_1(X)$. Since $R[X]$ is an integral domain, it follows that $X^{n-1} = f(X)g_1(X)$. By induction hypothesis, the polynomials $f(X)$ and $g_1(X)$ are both monomials.

If $f(0) = 0$, we will arrive at the equation $X^{n-1} = f_1(X)g(X)$. By induction hypothesis, we will get $f_1(X)$ to be a monomial and hence $f(X)$ will be a monomial. This completes the proof. $\square$

The above lemma is false, if the ring R is not an integral domain as the following example shows.

Example 3.9.44. Let $R = \mathbb{Z}_4[X]$ and $f(X) = X + 2$. We can write $f(X)^2 = (X+2)^2 = X^2$ in $\mathbb{Z}_4[X]$. Thus the polynomial $f(X)$ divides the monomial X^2 but $f(X)$ is not a monomial.

Let $f(X) = \sum_{i=0}^{n} a_i X^i$ be a reducible polynomial in $\mathbb{Z}[X]$ and let $f(X) = g(X)h(X)$ where $g(X) = \sum_{j=0}^{r} b_j X^j$ and $h(X) = \sum_{l=0}^{s} c_l X^l$ are nonconstant polynomials in $\mathbb{Z}[X]$. We multiply $g(X)$ and $h(X)$ and write it as

$$f(X) = b_0c_0 + (b_0c_1 + b_1c_0)X + \cdots + b_rc_sX^{r+s}.$$

Let p be a prime number such that p does not divide a_n. Then p does not divide b_r and c_s.

If p divides a_i for $0 \leq i \leq (n-1)$, then

$$\overline{f(X)} = \bar{a}_n X^n = \overline{g(X)}\,\overline{h(X)}$$

and $\overline{f(X)}$ is a monomial in $\mathbb{Z}_p[X]$. The polynomials $\overline{g(X)}$ and $\overline{h(X)}$ divide the polynomial $\overline{f(X)}$. Hence it follows from Lemma 3.9.43 that the polynomials $\overline{g(X)}$ and $\overline{h(X)}$ are also monomials.

This means that the prime number p divides the coefficients b_i for $0 \leq i \leq (r-1)$ and c_j for $0 \leq j \leq (s-1)$. Therefore p^2 divides $b_0c_0 = a_0$. If we now assume that p^2 does not divide a_0, then $f(X)$ can not be reducible.

This proves one of the important result in checking the irreducibility of a polynomial in $\mathbb{Z}[X]$.

Theorem 3.9.45 (Eisenstein irreducibility criterion). *Let $f(X) = a_0 + a_1X + \cdots + a_nX^n$ be a nonconstant polynomial in $\mathbb{Z}[X]$. Let p be a prime number such that*

1. *p divides a_i for $0 \leq i \leq (n-1)$,*
2. *p does not divide a_n and*
3. *p^2 does not divide a_0.*

Then $f(X)$ is irreducible in $\mathbb{Z}[X]$.

Further if the polynomial is primitive then it is also irreducible in $\mathbb{Q}[X]$.

Remark 3.9.46. The result is true in a UFD as can be seen from the proof.

We will give a proof of this result over a UFD.

Suppose $f(X)$ is reducible in $R[X]$ and $f(X) = f_1(X)f_2(X)$ where $f_1(X) := b_0 + b_1X + b_2X^2 + \ldots + b_rX^r$ and $f_2(X) := c_0 + c_1X + c_2X^2 + \ldots + c_sX^s$ are polynomials in $R[X]$. Then

$$f(X) = f_1(X)f_2(X) = b_0c_0 + (b_0c_1 + b_1c_0)X + \ldots + a_rb_sX^{r+s}.$$

- The prime p does not divide $a_n = b_rc_s$ shows that p does not divide b_r and c_s.
- The prime p divides $a_0 = b_0c_0$ means that p divides either b_0 or c_0.

Let us assume that p divides b_0. We will now use our hypothesis that p divides a_i for $1 \leq i \leq n-1$ to prove our result.

Since p divides $a_1 = b_0c_1 + b_1c_0$, it follows that p divides $a_1 - b_0c_1 = b_1c_0$. If p does not divide c_0, then p divides b_1. We use this to conclude that p divides $b_2c_0 = a_2 - (b_1c_1 + b_0c_2)$. Hence p divides b_2. Proceeding like this, we see that p divides $b_rc_0 = (a_r - (b_0c_r + b_1c_{r-1} + \ldots + b_{r-1}c_1))$. Since p does not divide c_0, it follows that p divides b_r and hence p divides $a_n = b_rc_s$. This is contradiction to our assumption that p does not divide a_n. Therefore p divides c_0. As a consequence $a_0 = b_0c_0$ is a multiple of p^2, a contradiction. Hence our first assumption that the polynomial $f(X)$ is reducible in $R[X]$ is wrong!

Examples 3.9.47. 1. The polynomial $f(X) = X^7 + 48X - 24$ is irreducible in $\mathbb{Q}[X]$.

Let $p = 3$. Then p does not divide $a_7 = 1$, divides $a_0 = -24, a_1 = 48$ and p^2 does not divide $a_0 = -24$. Therefore $f(X)$ is irreducible in $\mathbb{Z}[X]$. Since $f(X)$ is a primitive polynomial, it is irreducible in $\mathbb{Q}[X]$ as well.

2. Let $p \geq 2$ be a prime number. Then the polynomial $f(X) = X^n - p$ is irreducible in $\mathbb{Z}[X]$ for $n \geq 2$.

3. Let p be a prime number and $\Phi_p(X) = \frac{X^p-1}{X-1} = 1 + X + X^2 + \ldots + X^{p-1}$ the p-th cyclotomic polynomial. The polynomial $\Phi_p(X)$ is irreducible over $\mathbb{Z}$.

   We can't apply Eisenstein's irreducibility criterion directly. Let $f(X) = \Phi(x+1)$. Then

$$\begin{aligned} f(X) &= \frac{(X+1)^p - 1}{(X+1)-1} \\ &= X^{p-1} + \binom{p}{1} X^{p-2} + \ldots + \binom{p}{p-2} X + \binom{p}{p-1} \\ &= X^{p-1} + \binom{p}{1} X^{p-2} + \ldots + \binom{p}{p-2} X + p. \end{aligned}$$

   The constant coefficient is exactly p and p^2 does not divide p. The coefficient of X^{p-1} is 1 not divisible by p and all other coefficients are divisible by p. Hence by Eisenstein's irreducibility criterion the polynomial $f(X)$ is irreducible in $\mathbb{Z}[X]$. Since it is primitive it is also irreducible over $\mathbb{Q}$.

Exercise 3.9.48. Let R be a commutative ring with identity and $f(X)$ be a nonconstant, monic polynomial of degree less than or equal to 3. Then $f(X)$ is irreducible in $R[X]$ iff $f(X)$ does not have a zero in R.

We will do for polynomials of degree 3 and leave the easy case to the readers.

Let $f(X) = a_0 + a_1X + a_2X^2 + a_3X^3$ be a monic polynomial of degree 3.

Assume that α in R is a root of the polynomial $f(X)$. Then there is a polynomial $q(X)$ of degree 2 such that $f(X) = (X - \alpha)q(X)$. This proves that $f(X)$ is reducible in $R[X]$.

For the converse part we assume $f(X)$ to be monic and assume that $f(X)$ is reducible in $R[X]$. Then we can write $f(X) = g(X)h(X)$ where $g(X)$ and $h(X)$ are nonconstant polynomials in $R[X]$. We may assume that $g(X) = (aX + b)$ and $h(X) = (cX^2 + dX + e)$. Since a_3 is a unit, it follows that a and c are units in R. Therefore $\alpha = -a^{-1}b$ is a root of $g(X)$. As a consequence $\alpha = -a^{-1}b$ is a root of $f(X) = g(X)h(X) = a(X + a^{-1}b)h(X)$ too.

Remark 3.9.49. 1. The converse part of the above result is not true if the polynomial is not monic.

For example let $R = \mathbb{Z}_6$ and $f(X) = 2X^3 + 1$ in $\mathbb{Z}_6[X]$. Then $f(X) = (2X+1)(4X^2+4X+1) = (2X+1)^3$ and none of the factors have a root in $\mathbb{Z}_6$.

Let $R = \mathbb{Z}$ and $f(X) = 2X^3 + X^2 + X - 1$ in $\mathbb{Z}[X]$. The $f(X)$ is reducible in $\mathbb{Z}[X]$ and $f(X) = (2X-1)(X^2+X+1)$ $\mathbb{Z}[X]$. But $f(X)$ does not have a root in $\mathbb{Z}$.

2. The above result is not valid if degree of $f(X) \geq 4$. For example, the polynomial $f(X) = (X^2+1)(X^2+2)$ is reducible in $\mathbb{R}[X]$ but it does not have a zero in $\mathbb{R}$.

Exercise 3.9.50. Let $f(X)$ be a polynomial of degree 3 in $\mathbb{R}[X]$. Then $f(X)$ is reducible in $\mathbb{R}[X]$.

Exercise 3.9.51. Show that the following polynomials are irreducible over $\mathbb{Q}$.

1. $X^7 + 48X - 24$.

2. $3X^{19} - 7X^{15} + 49X^{10} - 28X^6 - 35$.

3. $2X^{17} - 18X^{12} + 24X^8 + 243X^6 - 36X^3 - 6$.

4. $X^6 + X^3 + 1$.

5. $X^4 + 1$.

Exercise 3.9.52. 1. Factorize the polynomial $f(X) = X^4 + 4$ in $\mathbb{Q}[X]$.

2. In which of the following rings, the polynomial $f(X) = X^3 + X^2 + X + 2$ is irreducible?

   (a) $\mathbb{R}[X]$

   (b) $\mathbb{C}[X]$

   (c) $\mathbb{Z}_p[X]$ for $p = 2, 3$ and 5

   (d) $\mathbb{Q}[X]$

3. Show that the polynomial $f(X) = 3X^2 + 4X + 3$ is reducible in $\mathbb{Z}_5[X]$.

Exercise 3.9.53. 1. Let $f(X) = aX^2 + bX + c$ in $\mathbb{R}[X]$ be an irreducible polynomial of degree 2. Show that the ring $\mathbb{R}[X]/(f(X))$ is isomorphic to the field of complex numbers.

2. Let $f_1(X) = X^2 + 1$ and $f_2(X) = X^2 + X + 1$. Show that the quotient rings $\mathbb{R}[X]/(f_1(X))$ and $\mathbb{R}[X]/(f_2(X))$ are isomorphic.

3. Let $f(X)$ be an irreducible polynomial in $\mathbb{R}[X]$. Identify the quotient ring $\mathbb{R}[X]/(f(X))$.

4. Let $f_1(X)$ and $f_2(X)$ be two irreducible polynomials in $\mathbb{R}[X]$. Are the two rings $\mathbb{R}[X]/(f_1(X))$ and $\mathbb{R}[X](f_2(X))$ isomorphic?

5. Let $f_1(X) = 1 + X + X^2$ and $f_2(X) = 1 + X + X^2 + X^3 + X^4$. Are the two rings $\mathbb{Q}[X]/(f_1(X))$ and $\mathbb{Q}[X]/(f_2(X))$ isomorphic?

3.10 Group Rings

Let R be a commutative ring with identity and G be a finite group. Now we construct a ring called the *Group ring* over the ring R and the group G.

This ring denoted by $R[G]$ is the set of formal finite sum $\sum_{g\in G} a_g g$; here, for each $g \in G$, the element a_g is in R. That is

$$R[G] := \left\{ \sum_{g\in G} a_g g : g \in G \text{ and } a_g \in R \right\}$$

and the addition and multiplication are defined by

$$\begin{aligned}\left(\sum_{g\in G} a_g g\right) + \left(\sum_{h\in G} b_h h\right) &:= \sum_{g\in G} (a_g + b_g)\, g \\ \left(\sum_{g\in G} a_g g\right)\left(\sum_{h\in G} b_h h\right) &:= \sum_{g,h\in G} (a_g b_h)\, gh\end{aligned}$$

for $\sum_{g\in G} a_g g$ and $\sum_{h\in G} b_h h$ in $R[G]$.

It is easy to show that addition is associative: If $\sum_{g\in G} a_g g$, $\sum_{h\in G} b_h h$ and $\sum_{k\in G} c_k k$ are in $R[G]$, then

$$\begin{aligned}\sum_{g\in G} a_g g + \left(\sum_{h\in G} b_h h + \sum_{k\in G} c_k k\right) &= \sum_{g\in G} a_g g + \sum_{h\in G} (b_h + c_h)h) \\ &= \sum_{g\in G} (a_g + (b_g + c_g))\, g \\ &= \sum_{g\in G} ((a_g + b_g) + c_g)\, g \\ &= \sum_{g\in G} (a_g + b_g)\, g + \sum_{k\in G} c_k k \\ &= \left(\sum_{g\in G} a_g g + \sum_{h\in G} b_h h\right) + \sum_{k\in G} c_k k.\end{aligned}$$

This proves the associativity law for addition. Other axioms are easy to verify: $\sum_{g\in G} 0\, g$ is the additive identity.

If $\sum_{g\in G} a_g g$ is an element in $R[G]$ then $\sum_{g\in G}(-a_g)g$ is the additive inverse of $\sum_{g\in G} a_g g$. Thus $(R[G], +)$ is an abelian group.

Similarly we can check the axioms for multiplication also. The element $1.e + \sum_{g\neq e\in G} 0.g$ is the multiplicative identity.

The ring $R[G]$ can also be thought of as the set of all functions from the group G taking values in the ring R. That is

$$R[G] := \{f : G \to R : f \text{ is a function }\}$$

with the addition and multiplication as described below.

If $f : G \to R$ is a function and if $f(g) := a_g$, then f can be represented by the tuple $(a_g)_{g\in G}$ and we denoted this as the formal sum $\sum_{g\in G} a_g g$. In this identification, the addition of two elements in $R[G]$ correspond to the sum of functions of $R[G]$.

Let $f_1 \leftrightarrow \sum_{g\in G} a_g g$ and $f_2 \leftrightarrow \sum_{h\in G} b_h h$, then the product was defined as

$$\left(\sum_{gng} a_g g\right)\left(\sum_{h\in G} b_h h\right) = \left(\sum_{g,h\in G} (a_g b_h) gh\right).$$

For two elements g and h in G, let $x = gh$. Then $h = g^{-1}x$. We use this and rewrite the product

$$\begin{aligned}\left(\sum_{gng} a_g g\right)\left(\sum_{h\in G} b_h h\right) &= \left(\sum_{g,h\in G} (a_g b_h) gh\right)\\ &= \sum_{x\in G}\left(\sum_{g\in G} a_g b_{g^{-1}x}\right) x.\end{aligned}$$

This shows that the this product is not same as the product of two functions. This product is called the *convolution* of the two functions f_1 and f_2 and it is denoted by $f_1 * f_2$. That is $f_1 * f_2 : G \to R$ is the function defined by $(f_1 * f_2)(x) := \sum_{g\in G} a_g b_{g^{-1}x}$.

Remark 3.10.1.

1. The group G is contained in $R[G]$ via the map $g \mapsto 1.g$. Every element $g \in G$ is a unit in $R[G]$. For $(1.g)(1.g^{-1}) = 1.(g.g^{-1}) = 1.e$.

2. Suppose $|G| \geq 2$. Let $g \in G$ be an element of order $k \geq 2$. Then $g^k = e$. Therefore

$$0 = 1.e - 1.g^k = (1.e - 1.g)(1.e + 1.g + 1.g^2 + \cdots + 1.g^{k-1}).$$

Thus $1 - g$ is a zero divisor in $R[G]$ and this proves that if G is a group of order $n \geq 2$, then the ring $R[G]$ has zero divisors.

3. If G is abelian then $R[G]$ is a commutative ring and conversely.

Chapter 4

Field Theory

Let $a(\neq 0)$ and b be two rational numbers and $f(X) = aX + b$ be a linear polynomial. Then the polynomial $f(X)$ has only one root namely $-\frac{b}{a}$. This root lies in $\mathbb{Q}$.

Let b and c be two rational numbers and $f(X) = X^2 + bX + c$ be a quadratic polynomial. We know that the roots of the polynomial $f(X)$ are $\frac{-b \pm \sqrt{b^2 - 4c}}{2}$. The roots are found by performing the algebraic operations of addition, subtraction, multiplication, division of the coefficients of the polynomial $f(X)$ and extracting roots.

Now we want to know if we can find all the roots of polynomials of degree bigger than 2 by the algebraic operations mentioned above.

In this chapter we will show that the roots of polynomials of degree 3 and 4 can be found by this method. However starting with polynomials of degree 5 and above we can't find the roots by this method. This is a very important result due to Galois and Abel.

In this chapter we prove the results due to Galois and Abel and this is done without going too much in to general field theory.

4.1 Field Extensions

4.1.1 Definition and Examples

Definition 4.1.1. A field $\mathbb{E}$ is said to be an extension of a field $\mathbb{F}$ if there exists a ring homomorphism $\sigma : \mathbb{F} \to \mathbb{E}$.

If a field $\mathbb{E}$ is an extension of a field $\mathbb{F}$, we denote it by $\mathbb{E}/\mathbb{F}$.

Exercise 4.1.2. Let $\mathbb{F}$ be a field and R be a ring and $\sigma : \mathbb{F} \to R$ be a homomorphism. Show that

1. $\sigma : \mathbb{F} \to R$ is one-one,

2. $\sigma(a)$ is a unit in R, if a in $\mathbb{F}$ is nonzero and

3. the subring $\sigma(\mathbb{F})$ is a field.

We have seen in Chapter 3 that kernel of a ring homomorphism is an ideal. Since $\mathbb{F}$ is a field, it has only two ideals, namely the full ring $\mathbb{F}$ and (0), the zero ideal. Therefore kernel of σ is the ideal (0) and σ is one-one.

Let $a \in \mathbb{F}$ be nonzero. Then $1 = \sigma(1) = \sigma(aa^{-1}) = \sigma(a)\sigma(a^{-1})$. Hence $\sigma(a)$ is a unit in R and $\sigma(\mathbb{F})$ is a field.

Remark 4.1.3. If a field $\mathbb{E}$ is an extension of a field $\mathbb{F}$ and $\sigma : \mathbb{F} \to \mathbb{E}$ is a homomorphism, we identify the elements of the field $\mathbb{F}$ with the elements in the image $\sigma(\mathbb{F})$. If x is an element in the field $\mathbb{F}$, we write the image $\sigma(x)$ also as x.

Example 4.1.4.

1. The field $\mathbb{R}$ of real numbers is an extension of the field $\mathbb{Q}$ of rational numbers.

2. The field $\mathbb{C}$ of complex numbers is an extension of the field of real numbers $\mathbb{R}$.

3. Let p be a prime number. Then $\mathbb{Q}[\sqrt{p}] := \{a + b\sqrt{p} : a, b \in \mathbb{Q}\}$ is a field containing the field of rational numbers $\mathbb{Q}$. Hence $\mathbb{Q}[\sqrt{p}]$ is an extension of $\mathbb{Q}$.

   We have already shown in Chapter 3 that $\mathbb{Q}[\sqrt{p}]$ is a field.

4. Let D be a square free integer and let $\mathbb{E} = \mathbb{Q}[\sqrt{D}]$. Then as in the above example $\mathbb{E}$ is a field it is an extension of $\mathbb{Q}$.

5. Let $\zeta \neq 1$ be a cube root of unity and let $\mathbb{E} = \mathbb{Q}[\zeta]$. Then $\mathbb{E}$ is an extension of $\mathbb{Q}$.

   We have already seen that $\mathbb{E}$ is a ring. Since $\zeta^3 = 1$ and $\zeta \neq 1$, it follows that $1 + \zeta + \zeta^2 = 0$.

   Let $\varphi : \mathbb{Q}[X] \to \mathbb{E}$ be the homomorphism defined by $\varphi(X) = \zeta$. This homomorphism is on to. The polynomial $f(X) = 1 + X + X^2$ is in $\ker(\varphi)$. Since the polynomial $f(X)$ is irreducible over $\mathbb{Q}$, the principal ideal generated by $f(X)$ is a maximal ideal in $\mathbb{Q}[X]$ and it is contained in $\ker(\varphi)$. Therefore $\ker(\varphi) = (f(X))$. By Fundamental theorem of homomorphism for rings $\mathbb{Q}[X]/(f(X))$ is isomorphic to $\mathbb{E}$. Since $\mathbb{Q}[X]/(f(X))$ is a field, the ring $\mathbb{E}$ is also a field and it is an extension of $\mathbb{Q}$.

6. Let $\mathbb{K}$ be a field and $\mathbb{E} := \mathbb{K}(X)$, the quotient field of the polynomial ring $\mathbb{K}[X]$. This field is also called the rational function field over $\mathbb{K}$. Then $\mathbb{K}(X)$ is an extension of the field $\mathbb{K}$.

7. Let $\mathbb{K}$ be a field and $\mathbb{E}$ be the ring of Laurent series over the field $\mathbb{K}$. We have shown in Theorem 3.3.27 of Chapter 3 that $\mathbb{E}$ is a field. Hence $\mathbb{E}$ is an extension of the field $\mathbb{K}$.

8. Let $\mathbb{K}$ be a field. Then the ring $\mathbb{K}[X]$ is a subring of $\mathbb{K}[\![X]\!]$. Hence the quotient field $\mathbb{K}((X))$ of the ring $\mathbb{K}[\![X]\!]$ is an extension of $\mathbb{K}(X)$.

Let $\mathbb{F}$ be a field and and $\mathbb{E}$ be an extension of $\mathbb{F}$.

Let x be an element in $\mathbb{F}$ and v be an element in the field $\mathbb{E}$. Then, since $\mathbb{F}$ is a subfield of $\mathbb{E}$, the product $x.v$ is also in $\mathbb{E}$. This defines a scalar multiplication of the field $\mathbb{F}$ on the abelian group $(\mathbb{E}, +)$. With this product the abelian group $(\mathbb{E}, +)$ becomes a vector space over the field $\mathbb{F}$.

We denote the dimension of the vector space $\mathbb{E}$ over $\mathbb{F}$ by $[\mathbb{E} : \mathbb{F}]$.

Definition 4.1.5. Let $\mathbb{E}$ be an extension of the field $\mathbb{F}$.

If the dimension $[\mathbb{E} : \mathbb{F}]$ is finite, then the extension $\mathbb{E}/\mathbb{F}$ is called a finite extension and $[\mathbb{E} : \mathbb{F}]$ is called the degree of the extension.

If the dimension $[\mathbb{E} : \mathbb{F}]$ is not finite, then the extension $\mathbb{E}/\mathbb{F}$ is called an infinite extension

Example 4.1.6.

1. The extension $\mathbb{C}/\mathbb{R}$ is a finite extension.

   If $z = x + iy$ is in $\mathbb{C}$, then $z = x \cdot 1 + y \cdot i$. Thus $\mathbb{C} = \mathbb{R}[i]$ and $[\mathbb{C} : \mathbb{R}] = 2$.

2. Let p be a prime number. Then the vector space $\mathbb{Q}[\sqrt{p}]$ is of dimension 2 over $\mathbb{Q}$. Hence $\mathbb{Q}[\sqrt{p}]/\mathbb{Q}$ is a finite extension.

   The field $\mathbb{Q}[\sqrt{p}] := \{a + b\sqrt{p} : a, b \in \mathbb{Q}\}$ has a basis $\{1, \sqrt{p}\}$ over $\mathbb{Q}$.

3. Let $\zeta \neq 1$ be a cube root of unity. Then $\mathbb{Q}[\zeta]$ is a vector space of dimension 2 over $\mathbb{Q}$. Hence $\mathbb{Q}[\zeta]/\mathbb{Q}$ is a finite extension.

   The set $\{1, \zeta\}$ is a basis for the vector space $\mathbb{Q}[\zeta]$ over $\mathbb{Q}$.

4. The extension $\mathbb{R}/\mathbb{Q}$ is not a finite extension.

   Hint, The field $\mathbb{Q}$ is countable and $\mathbb{R}$ is uncountable.

5. Let $\mathbb{K}$ be a field. Then the extension $\mathbb{K}(X)/\mathbb{K}$ is not finite.

   For every $n \geq 2$, the set $\{1, X, X^2, \ldots, X^n\}$ is linearly independent over $\mathbb{K}$.

One of the first important and useful property about the extension of fields is the following proposition.

Proposition 4.1.7 (Tower Law). *Let $\mathbb{F}$, $\mathbb{E}$ and $\mathbb{K}$ be fields such that $\mathbb{F} \subseteq \mathbb{E} \subseteq \mathbb{K}$. Then $[\mathbb{K} : \mathbb{F}] = [\mathbb{K} : \mathbb{E}][\mathbb{E} : \mathbb{F}]$.*

Proof. Assume that one of the extensions $\mathbb{K}/\mathbb{E}$ or $\mathbb{E}/\mathbb{F}$ is not finite, say $\mathbb{K}/\mathbb{E}$.

Then, for every $n \geq 2$, there is a subset $S = \{v_1, v_2, v_3, \ldots, v_n\}$ in $\mathbb{K}$, that is linearly independent over $\mathbb{E}$. Since $\mathbb{F}$ is a subfield of $\mathbb{E}$, the set S is linearly independent over $\mathbb{F}$ as well. This proves that the extension $\mathbb{K}/\mathbb{F}$ is not finite. Hence the equality $[\mathbb{K} : \mathbb{F}] = [\mathbb{K} : \mathbb{E}][\mathbb{E} : \mathbb{F}]$ is valid.

Similarly we can show that $\mathbb{K}/\mathbb{F}$ is not a finite extension if $\mathbb{E}/\mathbb{F}$ is not finite.

We will now assume that both the extensions $\mathbb{K}/\mathbb{E}$ and $\mathbb{E}/\mathbb{F}$ are finite. Let $[\mathbb{K} : \mathbb{E}] = m$ and $[\mathbb{E} : \mathbb{F}] = n$.

Let $\{x_1, x_2, \ldots, x_m\}$ be a basis of $\mathbb{K}$ over $\mathbb{E}$ and $\{y_1, y_2, \ldots, y_n\}$ be a basis of $\mathbb{E}$ over $\mathbb{F}$. We claim that $\{x_i y_j : 1 \leq i \leq m \text{ and } 1 \leq j \leq n\}$ is a basis of the vector space $\mathbb{K}$ over the field $\mathbb{F}$.

Let v be an element of the vector space $\mathbb{K}$. Since $\{x_1, x_2, \ldots, x_m\}$ is a basis of $\mathbb{K}$ over the field $\mathbb{E}$, there exist scalars $\alpha_1, \alpha_2, \ldots, \alpha_m$ in $\mathbb{E}$ such that $v = \sum_{i=1}^m \alpha_i x_i$. Since $\{y_1, y_2, \ldots, y_n\}$ is a basis of $\mathbb{E}$ over $\mathbb{F}$, for each α_i, there exist scalars β_{ji} in the field $\mathbb{F}$ such that $\alpha_i = \sum_{j=1}^n \beta_{ji} y_j$. Hence

$$v = \sum_{i=1}^{m} \alpha_i x_i = \sum_{i,j} \beta_{ji} y_j x_i.$$

Thus the set $\{x_i y_j : 1 \leq i \leq m \text{ and } 1 \leq j \leq n\}$ generates the vector space $\mathbb{K}$ over $\mathbb{F}$.

Let θ_{ji} for $1 \leq i \leq m$ and $1 \leq j \leq n$ be scalars in $\mathbb{F}$ such that the linear combination $\sum_{i,j} \theta_{ji} x_i y_j = 0$. We rewrite this sum as $\sum_{i=1}^m \left(\sum_{j=1}^n \theta_{ji} y_j \right) x_i = 0$. The elements $\sum_{j=1}^n \theta_{ji} y_j$ are in $\mathbb{E}$. Since the set $\{x_i : 1 \leq i \leq m\}$ is a basis of the vector space $\mathbb{K}/\mathbb{F}$ we conclude that $\sum_{j=1}^n \theta_{ji} y_j = 0$ for each $i = 1, 2, \ldots, n$. Since $\{y_j : 1 \leq j \leq n\}$ is a basis of $\mathbb{E}$ over $\mathbb{F}$, we conclude that $\theta_{ji} = 0$ for each $j = 1, 2, \ldots, m$. Hence $\theta_{ji} = 0$ for all $i = 1, 2, \ldots, n$ and $j = 1, 2, \ldots, m$ and this proves that the set $\{x_i y_j : 1 \leq i \leq m \text{ and } 1 \leq j \leq n\}$ is linearly independent over $\mathbb{F}$. Thus we have shown that $[\mathbb{K} : \mathbb{F}] = mn = [\mathbb{K} : \mathbb{E}][\mathbb{E} : \mathbb{F}]$. □

Definition 4.1.8. Let $\mathbb{K}$ be an extension of a field $\mathbb{F}$ and S be a subset of $\mathbb{K}$. The smallest subring containing $\mathbb{F}$ and the set S, denoted by $\mathbb{F}[S]$, is called the subring generated by the set S over $\mathbb{F}$.

The smallest field containing the set S and $\mathbb{F}$ in $\mathbb{K}$ is called the subfield generated by S and it is denoted by $F(S)$.

Example 4.1.9.

1. Let $\mathbb{K} = \mathbb{R}$ and $\mathbb{F} = \mathbb{Q}$. If $S = \{\sqrt{2}\}$, then the subring generated by S is the ring $\mathbb{Q}[S] = \{a + b\sqrt{2} : a, b \in \mathbb{Q}\}$. Notice that the subring $\mathbb{Q}[\sqrt{2}]$ is also a field(why?) and hence $\mathbb{Q}[\sqrt{2}] = \mathbb{Q}(\sqrt{2})$.

2. Let $\mathbb{K} = \mathbb{R}$ and $\mathbb{F} = \mathbb{Q}$. Let $S = \{\pi\}$. Then the subring $\mathbb{Q}[S]$ is $\mathbb{Q}[\pi] = \{a_0 + a_1\pi + \cdots + a_n\pi^n : a_i \in \mathbb{Q} \text{ for } 0 \leq i \leq n \text{ and } n \in \mathbb{N} \cup \{0\}\}$ and the subfield $\mathbb{Q}(\pi)$ is the quotient ring of $\mathbb{Q}[\pi]$.

3. Let $\mathbb{F}$ be a field and $\mathbb{K} = \mathbb{F}(X)$. If $S = \{X\}$, then $\mathbb{F}[S]$ is the polynomial ring $\mathbb{F}[X]$ and $\mathbb{F}(S)$ is the field $\mathbb{K}$.

Exercise 4.1.10.

1. Let $n \geq 2$ and $\alpha = \sqrt[n]{2}$. Is the subring $\mathbb{Q}[\alpha]$, generated by α a subfield of $\mathbb{R}$?

2. Let α be as in the previous exercise and $T := \{\alpha^k : 1 \leq k \leq n-1\}$.
   (a) Is the subring $\mathbb{Q}[T]$ a field?
   (b) Is the subring $\mathbb{Q}[\alpha^2]$ a field? Is it a subring of $\mathbb{Q}[\alpha]$?

3. Let $S = \{\sqrt{2}, \sqrt{3}\}$ and $T = \{\sqrt{2} + \sqrt{3}\}$. Show that $\mathbb{Q}[S] = \mathbb{Q}[T]$.

   It is easy to see that $\mathbb{Q}[T]$ is contained in $\mathbb{Q}[S]$.
   We will now show that $\sqrt{2}$ and $\sqrt{3}$ are in $\mathbb{Q}[T]$ as follows:

   - $\sqrt{2}\sqrt{3} = \frac{(\sqrt{2}+\sqrt{3})^2-5}{2}$. Thus $\sqrt{2}\sqrt{3}$ is in $\mathbb{Q}[T]$.

- $\sqrt{2}\sqrt{3}(\sqrt{2}+\sqrt{3}) = 3\sqrt{2}+2\sqrt{3}$ is in $\mathbb{Q}[T]$. As a consequence
- $\sqrt{2} = 3\sqrt{2} + 2\sqrt{3} - 2(\sqrt{2} + \sqrt{3})$ is in $\mathbb{Q}[T]$. Hence $\sqrt{3}$ is also in $\mathbb{Q}[T]$ proving that $\mathbb{Q}[S] = \mathbb{Q}[T]$.

Exercise 4.1.11. Let $\mathbb{K}$ be an extension of a field $\mathbb{F}$. If S is a subset of $\mathbb{K}$, then show that $\mathbb{F}(S)$ is the quotient field of the ring $\mathbb{F}[S]$.

Definition 4.1.12. A field extension $\mathbb{K}/\mathbb{F}$ is called finitely generated if there exists a finite subset S in $\mathbb{K}$ such that $\mathbb{K} = \mathbb{F}(S)$.

Example 4.1.13. 1. The extension $\mathbb{C}/\mathbb{R}$ is finitely generated.

2. The extension $\mathbb{Q}[\sqrt{2}]/\mathbb{Q}$ is finitely generated.
3. Let ζ be a cube root of unity. Let $\mathbb{K} = \mathbb{Q}[\zeta]$. We have seen earlier that $\mathbb{K}/\mathbb{Q}$ is finitely generated.
4. The extension $\mathbb{R}/\mathbb{Q}$ is not finitely generated.

Exercise 4.1.14. Let $\mathbb{K}/\mathbb{F}$ be an extension. If $[\mathbb{K} : \mathbb{F}] < \infty$, show that $\mathbb{K}/\mathbb{F}$ is finitely generated.

Exercise 4.1.15. Let $\mathbb{F}$ be a field and $\mathbb{K} = \mathbb{F}(X)$ be the rational function field over $\mathbb{F}$. Show that the extension $\mathbb{K}/\mathbb{F}$ is a finitely generated extension but not a finite extension.

4.1.2 Algebraic Extensions

Now we introduce the notion of algebraic extension of a field.

Definition 4.1.16 (Algebraic Elements). Let $\mathbb{K}/\mathbb{F}$ be an extension. An element $\alpha \in \mathbb{K}$ is called algebraic over $\mathbb{F}$ if there exists a polynomial $f(X)$ in $\mathbb{F}[X]$ such that $f(\alpha) = 0$. That is the element α is a root of a polynomial in $\mathbb{F}[X]$.

If there is no polynomial $f(X)$ in $\mathbb{F}[X]$ such that $f(\alpha) = 0$, then α is called transcendental over the field $\mathbb{F}$.

Definition 4.1.17 (Algebraic Extension). An extension $\mathbb{K}/\mathbb{F}$ is called an algebraic extension, if every element α in $\mathbb{K}$ is algebraic over the field $\mathbb{F}$.

Example 4.1.18. 1. Let $\mathbb{F}$ be a field and $\mathbb{K} = \mathbb{F}$. Then the extension $\mathbb{K}/\mathbb{F}$ is algebraic.

Every element $\alpha \in \mathbb{F}$ is root of the polynomial $f(X) = X - \alpha$ in $\mathbb{F}[X]$.

2. Let $\mathbb{K} = \mathbb{C}$ and $\mathbb{F} = \mathbb{R}$. Then every element in $\mathbb{C}$ is algebraic over $\mathbb{R}$.

   Let $\alpha = a + ib$ be a nonzero complex number. Then α is a root of the polynomial $f(X) = X^2 - 2aX + (a^2 + b^2)$ in $\mathbb{R}[X]$

3. The element π in $\mathbb{R}$ is transcendental over $\mathbb{Q}$. We will not prove this here. For a proof we refer to [1] or [5].
4. Let $\mathbb{K}$ be a field. Then the element X in $\mathbb{K}(X)$ is transcendental over $\mathbb{K}$.

An important class of algebraic extensions are finite extensions as shown in the following proposition.

Proposition 4.1.19. *Let $\mathbb{K}/\mathbb{F}$ be a finite extension. Then it is an algebraic extension.*

Proof. Let $[\mathbb{K}:\mathbb{F}] = d$ be the degree of the extension. Let α be an element of $\mathbb{K}$. Then the set $\{1, \alpha, \alpha^2, \ldots, \alpha^d\}$ is linearly dependent. Hence there exist scalars $c_0, c_1, c_2, \ldots, c_d$ not all zero in $\mathbb{F}$ such that $c_0 + c_1\alpha + c_2\alpha^2 + \cdots + c_d\alpha^d = 0$. This proves that α is a root of the polynomial $c_0 + c_1X + \cdots + c_dX^d$ in $\mathbb{F}[X]$ and the proof is complete. □

Let $\mathbb{K}/\mathbb{F}$ be an extension and α be an element in $\mathbb{K}$. Then there exists a unique ring homomorphism $e_\alpha : \mathbb{F}[X] \to \mathbb{K}$ such that $e_\alpha(X) = \alpha$. This homomorphism is called the evaluation homomorphism and the image is the subring $\mathbb{F}[\alpha]$ in $\mathbb{K}$.

Let $I = \ker(e_\alpha)$, the kernel of the homomorphism e_α.

Assume e_α is one-one. Hence $e_\alpha(f(X)) = f(\alpha) \neq 0$ for every polynomial $f(X) \in \mathbb{F}[X]$. That is α is transcendental over the field $\mathbb{F}$.

Conversely suppose that α is transcendental over the field $\mathbb{F}$. This means that $f(\alpha) \neq 0$ for every polynomial $f(X) \in \mathbb{F}[X]$. Hence the homomorphism $e_\alpha : \mathbb{F}[X] \to \mathbb{K}$ is one one.

We have proved that the homomorphism e_α is one to one iff the element α in $\mathbb{K}$ is transcendental over $\mathbb{F}$.

Let us now assume that e_α is not one to one. Then by definition there exists a polynomial $f(X) \in \mathbb{F}[X]$ such that $e_\alpha(f(X)) = f(\alpha) = 0$. Thus α is algebraic over $\mathbb{F}$

Conversely suppose α is algebraic over $\mathbb{F}$. Then there exists a polynomial $f(X)$ in $\mathbb{F}[X]$ such that $f(\alpha) = 0$. Hence e_α is not one one.

This proves that the kernel I nonzero iff the element α is algebraic over $\mathbb{F}$.

Let us now assume that the element α in $\mathbb{K}$ is algebraic over $\mathbb{F}$ and let $I = \ker(e_\alpha)$. Since $\mathbb{F}[X]$ is a PID, there exists a polynomial $f(X)$ in $\mathbb{F}[X]$ such that $I = (f(X))$ and $f(\alpha) = 0$.

Let $g(X)$ be a nonzero polynomial in the kernel of e_α. Then there exists a nonzero polynomial $q(X)$ in $\mathbb{F}[X]$ such that $g(X) = f(X)q(X)$. Therefore $\deg(f(X)) \leq \deg(g(X))$ for all polynomials $g(X)$ in $\ker(e_\alpha)$. We may assume that the leading coefficient of the polynomial $f(X)$ is 1. Then $f(X)$ is the unique monic polynomial in $\mathbb{F}[X]$ such that $\ker(e_\alpha) = (f(X))$.

We will now show that the polynomial $f(X)$ is irreducible over $\mathbb{F}$. Suppose $f(X) = f_1(X)f_2(X)$. Then $0 = f(\alpha) = f_1(\alpha)f_2(\alpha)$. Therefore either $f_1(\alpha) = 0$ or $f_2(\alpha) = 0$. This shows that the polynomial $f_1(X)$ or $f_2(X)$ is in the kernel of e_α.

If $f_1(X)$ is in the kernel of e_α, then $f_1(X) = q(X)f(X)$. If we substitute this in the equation $f(X) = f_1(X)f_2(X)$ we get $f(X) = f(X)q(X)f_2(X)$. This proves that $q(X)f_2(X) = 1$ and $f_2(X)$ is a unit.

Similarly we can show that $f_1(X)$ is a unit, if $f_2(X)$ is in the kernel of e_α. Therefore the polynomial $f(X)$ is irreducible over $\mathbb{F}$ and it follows from Proposition 3.9.20 that the ideal $\ker(e_\alpha) = (f(X))$ is a maximal ideal in $\mathbb{F}[X]$. As a consequence the ring $\mathbb{F}[X]/(f(X))$ is a field and by fundamental theorem of homomorphism, $\mathbb{F}[X]/(f(X))$ is isomorphic to the ring $\mathbb{F}[\alpha]$. This proves that $\mathbb{F}[\alpha]$ is a field.

We denote the polynomial $f(X)$ by $\min(\alpha, \mathbb{F})(X)$ and call it the *minimal polynomial* of α over the field $\mathbb{F}$.

Suppose $g(X)$ is a polynomial in $\mathbb{F}[X]$ such that α is a root of the polynomial $g(X)$. Then the polynomial $g(X)$ is in the principal ideal $(\min(\alpha, \mathbb{F})(X))$ and hence there exists a polynomial $h(X)$ such that $g(X) = h(X)\min(\alpha, \mathbb{F})(X)$. This proves that $\min(\alpha, \mathbb{F})(X)$ divides the polynomial $g(X)$.

We summarize the discussion above in the following theorem.

Theorem 4.1.20. *Let $\mathbb{K}/\mathbb{F}$ be a field extension and let $\alpha \in \mathbb{K}$ be algebraic over $\mathbb{F}$. Then*

1. *there exists a unique monic irreducible polynomial $p(X)$ with leading coefficient 1 in $\mathbb{F}[X]$ such that α is a root of $p(X)$.*

2. *α is a root of a polynomial $f(X) \in \mathbb{F}[X]$ iff the polynomial $p(X)$ divides the polynomial $f(X)$ in $\mathbb{F}[X]$.*

3. *the field $\frac{\mathbb{F}[X]}{(p(X))}$ is isomorphic to the subfield $\mathbb{F}[\alpha]$ of $\mathbb{K}$.*

Remark 4.1.21.
1. The unique minimal polynomial $p(X)$ in the theorem is denoted as $\min(\alpha, \mathbb{F})(X)$.

2. Let us now observe that, if an element α in $\mathbb{K}$ is algebraic over the field $\mathbb{F}$, then the ring $\mathbb{F}[\alpha]$ is a field and hence the smallest field containing α and $\mathbb{F}$ is $\mathbb{F}(\alpha)$.

Corollary 4.1.22. *Let $\mathbb{K}/\mathbb{F}$ be a field extension. If α in $\mathbb{K}$ is algebraic over $\mathbb{F}$, then $\mathbb{F}[\alpha]/\mathbb{F}$ is a finite extension.*

Proof. Let α in $\mathbb{K}$ be algebraic over $\mathbb{F}$ and let the minimal polynomial of α be $\min(\alpha, \mathbb{F})(X) := c_0 + c_1X + c_2X^2 + \cdots + X^d$ in $\mathbb{F}[X]$. Therefore $\min(\alpha, \mathbb{F})(\alpha) = \alpha^d + c_{d-1}\alpha^{d-1} + c_{d-2}\alpha^{d-2} + \cdots + c_1\alpha + c_0 = 0$. We can write this as $\alpha^d = -(c_{d-1}\alpha^{d-1} + c_{d-2}\alpha^{d-2} + \cdots + c_1\alpha + c_0)$. Using this we can write α^{d+r} as a linear combination of elements of the set $B := \{1, \alpha, \ldots \alpha^{d-1}\}$ for every $r \geq 0$. As a consequence every element in $\mathbb{F}[\alpha]$ can be written as a linear combination of elements of B with coefficients from $\mathbb{F}$. Hence the dimension of the extension $\mathbb{F}[\alpha]/\mathbb{F}$ is at most d.

We can show that the set $B := \{1, \alpha, \alpha^2, \ldots, \alpha^{d-1}\}$ is linearly independent $\mathbb{F}(\alpha)$ over $\mathbb{F}$. This will show that B is a basis $\mathbb{F}[\alpha]$ over $\mathbb{F}$.

Let a_0, $a_1, a_2, \ldots, a_{d-1}$ in $\mathbb{F}$ be such that $a_{d-1}\alpha^{d-1} + a_{d-2}\alpha^{d-2} + \cdots + a_1\alpha + a_0 = 0$. Hence the polynomial $g(X) = a_{d-1}X^{d-1} + a_{d-2}X^{d-2} + \cdots + a_1X + a_0$ is in the ideal $(\min(\alpha, \mathbb{F})(X))$. If g is a non zero polynomial, then $d = \deg(f(X)) \leq \deg(g(X)) \leq d - 1$, a contradiction. This proves that $a_i = 0$ for all $i = 0, 1, 2, \ldots, d-1$ and the set B is linearly independent in $\mathbb{F}[\alpha]$. □

Corollary 4.1.23. *Let α in $\mathbb{K}$ be algebraic over $\mathbb{F}$. Then every element β in $\mathbb{F}[\alpha]$ is algebraic over $\mathbb{F}$.*

Proof. Let $[\mathbb{F}[\alpha] : \mathbb{F}] = d$ and $\beta \neq 0$ be an element of $\mathbb{F}[\alpha]$. Since $\mathbb{F}[\alpha]$ is a field, $\beta^j \neq 0$ for every $j \geq 0$. Therefore the set $S = \{1, \beta, \ldots, \beta^d\}$ is linearly dependent over $\mathbb{F}$. Let b_0, $b_1, \ldots,$ b_d be scalars in $\mathbb{F}$ not all zero such that $b_0 + b_1\beta + \cdots + b_d\beta^d = 0$. This proves that β is a root of the polynomial $g(X) = b_0 + b_1X + \cdots + b_dX^d$ in $\mathbb{F}[X]$. □

Corollary 4.1.24. *Let $\mathbb{F} \subseteq \mathbb{K} \subseteq \mathbb{L}$ be extension of fields. If an element α in $\mathbb{L}$ is algebraic over the field $\mathbb{F}$, then*

1. *α is algebraic over $\mathbb{K}$ and*
2. *the polynomial $\min(\alpha, \mathbb{K})(X)$ divides the polynomial $\min(\alpha, \mathbb{F})(X)$ in $\mathbb{K}$.*

Proof. Let α be an element in $\mathbb{L}$ and $p(X)$ in $\mathbb{F}[X]$ be such that $p(\alpha) = 0$. Since $\mathbb{F} \subseteq \mathbb{K}$, the polynomial $p(X)$ is in $\mathbb{K}[X]$ also. Thus α in $\mathbb{L}$ is algebraic over $\mathbb{K}$.

The minimal polynomial $\min(\alpha, \mathbb{F})(X) \in \mathbb{F}[X]$ of α is also in $\mathbb{K}[X]$ and $\min(\alpha, \mathbb{F})(X)$ is in the principal ideal $(\min(\alpha, \mathbb{K})(X))$ in $\mathbb{K}[X]$. This proves that the polynomial $\min(\alpha, \mathbb{F})(X)$ divides $\min(\alpha, \mathbb{K})(X)$ in $\mathbb{K}[X]$. □

Example 4.1.25. Let $\alpha = \sqrt[4]{2}$ be the positive 4-th root of 2 and let $\mathbb{L} = \mathbb{Q}[\alpha]$. Then α is a root of the polynomial $X^4 - 2$ in $\mathbb{Q}[X]$ and $\min(\alpha, \mathbb{Q})(X) = X^4 - 2$. On the other hand we can factorize $X^4 - 2$ as $X^4 - 2 = (X^2 - \sqrt{2})(X^2 + \sqrt{2})$ in $\mathbb{K}[X]$ where $\mathbb{K} = \mathbb{Q}[\sqrt{2}]$. Now we see that α is a root of $X^2 - \sqrt{2}$. It is an easy exercise to show that $X^2 - \sqrt{2}$ is irreducible in $\mathbb{K}[X]$. Thus $\min(\alpha, \mathbb{K})(X) = X^2 - \sqrt{2}$ and it divides $\min(\alpha, \mathbb{Q})(X)$ in $\mathbb{K}$.

Definition 4.1.26. A complex number $\omega \neq 1$ is called a n-th root of unity if $\omega^n = 1$. In this cases ω is a root of the polynomial $1 + X + X^2 + \cdots + X^{n-1}$.

The complex number ω is called a primitive n-th root of unity, $\omega^k \neq 1$ for $0 < k < n$.

Example 4.1.27. Let ω be a primitive n-th root of unity. By definition $\omega \in \mathbb{C}$ is algebraic over $\mathbb{Q}$. Hence $\mathbb{Q}[\omega]$ is a field and $\mathbb{Q}[\omega]$ is a finite extension of $\mathbb{Q}$.

Let $\mathbb{K}/\mathbb{F}$ be a field extension. We have seen above that if an element α in $\mathbb{K}$ is algebraic over $\mathbb{F}$, then $\mathbb{F}(\alpha)/\mathbb{F}$ is a finite extension and hence an algebraic extension. We will now show that if α and $\beta(\neq 0)$ in $\mathbb{K}$ are algebraic over $\mathbb{F}$, then $\alpha \pm \beta$, $\alpha\beta$ and $\frac{\alpha}{\beta}$ are algebraic over $\mathbb{F}$.

Let $\mathbb{K}/\mathbb{F}$ be a field extension. For α and β in $\mathbb{K}$, we denote by $(\mathbb{F}(\alpha))(\beta)$ to be the subfield generated by $\mathbb{F}(\alpha)$ and β in $\mathbb{K}$. Then we have the following lemma.

Lemma 4.1.28. *Let $\mathbb{K}/\mathbb{F}$ be a field extension. Then*

$$(\mathbb{F}(\alpha))(\beta) = \mathbb{F}(\alpha, \beta)$$

for any two elements α and β in $\mathbb{K}$.

Proof. The field $\mathbb{F}(\alpha, \beta)$ contains the field $\mathbb{F}$ and α. Hence the field $\mathbb{F}(\alpha)$ is contained in $\mathbb{F}(\alpha, \beta)$. Since the element β is also in $\mathbb{F}(\alpha, \beta)$, it follows that the field $(\mathbb{F}(\alpha))(\beta)$ is contained in the field $\mathbb{F}(\alpha, \beta)$.

Now the field $(\mathbb{F}(\alpha))(\beta)$ contains the field $\mathbb{F}$ and the elements α and β. But by definition $\mathbb{F}(\alpha, \beta)$ is the smallest field containing the field $\mathbb{F}$ and the elements α and β. Thus $\mathbb{F}(\alpha, \beta) \subseteq (\mathbb{F}(\alpha))(\beta)$ and the proof is complete. □

Proposition 4.1.29. *Let $\mathbb{K}/\mathbb{F}$ be a field extension. If two elements α and β in $\mathbb{K}$ are algebraic over $\mathbb{F}$, then*

1. *the element β is algebraic over $\mathbb{F}(\alpha)$ and*
2. $[\mathbb{F}(\alpha, \beta) : \mathbb{F}] \leq deg(\min(\alpha, \mathbb{F})(X)).deg(\min(\beta, \mathbb{F})(X))$.

Proof. Let α and β be two elements in $\mathbb{K}$ which are algebraic over $\mathbb{F}$.

Let $\min(\beta, \mathbb{F})(X)$ be the minimal polynomial of β in $\mathbb{F}$. Since $\mathbb{F}$ is contained in $\mathbb{F}(\alpha)$, the polynomial $\min(\beta, \mathbb{F})(X)$ is in $\mathbb{F}(\alpha)[X]$ and β is a root of $\min(\beta, \mathbb{F})$ in $\mathbb{F}(\alpha)(X)$ too. Hence the $\mathbb{F}(\alpha, \beta)/\mathbb{F}(\alpha)$ is a finite extension. Further, since $\min(\beta, \mathbb{F}(\alpha))(X)$ divides $\min(\alpha, \mathbb{F})(X)$ in $\mathbb{F}[\alpha][X]$, it follows that

$$[\mathbb{F}(\alpha, \beta) : \mathbb{F}(\alpha)] = \deg(\min(\beta, \mathbb{F}(\alpha))(X)) \leq \deg(\min(\beta, \mathbb{F})(X)).$$

We now apply tower law for the extensions $\mathbb{F} \subseteq \mathbb{F}(\alpha) \subseteq \mathbb{F}(\alpha, \beta)$ to get

$$\begin{aligned} [\mathbb{F}(\alpha, \beta) : \mathbb{F}] &= [\mathbb{F}(\alpha, \beta) : \mathbb{F}(\alpha)][\mathbb{F}(\alpha) : \mathbb{F}] \\ &= \deg(\min(\beta, \mathbb{F}(\alpha))(X))\deg(\min(\alpha, \mathbb{F})(X)) \\ &\leq \deg(\min(\beta, \mathbb{F})(X))\deg(\min(\alpha, \mathbb{F})(X)). \end{aligned}$$

This completes the proof. □

Corollary 4.1.30. *Let $\mathbb{K}/\mathbb{F}$ be an extension. If α and $\beta(\neq 0)$ in $\mathbb{K}$ are algebraic over $\mathbb{F}$, then $\alpha \pm \beta$, $\alpha\beta$ and $\frac{\alpha}{\beta}$ are algebraic over $\mathbb{F}$.*

Proof. We have shown above that the extension $\mathbb{F}(\alpha, \beta)/\mathbb{F}$ is a finite extension. Therefore the elements $\alpha \pm \beta$, $\alpha\beta$ and $\frac{\alpha}{\beta}$ are algebraic over $\mathbb{F}$. □

Remark 4.1.31. The inequality in the proposition could be equal in some cases and strict in some other cases.

Example 4.1.32. 1. Let $\alpha = \sqrt{2}$ and $\beta = \sqrt{3}$. Then the minimal polynomials of $\sqrt{2}$ and $\sqrt{3}$ over $\mathbb{Q}$ are $\min(\sqrt{2}, \mathbb{Q})(X) = X^2 - 2$ and $\min(\sqrt{3}, \mathbb{Q})(X) = X^2 - 3$ respectively.

The real number $\sqrt{3}$ is not in $\mathbb{Q}[\sqrt{2}]$. Hence the polynomial $X^2 - 3$ is irreducible over $\mathbb{Q}[\sqrt{2}]$ also. Therefore

$$[\mathbb{Q}[\sqrt{2}, \sqrt{3}] : \mathbb{Q}[\sqrt{2}]] = 2 = \deg\min(\sqrt{2}, \mathbb{Q})(X).$$

Now we use Tower law to get

$$\begin{aligned} \mathbb{Q}[\sqrt{2}, \sqrt{3}] : \mathbb{Q}] &= \mathbb{Q}[\sqrt{2}, \sqrt{3}] : \mathbb{Q}[\sqrt{2}]][\mathbb{Q}[\sqrt{2}] : \mathbb{Q}] \\ &= \deg(\min(\sqrt{3}, \mathbb{Q}[\sqrt{2}])(X) \deg(\min(\sqrt{2}, \mathbb{Q})(X) \\ &= \deg(\min(\sqrt{3}, \mathbb{Q})(X) \deg(\min(\sqrt{2}, \mathbb{Q})(X) \\ &= 4. \end{aligned}$$

Now we prove the following theorem.

Theorem 4.1.33. *Let $\mathbb{L}/\mathbb{F}$ be an extension of a field $\mathbb{F}$ and let*

$$\mathbb{K} := \{\alpha \in \mathbb{L} : \alpha \text{ is algebraic over } \mathbb{F}\}.$$

Then

1. *$\mathbb{K}$ is a field and*

2. *$\mathbb{K}$ is algebraic over $\mathbb{F}$.*

Proof. Let α and $\beta \neq 0$ be two elements in $\mathbb{K}$. Then we know that $\alpha \pm \beta$, $\alpha\beta$ and $\frac{\alpha}{\beta}$ are all algebraic over $\mathbb{F}$. This proves that $\mathbb{K}$ is a field and by definition $\mathbb{K}$ is also algebraic over $\mathbb{F}$. □

Remark 4.1.34. The extension $\mathbb{K}/\mathbb{F}$ may not be a finite extension. For example let

$$\mathbb{K} := \{\alpha \in \mathbb{C} : \alpha \text{ is algebraic over} \mathbb{Q}\}.$$

Now given any integer $n \geq 3$, let ζ_n be a primitive n-th root of unity. Then $\mathbb{Q}[\zeta_n] = \mathbb{Q}(\zeta_n)$ is in $\mathbb{K}$ and $[\mathbb{Q}(\zeta_n) : \mathbb{Q}] = n - 1$ and by tower law $[\mathbb{K}, \mathbb{Q}] \geq n$. This shows that $\mathbb{K}/\mathbb{Q}$ is not a finite extension.

This field $\mathbb{K}$ is also denoted by $\overline{\mathbb{Q}}$.

The real numbers e and π are transcendental over $\mathbb{Q}$. In particular they are not in $\mathbb{K}$. Hence $\mathbb{K}$ is a proper subfield of $\mathbb{C}$.

Exercise 4.1.35. Let $S = \{\sqrt{p} : p \text{ is a prime number}\}$ and $\mathbb{K} = \mathbb{Q}(S)$ be the field generated by the set S. Show that the extension $\mathbb{K}/\mathbb{Q}$ is algebraic and it is not a finite extension.

We will first show that the extension is not a finite extension.

Let $n \geq 2$ and $p_1, p_2, \ldots, p_n$ be prime numbers. Then we use tower law to conclude that $[\mathbb{Q}[\sqrt{p_1}, \sqrt{p_2}, \ldots, \sqrt{p_n}] : \mathbb{Q}] = 2^n$. Hence the extension $\mathbb{K}/\mathbb{Q}$ is not a finite extension.

Let α be an element of $\mathbb{Q}(S)$. Then $\alpha := \frac{a_1 s_1^{r_1} + a_2 s_2^{r_2} + \cdots + a_k s_k^{r_k}}{b_1 s_1^{l_1} + b_2 s_2^{l_2} + \cdots + b_k s_k^{l_k}}$ where the coefficients a_i, b_i are in $\mathbb{Q}$, $s_i = \sqrt{p_i}$ and $0 \leq r_i, l_i \leq 1$ are integers.

The elements $a_1 s_1^{r_1} + a_2 s_2^{r_2} + \cdots + a_k s_k^{r_k}$ and $b_1 s_1^{l_1} + b_2 s_2^{l_2} + \cdots + b_k s_k^{l_k}$ are in the field $\mathbb{E} = \mathbb{Q}[\sqrt{p_1}, \sqrt{p_2}, \ldots, \sqrt{p_k}]$. Therefore the element α is in $\mathbb{E}$. Since $\mathbb{E}/\mathbb{Q}$ is a finite extension, it is algebraic. Hence α is algebraic over $\mathbb{Q}$. As a consequence $\mathbb{K}/\mathbb{Q}$ is an algebraic extension.

Exercise 4.1.36. Let $\mathbb{F} = \mathbb{Q}$. Find the dimension $[\mathbb{F}[\alpha, \beta] : \mathbb{F}]$ if

1. $\{\alpha, \beta\} = \{i, \sqrt{2}\}$

2. $\{\alpha, \beta\} = \{i, \sqrt[3]{2}\}$

3. $\{\alpha, \beta\} = \{\sqrt{2}, \sqrt{3}\}$

Now we ask the following question.

Question 4.1.37. Let $\mathbb{F}$ be a field and $f(X)$ be a non-constant polynomial in $\mathbb{F}[X]$. Does there exist a field extension $\mathbb{K}/\mathbb{F}$ such that the polynomial $f(X)$ has a root in $\mathbb{K}$?

> If $f(X)$ is a linear polynomial, say, $f(X) = aX + b$. Then $-\frac{b}{a}$ is the unique root of $f(X)$ in $\mathbb{F}$.
>
> If the polynomial $f(X)$ is of degree bigger than or equal to 2, then it may not have a root in $\mathbb{F}$. For example the polynomial $X^2 + X + 1$ does not have a root in $\mathbb{Q}$ and $\mathbb{R}$. However it has a root in $\mathbb{C}$. In fact all the roots of $X^2 + X + 1$ are in $\mathbb{C}$.
>
> Let $f(X) = X^3 - 2$. Then $\sqrt[3]{2}$ is a root in $\mathbb{R}$ not in $\mathbb{Q}$. The other two roots are not in $\mathbb{R}$.

We will answer the question above in affirmative by showing that given a field $\mathbb{F}$ and a polynomial $f(X)$ in $\mathbb{F}[X]$, there exists an extension $\mathbb{K}$ of the field $\mathbb{F}$ such that the polynomial $f(X)$ has a root α in $\mathbb{K}$.

Let $\mathbb{F}$ be a field and $f(X)$ be a non-constant polynomial in $\mathbb{F}[X]$. Suppose there is a field extension $\mathbb{K}/\mathbb{F}$ such that $f(X)$ has a root α in $\mathbb{K}$.

Let $f(X) = f_1(X)f_2(X)\cdots f_r(X)$ be the irreducible factorization of the polynomial $f(X)$. Since $f(\alpha) = 0$, we see that α is a root of one of the irreducible polynomial $f_i(X)$. Hence we may assume that $f(X)$ is irreducible and ask if there is an extension $\mathbb{K}$ over $\mathbb{F}$ such that $f(X)$ has a root in $\mathbb{K}$.

Suppose there is an extension $\mathbb{K}$ over $\mathbb{F}$ which has a root α of $f(X)$. Let $m_{\alpha,\mathbb{F}}(X) := \min(\alpha, \mathbb{F})(X)$ be the minimal polynomial of α in $\mathbb{F}[X]$. We know that the polynomial $m_{\alpha,\mathbb{F}}(X)$ is the unique monic irreducible polynomial such that $m_{\alpha,\mathbb{F}}(\alpha) = 0$. Since the polynomial $f(X)$ is irreducible and α is a root of $f(X)$, it follows that $f(X) = cm_{\alpha,\mathbb{F}}(X)$ for some constant $c \in \mathbb{F}$. Therefore the principal ideals $(f(X))$ and $(m_{\alpha,\mathbb{F}}(X))$ are same. Now we use the isomorphism between the fields $\frac{\mathbb{F}[X]}{(m_{\alpha,\mathbb{F}}(X))}$ and $\mathbb{F}[\alpha]$ to conclude that the field $\frac{\mathbb{F}[X]}{(f(X))} = \frac{\mathbb{F}[X]}{(m_{\alpha,\mathbb{F}}(X))} = \mathbb{F}[\alpha]$.

This tells us that the irreducible polynomial $f(X)$ has a root in the field $\frac{\mathbb{F}[X]}{(f(X))}$ and we will prove this in the following theorem.

Theorem 4.1.38 (Kronecker's Theorem). *Let $\mathbb{F}$ be a field and $f(X)$ be an irreducible polynomial in $\mathbb{F}[X]$. Then there exists a field extension $\mathbb{K}/\mathbb{F}$ such that the polynomial $f(X)$ has a root in $\mathbb{K}$.*

Further if there exists a field extension $\mathbb{L}/\mathbb{F}$ containing a root β of the polynomial $f(X)$, there exists a homomorphism $\varphi : \mathbb{K} \to \mathbb{L}$ such that $\varphi(\alpha) = \beta$. In particular $\mathbb{L}$ is an extension of $\mathbb{K}$.

Proof. Let $\mathbb{F}$ be a field and $f(X) = a_0 + a_1X + a_2X^2 + \cdots + a_dX^d$ be an irreducible polynomial in $\mathbb{F}[X]$. Since $\mathbb{F}[X]$ is a PID and the polynomial $f(X)$ is irreducible in $\mathbb{F}[X]$, the principal ideal $I := (f(X))$ is a maximal ideal in $\mathbb{F}[X]$. Hence the quotient ring $\dfrac{\mathbb{F}[X]}{I}$ is a field. We denote this field by $\mathbb{K}$.

There is a natural inclusion homomorphism i of $\mathbb{F}$ in to the polynomial ring $\mathbb{F}[X]$ and a homomorphism θ of $\mathbb{F}[X]$ on to the field $\mathbb{K} = \dfrac{\mathbb{F}[X]}{I}$. The composition $\theta \circ i : \mathbb{F} \to \mathbb{K}$ is a ring homomorphism and we have shown that $\mathbb{K}$ is an extension of the field $\mathbb{F}$.

Let $\alpha := X + I$. Then $\theta(X) = \alpha$ and $\theta(X^k) = \alpha^k = X^k + I$ for $k \geq 1$. Therefore

$$\begin{aligned} f(\alpha) &= \sum_{k=1}^{d} a_k (X+I)^k \\ &= \sum_{k=1}^{d} a_k (X^k + I) \\ &= \sum_{k=1}^{d} a_k X^k + I \\ &= I \\ &= 0. \end{aligned}$$

This proves that α in $\mathbb{K}$ is a root of the polynomial $f(X)$.

Now suppose there is an extension $\mathbb{L}/\mathbb{F}$ such that the polynomial $f(X)$ has a root β in $\mathbb{L}$. We will prove the result by constructing a homomorphism $\varphi : \mathbb{F}[X] \to \mathbb{L}$ such that $\ker(\varphi) = I$.

The map $\varphi : \mathbb{F}[X] \to \mathbb{L}$ defined by

$$\varphi(b_0 + b_1 X + b_2 X^2 + \cdots + b_r X^r) = b_0 + b_1\beta + b_2\beta^2 + \cdots + b_r\beta^r$$

for $b_0 + b_1 X + b_2 X^2 + \cdots + b_r X^r \in \mathbb{F}[X]$ is a ring homomorphism. Since β is a root of $f(X)$, the polynomial $f(X)$ is in the kernel of φ. If $\min_{\beta,\mathbb{F}}(X)$ denotes the minimal polynomial of β over $\mathbb{F}$, then from the discussion preceding the statement of the theorem we conclude that $f(X) = c\min_{\beta,\mathbb{F}}(X)$ for some constant $c \in \mathbb{F}$. Hence the principal ideal $I = (\min_{\beta,\mathbb{F}}(X))$. As a consequence

$$\mathbb{F}[\alpha] \simeq \frac{\mathbb{F}[X]}{I} = \frac{\mathbb{F}[X]}{(\min_{\beta,\mathbb{F}}(X))} \simeq \mathbb{F}[\beta].$$

Since $\mathbb{F}[\beta]$ is a subfield of $\mathbb{L}$ the result follows. □

Given an irreducible polynomial $f(X)$ in $\mathbb{F}[X]$, Kronecker's theorem gives us a method of constructing a field extension $\mathbb{K}$ of $\mathbb{F}$ containing a root of the polynomial $f(X)$. However this extension $\mathbb{K}$ may not contain all the roots of the polynomial $f(X)$.

Example 4.1.39. 1. Let $f(X) = X^2 + 1$ in $\mathbb{R}[X]$. Then $\frac{\mathbb{R}[X]}{(X^2+1)}$ is isomorphic to $\mathbb{C}$.

The two roots $\pm i$ are in $\mathbb{C}$.

2. Let $f(X) = X^2 - 2$ in $\mathbb{Q}[X]$. Then $f(X)$ has a root in $\frac{\mathbb{Q}[X]}{(X^2-2)} = \mathbb{Q}[\sqrt{2}]$.

In this example too the extension contains all the roots of the polynomial.

3. Let $f(X) = X^3 - 2$ in $\mathbb{Q}[X]$ and $\sqrt[3]{2}$ be the real root of the polynomial $f(X)$. If $\mathbb{K} = \mathbb{Q}(\sqrt[3]{2})$, then $f(X)$ has only one root in the field $\mathbb{K}$. The other two roots are complex roots and they are not in $\mathbb{K}$.

Corollary 4.1.40. *Let $\mathbb{F}$ be a field and $f(X)$ be a non-constant polynomial in $\mathbb{F}[X]$. Then there exists a field extension $\mathbb{K}$ of $\mathbb{F}$ such that all the roots of the polynomial $f(X)$ are in the field $\mathbb{K}$.*

Proof. We prove the result by induction on the degree of the polynomial $f(X)$ over any field $\mathbb{F}$.

Suppose $f(X) = aX + b$ is a polynomial of degree 1. Then $-\frac{b}{a}$ is the only root of the polynomial $f(X)$ and it is in $\mathbb{F}$. Hence $\mathbb{K} = \mathbb{F}$ in this case.

Let us now assume the result to be true for all polynomials of degree less than d over any field.

Let $f(X)$ be a polynomial of degree d and $g(X)$ be an irreducible factor of $f(X)$. Then by Kronecker's Theorem 4.1.38 there exists an extension $\mathbb{K}_1$ of $\mathbb{F}$ such that the polynomial $g(X)$ has a root, say, α in $\mathbb{K}_1$. Since α is a root of the polynomial $g(X)$, it is a root of $f(X)$ too. Hence the polynomial $X - \alpha$ divides the polynomial $f(X)$ in $\mathbb{K}_1[X]$ and we write $f(X) = (X-\alpha)f_1(X)$ where $f_1(X)$ is in $\mathbb{K}_1[X]$.

Now the polynomial $f_1(X)$ is of degree $d-1$ and by induction hypothesis, there exists an extension $\mathbb{K}$ of $\mathbb{K}_1$ such that all the roots of $f_1(X)$ are in $\mathbb{K}$. Hence the extension $\mathbb{K}$ contains all the roots of $f(X)$. □

Corollary 4.1.41. *Let $\mathbb{F}$ be a field and $f(X)$ be a polynomial in $\mathbb{F}[X]$ of degree $n \geq 2$. Then there exists a field $\mathbb{K}$ containing all the roots of $f(X)$ such that $[\mathbb{K} : \mathbb{F}] \leq n!$.*

Proof. We will prove this result by induction on the degree n of the polynomial $f(X)$.

Let $f(X) = aX + b$ be a polynomial of degree 1. Then $-\frac{b}{a} \in \mathbb{F}$ is the only root of the polynomial and hence $\mathbb{K} = \mathbb{F}$.

Let $f(X)$ be a polynomial of degree n. By Kronecker's theorem 4.1.38 there exists an extension $\mathbb{K}_1$ of $\mathbb{F}$ such that the polynomial $f(X)$ has a root, say α in $\mathbb{K}_1$ and $[\mathbb{K}_1 : \mathbb{F}] \leq \deg(f(X))$. We write the polynomial $f(X)$ in $\mathbb{K}_1(X)$ as $f(X) = (X - \alpha)f_1(X)$ for some polynomial $f_1(X)$ in $\mathbb{K}_1(X)$. By induction hypothesis, there exists a field extension $\mathbb{K}$ of $\mathbb{K}_1$ such that $f_1(X)$ has all its roots in $\mathbb{K}$ and $[\mathbb{K} : \mathbb{K}_1] \leq (n-1)!$. Now the polynomial $f(X)$ has all its roots in $\mathbb{K}$ and since $[\mathbb{K}_1 : \mathbb{F}] = \deg(f(X)) \leq n$ it follows that

$$[\mathbb{K} : \mathbb{F}] = [\mathbb{K} : \mathbb{K}_1][\mathbb{K}_1 : \mathbb{F}] \leq n!.$$

This completes the proof. □

Example 4.1.42. 1. Let $f(X) = X^p - 1$ for a prime number $p \neq 2$ and ζ be a primitive p-th root of unity. Then the extension field $\mathbb{K} := \mathbb{Q}[\zeta]$ contains all the roots of the polynomial $f(X)$.

A complex number ζ in $\mathbb{C}$ is called a primitive p-th root of unity, if $\zeta^k \neq 1$ for $0 < k < p$.

Let ζ be a primitive p-th root of unity and $\mathbb{K} = \mathbb{Q}[\zeta]$. Then for all $i \leq p-1$ the powers ζ^i of ζ are in $\mathbb{K}$ and $(\zeta^i)^p = (\zeta^p)^i = 1$. Thus ζ^is are all roots of the polynomial $f(X)$ and these are the only roots of $f(X)$. As a consequence the field $\mathbb{K}$ contains all the roots of the polynomial $f(X)$.

Further observe that the degree of the extension is $p - 1$.

2. Let $f(X) = (X^2+2)(X^2+3)$ over $\mathbb{Q}[X]$. Then the field $\mathbb{Q}[i\sqrt{2}, i\sqrt{3}]$ contains all the roots of the polynomial $f(X)$ over $\mathbb{Q}$.

3. Let $f(X) = X^4 + 1$ over $\mathbb{Q}$. We show that there exists a field $\mathbb{K}$ that contains all the roots of the polynomial $f(X)$ and $[\mathbb{K} : \mathbb{Q}] = 4$.

   The polynomial $f(X)$ is irreducible over $\mathbb{Q}$. We factorize $f(X) = (X^2+i)(X^2-i)$ over $\mathbb{C}$. The roots of the polynomial X^2+i are $\frac{1\pm i}{\sqrt{2}}$. They are also roots for the polynomial $X^2 - i$.
   Let $\alpha = \frac{1+i}{\sqrt{2}}$ and $\beta = \frac{1-i}{\sqrt{2}}$. Then $\alpha+\beta = \sqrt{2}$ and $\alpha - \beta = i\sqrt{2} = i(\alpha+\beta)$. Thus $\mathbb{K} = \mathbb{Q}[i, \sqrt{2}]$. (Why?) It follows from tower law that

   $$[\mathbb{K} : \mathbb{Q}] = [\mathbb{K} : \mathbb{Q}[i]][\mathbb{Q}[i] : \mathbb{Q}] = 4.$$

4. Let $f(X) = X^3 - 2$ in $\mathbb{Q}[X]$. If $\zeta(\neq 1)$ is a cube root of unity and $\sqrt[3]{2}$ is a positive real root of $f(X)$, then $\mathbb{K} = \mathbb{Q}[\zeta, \sqrt[3]{2}]$ and $[\mathbb{K} : \mathbb{Q}] = 3! = 6$.

   Let $\alpha = \sqrt[3]{2}$ be the positive real root of $f(X)$. If $\zeta(\neq 1)$ is a cube root of unity then α, $\zeta\alpha$ and $\zeta^2\alpha$ are the roots of the polynomial $f(X)$. Hence $\mathbb{K} = \mathbb{Q}[\zeta, \alpha]$.
   The degree of the extension is 6. It is a easy to check that the set $\{1, \zeta, \zeta^2, \alpha, \zeta\,\alpha, \zeta^2\alpha\}$ is a basis of $\mathbb{K}$ over $\mathbb{Q}$.
   This example shows that the constant $n!$ in the corollary can not be improved upon.

Exercise 4.1.43. 1. Let p be a prime number and $f(X) = X^p - 2$. Find an extension of the field $\mathbb{K}$ of $\mathbb{Q}$ containing all the roots of $f(X)$.

2. Let $f(X) = X^4 - 10X^2 + 1$ in $\mathbb{Q}[X]$. Find an extension $\mathbb{K}/\mathbb{Q}$ which contains all the roots of $f(X)$ and find $[\mathbb{K} : \mathbb{Q}]$.

Exercise 4.1.44. Let $\mathbb{F}$ be a field. Let $f_1(X)$ and $f_2(X)$ be two polynomials in $\mathbb{F}[X]$. Show that there exists a field $\mathbb{K}$ of $\mathbb{F}$ such that the field $\mathbb{K}$ has all the roots of the two polynomials f_1 and f_2.

Exercise 4.1.45. Let $f(X) = X^2 - 2$ and $g(X) = X^2 - 3$. Find a field $\mathbb{K}$ containing all the roots of these two polynomials $f(X)$ and $g(X)$.

4.1.3 Splitting Fields

Given a non-constant polynomial $f(X)$ over a field $\mathbb{F}$, we saw in previous section that there exists an extension $\mathbb{K}$ of $\mathbb{F}$ such that $\mathbb{K}$ has all the roots of f. In this section we will study those extensions which are minimal with this property.

Definition 4.1.46 (Splitting Field). 1. Let $\mathbb{K}$ be an extension of a field $\mathbb{F}$ and $f(X)$ be a polynomial in $\mathbb{F}[X]$. We say that the polynomial splits in the field $\mathbb{K}$, if all the roots of the polynomial $f(X)$ are in $\mathbb{K}$. That is there exist $\alpha_1, \alpha_2, \ldots, \alpha_r$ in $\mathbb{K}$ such that $f(X) = c\prod_{i=1}^{r}(X - \alpha_i)$.

2. Let $\mathbb{F}$ be a field and $f(X)$ be a polynomial in $\mathbb{F}[X]$. An extension $\mathbb{K}/\mathbb{F}$ is called a splitting field of the polynomial $f(X)$

   (a) if the polynomial $f(X)$ splits in $\mathbb{K}$ and

(b) if $\mathbb{L}$ is a proper subfield of $\mathbb{K}$ then the polynomial $f(X)$ does not split in $\mathbb{L}$.

Example 4.1.47. 1. Let $k \in \mathbb{N}$ be square free and $f(X) = X^2 - k$ in $\mathbb{Q}[X]$. Then $\mathbb{Q}[\sqrt{k}]$ is a splitting field of the polynomial $f(X)$ over $\mathbb{Q}$.

2. Let $f(X) = X^3 - 2$ in $\mathbb{Q}[X]$. If $\sqrt[3]{2}$ is the real cube root of 2 and $\mathbb{K} = \mathbb{Q}[\sqrt[3]{2}]$, then $\mathbb{K}$ is not a splitting field of the polynomial $f(X)$. For the field $\mathbb{K}$ does not contain the other two complex roots.

3. Let $f(X) = X^3 - 2$ and $\zeta \neq 1$ be a cube root of unity. Then $\alpha = \sqrt[3]{2}\,\zeta$ is a root of the polynomial $f(X)$ and hence the extension $\mathbb{K} = \mathbb{Q}[\sqrt[3]{2}\,\zeta]$ contains a root of the polynomial $f(X)$. We will now show that the real root $\alpha := \sqrt[3]{2}$ of $f(X) = X^3 - 2$ is not in $\mathbb{K}$ and conclude that $\mathbb{K}$ is not a splitting field of the polynomial $f(X)$.

   Suppose α is in $\mathbb{K}$. Then $\zeta = \frac{\alpha\zeta}{\alpha}$ will be in $\mathbb{K}$ and hence $\alpha\zeta^2$ will also be in $\mathbb{K}$.

   Let a, b, c in $\mathbb{Q}$ be such that $a + b\alpha\zeta + c\alpha^2\zeta^2 = \alpha\zeta^2$. If we now multiply this equation by ζ, we get

$$a\zeta + b\alpha\zeta^2 + c\alpha^2 = \alpha\zeta^3 = \alpha$$

   and hence

$$a\zeta + b\alpha\overline{\zeta} = \alpha - c\alpha^2.$$

   Since $\alpha - c\alpha^2$ is a real number, we see that $a\zeta + b\alpha\overline{\zeta}$ is a real number.

   Let $\zeta = u + iv$. Then $a\zeta + b\alpha\zeta = (a + b\alpha)\,u + i(a - b\alpha)\,v$. Since $v \neq 0$, we have $a - b\alpha = 0$. This means that $\sqrt[3]{2} = \alpha = \frac{a}{b}$ is a rational number, a contradiction.

   Thus we have shown that $\mathbb{K}$ is not a splitting field of $f(X)$.

   We can show that the field $\mathbb{Q}[\sqrt[3]{2}, \zeta]$ is a splitting field of $f(X)$ over $\mathbb{Q}$.

4. Let $n \geq 2$ and $f(X) = 1 + X + X^2 + \cdots + X^{n-1}$. If ζ is a primitive n-th root of unity, then $\mathbb{K} = \mathbb{Q}[\zeta]$ is a splitting field of the polynomial $f(X)$.

5. Let p be a prime number and $f(X) = X^p - 2$. If ζ is a primitive p-th root of unity, then $\sqrt[p]{2}$ is a root of the polynomial $f(X)$ and $\mathbb{K} = \mathbb{Q}[\sqrt[p]{2}\zeta]$ does not contain one real root of the polynomial $f(X)$. Hence it is not a splitting field of the polynomial $f(X)$. The field $\mathbb{Q}[\sqrt[p]{2}, \zeta]$ is a splitting field of $X^p - 2$ over $\mathbb{Q}$.

The following theorem proves the existence of splitting field of a polynomial over a field.

Theorem 4.1.48. *Let $\mathbb{F}$ be a field and $f(X)$ be a non-constant polynomial. Then there exists a splitting field $\mathbb{K}$ of the polynomial over $\mathbb{F}$ and $[\mathbb{K} : \mathbb{F}] < \infty$.*

Proof. Let $f(X)$ be a non-constant polynomial of degree d in $\mathbb{F}[X]$. By Corollary 4.1.40 there exists an extension $\mathbb{L}$ such that the field $\mathbb{L}$ contains all the roots of the polynomial $f(X)$. Let $\alpha_1, \alpha_2, \ldots, \alpha_d$ be the distinct roots of the polynomial $f(X)$ and let $\mathbb{K} = \mathbb{F}(\alpha_1, \alpha_2, \ldots, \alpha_d) \subseteq \mathbb{L}$.

We claim that $\mathbb{K}$ is a splitting field of $f(X)$ over $\mathbb{F}$. Suppose $\mathbb{E}$ is an extension of the field $\mathbb{F}$ such that $f(X)$ splits in $\mathbb{E}$. Then all the roots α_i of $f(X)$ in $\mathbb{L}$ are in $\mathbb{E}$. As a consequence the field $\mathbb{K} = \mathbb{F}(\alpha_1, \alpha_2, \ldots, \alpha_d) \subseteq \mathbb{E}$ and $\mathbb{E} \subseteq \mathbb{K}$. This shows that $\mathbb{K}$ is a splitting field of $f(X)$ over $\mathbb{F}$. □

Example 4.1.49.

1. Let $f(X) = X^2 - 2$ in $\mathbb{Q}[X]$. Then the polynomial $f(X)$ splits in $\mathbb{R}$ and the roots are $\pm\sqrt{2}$. Hence a splitting field of $f(X)$ is $\mathbb{Q}[\sqrt{2}]$.

2. Let $f(X) = X^2 + 1$ in $\mathbb{Q}[X]$. Then the polynomial $f(X)$ splits in $\mathbb{C}$ with roots $\pm i$. Thus a splitting field of $f(X)$ over $\mathbb{Q}$ is $\mathbb{Q}[i]$.

   This polynomial is irreducible over $\mathbb{R}$ also and the splitting of $f(X)$ over $\mathbb{R}$ is $\mathbb{C}$!

3. Let $f(X) = X^3 - 1$ over the field $\mathbb{Q}$. This polynomial is not irreducible over $\mathbb{Q}$ and we write as $f(X) = (X-1)(X^2+X+1)$ in to irreducible factors. Let $\zeta \neq 1$ in $\mathbb{C}$ be a root of $X^2 + X + 1$. Then $\zeta^2 = \overline{\zeta}$, the complex conjugate of is another root of $X^2 + X + 1$. Hence a splitting field of the polynomial $X^3 - 1$ over $\mathbb{Q}$ is $\mathbb{Q}(\zeta) = \mathbb{Q}(i\sqrt{3})$.

Exercise 4.1.50. Let $f(X)$ be an irreducible polynomial over $\mathbb{R}$. Show that any splitting field of f over $\mathbb{R}$ is isomorphic to $\mathbb{R}$ or $\mathbb{C}$.

We will now prove that splitting field of a polynomial is unique up to isomorphism.

Lemma 4.1.51. *Let $\theta : R \to S$ be a homomorphism between two rings R and S. Then there exists a homomorphism $\widetilde{\theta} : R[X] \to S[X]$ such that $\widetilde{\theta}|_R = \theta$.*

Further if θ is an isomorphism then $\widetilde{\theta}$ is also an isomorphism.

Proof. Define $\widetilde{\theta} : R[X] \to S[X]$ by

$$\widetilde{\theta}(a_0 + a_1X + a_2X^2 + \cdots + a_dX^d) := \theta(a_0)X + \theta(a_1)X + \cdots + \theta(a_d)X^d$$

for $a_0 + a_1X + a_2X^2 + \cdots + a_dX^d$ in $R[X]$. Then it is easy to show that this map $\widetilde{\theta}$ is a homomorphism such that $\widetilde{\theta}|_R = \theta$ and $\widetilde{\theta}$ is an isomorphism if θ is also an isomorphism. □

Theorem 4.1.52. *Let $\mathbb{F}$ and $\mathbb{F}'$ be two fields and $\theta : \mathbb{F} \to \mathbb{F}'$ be an isomorphism. Let $\widetilde{\theta} : \mathbb{F}[X] \to \mathbb{F}'[X]$ be the ring homomorphism such that $\widetilde{\theta}|_{\mathbb{F}} = \theta$ as in Lemma 4.1.51. Let $p(X) \in \mathbb{F}[X]$ be an irreducible polynomial and α be a root of $p(X)$. If β is a root of $p'(X) = \widetilde{\theta}(p(X))$, then there exists an isomorphism $\overline{\theta} : \mathbb{F}(\alpha) \to \mathbb{F}'(\beta)$ such that $\overline{\theta}(\alpha) = \beta$.*

Proof. Let $\theta : \mathbb{F} \to \mathbb{F}'$ be an isomorphism of fields and $\widetilde{\theta} : \mathbb{F}[X] \to \mathbb{F}'[X]$ be the isomorphism defined by $a_0 + a_1X + \cdots + a_dX^d \mapsto \theta(a_0) + \theta(a_1)X + \cdots + \theta(a_d)X^d$. If we now compose this isomorphism $\widetilde{\theta}$ with the canonical map $\mathbb{F}'[X] \to \frac{\mathbb{F}'[X]}{(p'(X))}$, we get an on to ring homomorphism from $\mathbb{F}[X] \to \frac{\mathbb{F}'[X]}{(p'(X))}$ with kernel $(p(X))$. Hence the fields $\frac{\mathbb{F}'[X]}{(p'(X))}$ and $\frac{\mathbb{F}[X]}{(p(X))}$ are isomorphic. Let us denote this isomorphism also by $\widetilde{\theta}$.

Let α be a root of the polynomial $p(X)$ over $\mathbb{F}$. Then we know that the evaluation map $e_\alpha : \frac{\mathbb{F}[X]}{(p(X))} \to \mathbb{F}[\alpha]$ is an isomorphism between the fields. Similarly, if β is a root of the polynomial $\widetilde{\theta}(p(X)) = p'(X)$, the map $e_\beta : \mathbb{F}'(\beta) \to \frac{\mathbb{F}'[X]}{((p'(X)))}$ is an isomorphism. Therefore the homomorphism $\overline{\theta} := e_\beta \circ \widetilde{\theta} \circ e_\alpha^{-1} : \mathbb{F}[\alpha] \to \mathbb{F}[\beta]$ is an isomorphism such that $\overline{\theta}(\alpha) = \beta$. □

Corollary 4.1.53. *Let $\mathbb{F}$ be a field $f(X)$ be an irreducible polynomial in $\mathbb{F}[X]$. Let $\mathbb{K}$ be a splitting field of $f(X)$. Let α be a root of $f(X)$ in $\mathbb{K}$. Then the number of injective homomorphisms from $\mathbb{F}[\alpha]$ to $\mathbb{K}$ is the number of distinct roots of the polynomial $f(X)$ in $\mathbb{K}$.*

Proof. Let $\mathbb{F}' = \mathbb{F}$ and θ be the identity map of the field $\mathbb{F}$. Let $\mathbb{K}$ be a splitting field of an irreducible polynomial $f(X)$ and α be a root of the polynomial. If $\beta \neq \alpha$ is a root of the polynomial $f(X)$, then we get a non trivial isomorphism $\overline{\theta} : \mathbb{F}[\alpha] \to \mathbb{F}[\beta]$ such that $\overline{\theta}(\alpha) = \beta$. If there are r distinct roots of the polynomial $f(X)$, then we get at least r injective homomorphisms from $\mathbb{F}[\alpha]$ to $\mathbb{K}$.

Conversely suppose there exists a homomorphism $\overline{\theta} : \mathbb{F}[\alpha] \to \mathbb{K}$ and let $\beta = \overline{\theta}(\alpha)$. If $\beta \neq \alpha$, then

$$f(\beta) = f(\overline{\theta}(\alpha)) = \overline{\theta}(f(\alpha)) = 0.$$

This proves that there are exactly r injective homomorphisms from $\mathbb{F}[\alpha]$ to $\mathbb{K}$. □

Corollary 4.1.54. *Let $\mathbb{F}$ be a field and $\mathbb{K}$ be a splitting field of an irreducible polynomial $f(X)$. If all the roots of $f(X)$ are distinct and if α is a root of $f(X)$, then the number of injective homomorphisms from $\mathbb{F}[\alpha]$ to $\mathbb{K}$ is equal to $[\mathbb{F}[\alpha] : \mathbb{F}]$.*

Proof. Let α be a root of $f(X)$. We have seen in Corollary 4.1.53 that the number of homomorphisms of $\mathbb{F}[\alpha]$ in to $\mathbb{K}$ is equal to the number of distinct roots of the polynomial f. Since all the roots of $f(X)$ are distinct, the number of roots is equal to the degree of the polynomial $f(X)$. However we know that $[\mathbb{F}[\alpha] : \mathbb{F}] = \deg(f(X))$ and this completes the proof. □

Theorem 4.1.55. *Let $\theta : \mathbb{F} \to \mathbb{F}'$ be an isomorphism between two fields $\mathbb{F}$ and $\mathbb{F}'$ and $\widetilde{\theta} : \mathbb{F}[X] \to \mathbb{F}'[X]$ be the isomorphism defined as in Lemma 4.1.51. Let $f(X)$ be a non-constant polynomial in $\mathbb{F}[X]$ and $\mathbb{K}$ be a splitting field of the polynomial $f(X)$ over $\mathbb{F}$. If $\mathbb{K}'$ is a splitting field of the polynomial $f'(X) = \widetilde{\theta}(f(X))$, then there exists an isomorphism $\overline{\theta} : \mathbb{K} \to \mathbb{K}'$ such that $\overline{\theta}\,|_{\mathbb{F}}= \theta$.*

Proof. We prove this theorem by induction on the degree of the polynomial $f(X)$ in $\mathbb{F}[X]$.

If $\deg(f(X)) = 1$, then $\mathbb{F}$ is a splitting field of $f(X)$ and therefore $\overline{\theta} = \theta$.

We will now assume the result to be true for any field $\mathbb{F}$, any isomorphism of $\mathbb{F}$ and any polynomial $f(X)$ of degree $m < n$.

Let $\mathbb{F}$ be a field and $f(X)$ be a polynomial of degree $n \geq 2$. If $f_1(X)$ is an irreducible factor of $f(X)$, then the polynomial $f_1'(X) = \widetilde{\theta}(f_1(X))$ is an irreducible factor of $f'(X)$.

Let $\alpha \in \mathbb{K}$ be a root of $f(X)$. If $\beta \in \mathbb{K}'$ is a root of $f'(X)$, then we know that there exists an isomorphism $\tau : \mathbb{F}[\alpha] \to \mathbb{F}'(\beta)$ such that $\tau\mid_{\mathbb{F}}= \theta$.

We also have $f(X) = (X - \alpha)f_1(X)$ where $f_1(X)$ is a polynomial of degree $n-1$ in $\mathbb{F}[\alpha][X]$. Similarly $f'(X) = (X - \beta)f_1'(X)$ where $f_1'(X) \in \mathbb{F}'(\beta)[X]$ is a polynomial of degree $n-1$.

Since $\mathbb{K}$ is a splitting field of $f(X)$ over $\mathbb{F}$ and $\mathbb{F} \subseteq \mathbb{F}[\alpha]$, it is a splitting field of $f(X)$ over $\mathbb{F}[\alpha]$ too. Similarly $\mathbb{K}'$ is a splitting field of $f'(X)$ over $\mathbb{F}'(\beta)$. Hence by induction hypothesis there exists an isomorphism $\overline{\theta} : \mathbb{K} \to \mathbb{K}'$ such that $\overline{\theta}\mid_{\mathbb{F}[\alpha]}= \tau$. Since τ extends the isomorphism $\theta : \mathbb{F} \to \mathbb{F}'$, it follows that $\overline{\theta}$ is an extension of θ as well. □

Corollary 4.1.56. *Let $\mathbb{F}$ be a field and $f(X)$ be a polynomial in $\mathbb{F}[X]$. If $\mathbb{K}$ and $\mathbb{K}'$ are splitting fields of $f(X)$ over $\mathbb{F}$ then $\mathbb{K}$ and $\mathbb{K}'$ are isomorphic.*

Proof. In Theorem 4.1.55, we let $\mathbb{F} = \mathbb{F}'$ and $\theta = Id$, the identity homomorphism. Then there exists an isomorphism $\overline{\theta} : \mathbb{K} \to \mathbb{K}'$ such that $\overline{\theta}$ is identity on $\mathbb{F}$. □

This corollary proves that splitting field of a polynomial over a field $\mathbb{F}$ is unique up to isomorphism. Hence we fix one splitting field $\mathbb{K}$ and call it the splitting field of $f(X)$ over $\mathbb{F}$.

Example 4.1.57.

1. Let $f(X) = X^2 - 2$ over $\mathbb{Q}$. Then $\mathbb{Q}(\sqrt{2})$ is a splitting field of the polynomial $f(X)$ over $\mathbb{Q}$ and we fix $\mathbb{Q}(\sqrt{2})$ and call it the splitting field of $f(X)$ over $\mathbb{Q}$.

2. Let $f(X) = (X^2-2)(X^2-3)$ over $\mathbb{Q}$. Then the splitting of the polynomial $f(X)$ over $\mathbb{Q}$ is $\mathbb{Q}(\sqrt{2}, \sqrt{3})$ and the degree of the extension is 4.

   The field $\mathbb{Q}(\sqrt{6})$ the splitting field of the polynomial $X^2 - 6$ over $\mathbb{Q}$ is a subfield of $\mathbb{Q}(\sqrt{2}, \sqrt{3})$.

3. Let $f(X) = X^3 - 1$ over $\mathbb{Q}$. Then $f(X) = (X - 1)(X^2 + X + 1)$ is the irreducible factorization of the polynomial $f(X)$. The root 1 of $X - 1$ is in $\mathbb{Q}$ itself. The remaining roots $\zeta = \frac{1+\sqrt{-3}}{2}$ and $\overline{\zeta} = \frac{1-\sqrt{-3}}{2}$ are not in $\mathbb{Q}$. Therefore the splitting field of $f(X)$ over $\mathbb{Q}$ is $\mathbb{Q}(\zeta, \overline{\zeta})$. Since $\frac{1}{2}$ is in $\mathbb{Q}$, it follows that $\mathbb{Q}[\zeta] = \mathbb{Q}[\sqrt{-3}]$. Hence $\mathbb{Q}[\zeta, \overline{\zeta}] = \mathbb{Q}[\sqrt{-3}]$ is the splitting field of the polynomial $X^3 - 1$ over $\mathbb{Q}$ and the degree of the extension is 2.

4. Let $f(X) = X^3 - 2$ over $\mathbb{Q}$. The roots of this polynomial are $\sqrt[3]{2}$, $\sqrt[3]{2}\,\zeta$ and $\sqrt[3]{2}\,\zeta^2$ where $\zeta \neq 1$ is a cube root of unity.

   Since $\sqrt{-3} = \zeta + \zeta^2$, the splitting field $\mathbb{Q}[\sqrt[3]{2}, \zeta] = \mathbb{Q}[\sqrt[3]{2}, \sqrt{-3}]$ and the degree of the extension is 6 over $\mathbb{Q}$.

5. Let $f(X) = X^3 - 2$ over $\mathbb{R}$. Then one of the roots $\sqrt[3]{2}$ is in $\mathbb{R}$ and hence the splitting field of $f(X)$ over $\mathbb{R}$ is $\mathbb{R}[\sqrt{-3}] = \mathbb{C}$ and the degree of the extension is 2.

6. Let $f(X) = X^4 + 4$ over $\mathbb{Q}$. Then $f(X) = X^4 + 4$ has a two irreducible factors $f_1(X) = X^2 + 2X + 2$ and $f_2(X) = X^2 - 2X + 2$ over $\mathbb{Q}$. That

is $f(X) = (X^2 + 2X + 2)(X^2 - 2X + 2)$. The roots of $f_1(X)$ and $f_2(X)$ are $\frac{-1 \pm i}{2}$ and $\frac{1 \pm i}{2}$ respectively. Therefore the splitting of f over $\mathbb{Q}$ is $\mathbb{Q}[i]$ and the degree of the extension is 2.

Exercise 4.1.58. Find the splitting filed $\mathbb{K}$ of the polynomial $f(X)$ over $\mathbb{F} = \mathbb{Q}$ and the dimension of the extension $[\mathbb{K} : \mathbb{F}]$ if

1. $f(X) = X^p - 1$ for a prime number p.
2. $f(X) = X^n - 1$ for $n \geq 2$.
3. $f(X) = X^p - 2$ for a prime number p.
4. $f(X) = X^n - 2$ for $n \geq 2$.
5. $f(X) = X^p - k$ where p is a prime number and $k \geq 2$.
6. $f(X) = X^n - k$ where k, $n \geq 2$.

Exercise 4.1.59. Let $\mathbb{F}$ be a field and $\mathbb{K}$ be the splitting field of a polynomial $f(X)$ over $\mathbb{F}$. If $\theta : \mathbb{K} \to \mathbb{K}$ is a homomorphism such that $\theta = Id$ on $\mathbb{F}$, then $\theta(\alpha)$ is also a root of the polynomial $f(X)$.

Let $f(X) = a_0 + a_1X + a_2X^2 + \cdots + a_nX^n$ where a_i's are in the field $\mathbb{F}$. Let $\theta : \mathbb{K} \to \mathbb{K}$ be a homomorphism such that θ is identity on $\mathbb{F}$. If α is a root of $f(X)$, then

$$\begin{aligned} \sum_{i=0}^{n} a_i\theta(\alpha)^i &= \sum_{i=0}^{n} a_i\theta(\alpha^i) \\ &= \sum_{i=0}^{n} \theta(a_i\alpha^i) \\ &= \theta(0) \\ &= 0. \end{aligned}$$

Theorem 4.1.60. *Let $\theta : \mathbb{F} \to \mathbb{F}'$ be an isomorphism between two fields $\mathbb{F}$ and $\mathbb{F}'$ and $\widetilde{\theta} : \mathbb{F}[X] \to \mathbb{F}'[X]$ be the isomorphism defined as in Lemma 4.1.51. Let $f(X)$ be a non-constant polynomial in $\mathbb{F}[X]$ and $\mathbb{K}$ be the splitting field of the polynomial $f(X)$ over $\mathbb{F}$. If $\mathbb{K}'$ is the splitting field of the polynomial $f'(X) = \widetilde{\theta}(f)$, then the number of isomorphisms $\overline{\theta} : \mathbb{K} \to \mathbb{K}'$ such that $\overline{\theta}\mid_{\mathbb{F}} = \theta$ is at most $[\mathbb{K} : \mathbb{F}]$. Further it is equal to $[\mathbb{K} : \mathbb{F}]$ iff the polynomial $f(X)$ has distinct roots.*

Proof. We will prove the result by induction on the degree of the extension $[\mathbb{K} : \mathbb{F}]$.

If $\mathbb{K} = \mathbb{F}$, the $\overline{\theta} = \theta$ is the only possible extension.

Now we assume $[\mathbb{K} : \mathbb{F}] > 1$. Let $f_1(X)$ be an irreducible factor of the polynomial $f(X)$ in $\mathbb{F}[X]$ and α be a root of the polynomial $f_1(X)$ in $\mathbb{K}$.

If $\overline{\theta} : \mathbb{K} \to \mathbb{K}'$ is an extension of the isomorphism $\theta : \mathbb{F} \to \mathbb{F}'$, then $\overline{\theta}$ restricts to an isomorphism θ_α of $\mathbb{F}[\alpha]$ to some subfield of $\mathbb{K}'$. The isomorphism θ_α is determined by $\overline{\theta}(\alpha)$. Further $\overline{\theta}(\alpha)$ is a root of the polynomial $f_1'(X) = \widetilde{\theta}(f_1(X))$.

Conversely given a root β of $f_1'(X)$ in $\mathbb{K}'$, there is an isomorphism $\theta_\alpha : \mathbb{F}[\alpha] \to \mathbb{F}'[\beta]$ which extends θ.

We know from Corollary 4.1.53 that the number of extensions of θ to θ_α is equal to the number of distinct roots of $f_1'(X)$. Since the number of distinct roots of $f_1(X)$ and $f_1'(X)$ are same it follows that the number of extensions of the isomorphism θ to θ_α is less than or equal to the degree of the extension $[\mathbb{F}[\alpha] : \mathbb{F}]$. (Why?)

Now $\mathbb{K}$ is the splitting field of $f(X)$ over $\mathbb{F}[\alpha]$, $\mathbb{K}'$ is the splitting field of $f'(X)$ over $\mathbb{F}'[\beta]$ and $[\mathbb{K} : \mathbb{F}[\alpha]] < [\mathbb{K} : \mathbb{F}]$. Therefore we apply induction hypothesis to extend each of these θ_α's and conclude that the number of extensions $\overline{\theta}$ of each of these θ_α is less than or equal to $[\mathbb{K} : \mathbb{F}[\alpha]]$.

Since $[\mathbb{K} : \mathbb{F}] = [\mathbb{K} : \mathbb{F}[\alpha]][\mathbb{F}[\alpha] : \mathbb{F}]$, it follows that the number of extensions $\overline{\theta}$ of the isomorphism θ are less than or equal to $[\mathbb{K} : \mathbb{F}]$.

Suppose all the roots of $f(X)$ are distinct. Then all the irreducible factors of $f(X)$ also have distinct roots. Therefore the number of extensions of θ to θ_α is equal to $\deg(f_1) = [\mathbb{F}[\alpha] : \mathbb{F}]$ and the number of extensions $\overline{\theta}$ of each of these θ_α is equal to $[\mathbb{K} : \mathbb{F}[\alpha]]$. This shows that the number of extensions of $\theta : \mathbb{F} \to \mathbb{F}'$ to $\overline{\theta} : \mathbb{K} \to \mathbb{K}'$ is equal to $[\mathbb{K} : \mathbb{F}]$.

The converse is easy. □

Let $\mathbb{F}$ be a field and let $\mathbb{K}$ be the splitting field of a polynomial $f(X)$ over $\mathbb{F}$. If we take θ to be the identity map from $\mathbb{F}$ on to itself, then the previous theorem tells us that the number of extensions of the identity map of $\mathbb{F}$ to an isomorphism of $\mathbb{K}$ is less than or equal to the degree of the extension $[\mathbb{K} : \mathbb{F}]$. We record this as a corollary.

Corollary 4.1.61. *Let $\mathbb{K}$ be the splitting of a polynomial $f(X)$ over a field $\mathbb{F}$. Then the number of isomorphisms of $\mathbb{K}$ which are identity on $\mathbb{F}$ are less than or equal to $[\mathbb{K} : \mathbb{F}]$ and equality holds iff the polynomial $f(X)$ has distinct roots.*

We shall now give a sufficient condition for an irreducible polynomial $f(X)$ in $\mathbb{F}[X]$ over a field $\mathbb{F}$ to have distinct roots in the splitting field $\mathbb{K}$ of $f(X)$.

If $f(X) = a_0 + a_1X + a_2X^2 + \cdots + a_nX^n$, we define the derivative of $f(X)$, denoted by $f'(X)$ to be the polynomial $f'(X) := a_1 + 2a_2X + \cdots + na_nX^{n-1}$. If $f(X)$ is in $\mathbb{F}[X]$, then $f'(X)$ is also in $\mathbb{F}[X]$ and $\deg(f'(X)) \leq \deg(f(X)) - 1$. It is easy to show that, if $f(X)$ and $g(X)$ are two polynomials in $\mathbb{F}[X]$, then $(f(X)+g(X))' = f'(X) + g'(X)$ and $(f(X)g(X))' = f'(X)g(X) + f(X)g'(X)$.

Let $\mathbb{F}$ be a field and $f(X) = a_0 + a_1X + a_2X^2 + \cdots + a_nX^n$ be a non constant irreducible polynomial in $\mathbb{F}[X]$. Let $\mathbb{K}$ be the splitting field of the polynomial $f(X)$. Let α in $\mathbb{K}$ be a root of order $k \geq 1$. Then we can write $f(X) = (X-\alpha)^k g(X)$ where $g(X)$ is in $\mathbb{K}[X]$. The product rule for the derivative tells us that $f'(X) = k(X-\alpha)^{k-1}gX) + (X-\alpha)^k g'(X)$. Assume that $k \geq 2$. Then $f'(\alpha) = 0$. That is α is a root of the polynomial $f'(X)$ also. Since $f(X)$ is the minimal polynomial of α and α is a root of $f'(X) \in \mathbb{F}[X]$, it follows that the polynomial $f(X)$ divides the polynomial $f'(X)$ in $\mathbb{F}[X]$. Therefore there exists a polynomial $q(X)$ in $\mathbb{F}[X]$ such that $f'(X) = f(X)q(X)$. Since the degree of $f'(X)$ is less than n, this leads to a contradiction if $f'(X) \neq 0$.

Assume that the characteristic of the field $\mathbb{F}$ is zero. Then $n \neq 0$ in $\mathbb{F}$. Hence $f'(X) \neq 0$ and this leads to a contradiction. Thus every irreducible polynomial in $\mathbb{F}[X]$ has distinct roots in its splitting field.

Assume that the characteristic of the field $\mathbb{F}$ is $p \neq 0$. Then the polynomial $f'(X) = a_1 + 2a_2X + \cdots + na_nX^{n-1} \equiv 0$ iff $ka_k = 0$ for $1 \leq k \leq n$. This is possible iff k is a multiple of p whenever $a_k \neq 0$ for $1 \leq k \leq n$. In this case we can write $f(X) = \sum_{k=0}^{n} a_kX^k$ as $f(X) = \sum_{k=0}^{rp} a_kX^k = \sum_{k=0}^{r} a_{kp}(X^p)^k$. This polynomial can be written as $f(X) = \varphi(X^p)$ where $\varphi(X) = \sum_{k=0}^{r} a_kX^k$.

Thus we have proved the following.

Proposition 4.1.62. *Let $\mathbb{F}$ be a field and $f(X)$ be an irreducible polynomial in $\mathbb{F}[X]$. Then we have the following.*

1. *If characteristic of the field $\mathbb{F}$ is zero, then all the roots of f are distinct.*
2. *If characteristic of the field $\mathbb{F}$ is non zero, then the polynomial f has repeated roots iff $f(X) = \varphi(X^p)$ for some polynomial $\varphi(X)$ in $\mathbb{F}[X]$.*

If the field $\mathbb{F}$ is finite, then we can show that every irreducible polynomial has distinct roots.

Proposition 4.1.63. *Let $\mathbb{F}$ be a finite field. Then every irreducible polynomial $f(X) \in \mathbb{F}[X]$ has distinct roots in its splitting field.*

Proof. Let $\mathbb{F}$ be a finite field of characteristic p and $f(X)$ be an irreducible polynomial in $\mathbb{F}[X]$. If the polynomial $f(X)$ has repeated roots in its splitiing field $\mathbb{K}$, then $f(X)$ is of the form $f(X) = a_0 + a_1X^p + a_2X^{2p} + \cdots + a_kX^{kp}$.

Since the field $\mathbb{F}$ is finite, there exist $b_i \in \mathbb{F}$ such that $b_i^p = a_i$ for $0 \leq i \leq k$. (Why?) Therefore

$$\begin{aligned} f(X) &= a_0 + a_1X^p + a_2X^{2p} + \cdots + a_kX^{kp} \\ &= b_0^p + b_1^pX^p + b_2^pX^{2p} + \cdots + b_k^pkX^{kp} \\ &= (b_0 + b_1X + b_2X + \cdots + b_kX^k)^p. \end{aligned}$$

This contradicts the irreducibility of the polynomial $f(X)$ in $\mathbb{F}[X]$. □

4.1.4 Normal Extensions

Definition 4.1.64 (Normal Extensions). An extension $\mathbb{K}/\mathbb{F}$ is called a normal extension if an irreducible polynomial $f(X)$ in $\mathbb{F}[X]$ has one root in $\mathbb{K}$, then all the roots of the polynomial $f(X)$ are also in $\mathbb{K}$.

Example 4.1.65. Let $\mathbb{F}$ be a field. Then every degree two extension $\mathbb{K}$ over $\mathbb{F}$ is a normal extension.

> Let $\mathbb{K}/\mathbb{F}$ be a degree two extension and $f(X) \in \mathbb{F}[X]$ be an irreducible polynomial of degree n.
>
> We show that the extension is normal by showing that any irreducible polynomial $f(X) \in \mathbb{F}[X]$ with a root in $\mathbb{K}$ must have degree at most 2.
>
> Let α in $\mathbb{K}$ be a root of the irreducible polynomial $f(X)$. Then the degree of the extension $[\mathbb{F}[\alpha] : \mathbb{F}]$ is $n = \deg(f(X))$. Since $\mathbb{F}[\alpha]$ is a subfield of $\mathbb{K}$, we see that $[\mathbb{F}[\alpha] : \mathbb{F}] \leq [\mathbb{K} : \mathbb{F}] = 2$.

Assume $n = 1$. Then $f(X) = aX + b$ is a linear polynomial and α is in $\mathbb{F}$. A linear polynomial has only one root over a field and we are through.

Suppose $n = 2$. Then $\mathbb{F}[\alpha] = \mathbb{K}$ and $f(X) = (X - \alpha)q(X)$ where $q(X) = aX + b \in \mathbb{K}[X]$. It is now easy to see that the other root $-\frac{b}{a}$ of $f(X)$ is also in $\mathbb{K}$.

The relation between the splitting field of a polynomial $f(X)$ over a field $\mathbb{F}$ and a normal extension of the field $\mathbb{F}$ is in the following theorem.

Theorem 4.1.66. *Let $\mathbb{K}/\mathbb{F}$ be a finite extension. Then the extension $\mathbb{K}/\mathbb{F}$ is a normal extension iff $\mathbb{K}$ is the splitting field of a polynomial $f(X)$ over the field $\mathbb{F}$.*

Proof. We will assume that the finite extension $\mathbb{K}/\mathbb{F}$ is normal and show that it is the splitting field of a polynomial $f(X)$ over $\mathbb{F}$.

If r is the degree of the extension $\mathbb{K}/\mathbb{F}$, then there exists a basis $\{\alpha_1, \alpha_2, \ldots, \alpha_r\}$ of $\mathbb{K}$ over $\mathbb{F}$. Note that each α_i algebraic over $\mathbb{F}$.

Let $f_i(X) = \min_{\alpha_i,\mathbb{F}}(X)$ be the minimal polynomial of α_i over the field $\mathbb{F}$ for $1 \leq i \leq r$. Let $f(X) = \prod_{i=1}^{r} f_i(X)$. We will show that $\mathbb{K}$ is the splitting field of the polynomial $f(X)$ over $\mathbb{F}$.

The extension $\mathbb{K}/\mathbb{F}$ is normal and one root α_i of $f_i(X)$ is in $\mathbb{K}$ for each i. Hence all the roots of each of these irreducible polynomials $f_i(X)$ are in $\mathbb{K}$. As a consequence all the roots of the polynomial $f(X)$ are in $\mathbb{K}$.

Let $\mathbb{L}$ be a subfield of $\mathbb{K}$ containing all the roots of the polynomial $f(X)$. In particular the roots α_i are in $\mathbb{L}$. Hence the field $\mathbb{K} = \mathbb{F}[\alpha_1, \alpha_2, \cdots, \alpha_r]$ is contained in $\mathbb{L}$. Thus $\mathbb{K}$ is the splitting field of the polynomial $f(X)$ over $\mathbb{F}$.

Conversely assume that the extension $\mathbb{K}/\mathbb{F}$ is the splitting field of a polynomial $f(X)$. Let $g(X)$ be an irreducible polynomial in $\mathbb{F}[X]$ and let α be a root of $g(X)$ in $\mathbb{K}$. We need to show that all the roots of $g(X)$ are in $\mathbb{K}$.

Let β be another root of $g(X)$ in the splitting field $\mathbb{L}$ of the polynomial $g(X)$. Then there exists an isomorphism $\theta : \mathbb{F}[\alpha] \to \mathbb{F}[\beta]$ such that $\theta(\alpha) = \beta$ and θ is identity on $\mathbb{F}$. We need to show that β is in $\mathbb{K}$.

Let $\mathbb{M}$ be the splitting field of the product polynomial $f(X)g(X)$ over $\mathbb{F}$. Then the fields $\mathbb{K}$ and $\mathbb{L}$ are subfields of the field $\mathbb{M}$. Now we can extend the isomorphism $\theta : \mathbb{F}[\alpha] \to \mathbb{F}[\beta]$ to an isomorphism $\theta_1 : \mathbb{K}[\alpha] \to \mathbb{K}[\beta]$ such that θ is identity on $\mathbb{F}$. Since α is in $\mathbb{K}$, the field $\mathbb{K}[\alpha] = \mathbb{K}$ and hence $\mathbb{K}$ is isomorphic to the field $\mathbb{K}[\beta]$. Now we use tower law to conclude

$$\begin{aligned}
[\mathbb{K}[\beta] : \mathbb{F}] &= [\mathbb{K}[\beta] : \mathbb{F}[\beta]][\mathbb{F}[\beta] : \mathbb{F}] \\
&= [\theta_1(\mathbb{K}[\alpha]) : \theta_1(\mathbb{F}[\alpha])][\theta_1(\mathbb{F}[\alpha]) : \mathbb{F}] \\
&= [\mathbb{K}[\alpha] : \mathbb{F}[\alpha]][\mathbb{F}[\alpha] : \mathbb{F}] \\
&= [\mathbb{K}[\alpha] : \mathbb{F}] \\
&= [\mathbb{K} : \mathbb{F}].
\end{aligned}$$

This shows that $[\mathbb{K}[\beta] : \mathbb{F}] = [\mathbb{K} : \mathbb{F}]$. Since $\mathbb{K}$ is a subfield of $\mathbb{K}[\beta]$, it follows that β is in $\mathbb{K}$ and the proof is complete. □

Examples 4.1.67. Let $\zeta \neq 1$ be a cube root of unity and $\sqrt[3]{2}$ be a positive real root of the polynomial $X^3 - 2$. Then $\mathbb{K} = \mathbb{Q}(\sqrt[3]{2}, \zeta)$ is the splitting field of the polynomial $X^3 - 2$ over $\mathbb{Q}$ and hence it is normal. But the the extension $\mathbb{Q}[\sqrt[3]{2}]/\mathbb{Q}$ is not a normal extension.

Exercise 4.1.68. 1. Show that the extension $\mathbb{Q}(\sqrt{-3})/\mathbb{Q}$ is normal.

2. Show that the extension $\mathbb{Q}(\sqrt[4]{2})/\mathbb{Q}$ is not normal.

Another important property about normal extensions is the following theorem.

Theorem 4.1.69. *Let $\mathbb{L}/\mathbb{F}$ be a finite normal extension. Let $\mathbb{K}$ be a subfield of $\mathbb{L}$ containing $\mathbb{F}$. Then $\mathbb{L}/\mathbb{K}$ is a normal extension.*

Proof. We will prove the theorem by showing that $\mathbb{L}$ is the splitting field of a polynomial $g(X)$ in $\mathbb{K}[X]$.

Since $\mathbb{L}/\mathbb{F}$ is a finite extension, the extension $\mathbb{L}/\mathbb{K}$ is also finite. Let $S := \{\alpha_i \in \mathbb{L} : 1 \leq i \leq r\}$ be a $\mathbb{K}$ basis of $\mathbb{L}$. Then $\mathbb{L} = \mathbb{K}[S]$ and each of the elements α_i are algebraic over $\mathbb{K}$ as well as over $\mathbb{F}$.

Let $g_i(X)$ be the minimal polynomial of α_i over $\mathbb{K}[X]$ and $f_i(X)$ be the minimal polynomial of α_i over $\mathbb{F}[X]$. Since $\mathbb{F}$ is a subfield of $\mathbb{K}$, the polynomial $g_i(X)$ divides $f_i(X)$ in $\mathbb{K}[X]$ and there exists $h_i(X) \in \mathbb{K}[X]$ such that $f_i(X) = g_i(X)h_i(X)$. The field $\mathbb{L}$ contains one root α_i of the irreducible polynomial $f_i(X)$ in $\mathbb{F}[X]$. Therefore the field $\mathbb{L}$, being normal extension of $\mathbb{F}$, contains all the roots of the polynomial $f_i(X)$. Since the roots of $g_i(X)$ in $\mathbb{L}$ are roots of $f_i(X)$, the field $\mathbb{L}$ has all the roots of the irreducible polynomial $g_i(X)$ in $\mathbb{K}[X]$.

Let $g(X) = \prod_{i=1}^{r} g_i(X)$. Then $\mathbb{L}$ contains all the roots of the polynomial $g(X)$. Hence the splitting field $\mathbb{E}$ of $g(X)$ is contained in $\mathbb{L}$. Since $\mathbb{K}$ is a subfield of $\mathbb{E}$ and the roots $\alpha_1, \alpha_2, \ldots, \alpha_r$ are in $\mathbb{E}$, it follows that $\mathbb{L} = \mathbb{K}[S] \subseteq \mathbb{E} \subseteq \mathbb{L}$. Thus $\mathbb{E} = \mathbb{L}$ and the proof is complete. □

Remark 4.1.70. If $\mathbb{L}/\mathbb{F}$ is a finite normal extension and $\mathbb{K}$ is a subfield of $\mathbb{L}$ containing $\mathbb{F}$, then the extension $\mathbb{K}/\mathbb{F}$ need not be normal. This can be seen from Example 4.1.67(1).

4.1.5 Galois Group

Given a field $\mathbb{K}$, we let $\mathrm{Aut}(\mathbb{K})$ to be the set of all automorphisms of the field $\mathbb{K}$. We know that this set is a group under the composition of functions.

Now we want to study the automorphisms of a field extension $\mathbb{K}/\mathbb{F}$ which extend the identity map on $\mathbb{F}$.

Definition 4.1.71. (Galois Group) For a subfield $\mathbb{F}$ of $\mathbb{K}$, we let

$$\mathrm{Gal}(\mathbb{K}/\mathbb{F}) := \{\sigma \in \mathrm{Aut}(\mathbb{K}) : \sigma(x) = x \text{ for all } x \in \mathbb{F}\}.$$

Then it is easy to see that $\mathrm{Gal}(\mathbb{K}/\mathbb{F})$ is a subgroup of $\mathrm{Aut}(\mathbb{K})$. The subgroup $\mathrm{Gal}(\mathbb{K}/\mathbb{F})$ is called the *Galois group* of the extension $\mathbb{K}/\mathbb{F}$.

If $\mathbb{K}$ is the splitting field of a polynomial $f(X)$ in $\mathbb{F}[X]$, the Galois group $\mathrm{Gal}(\mathbb{K}/\mathbb{F})$ is denoted by $\mathrm{Gal}(f)$ and is also called the Galois group of the polynomial $f(X)$.

Example 4.1.72. 1. Let $\mathbb{K} = \mathbb{Q}[\sqrt{2}]$ and $\mathbb{F} = \mathbb{Q}$. Let $\sigma : \mathbb{K} \to \mathbb{K}$ be an isomorphism such that $\sigma(r) = r$ for all $r \in \mathbb{Q}$. Then

$$2 = \sigma(2) = \sigma(\sqrt{2})^2) = (\sigma(\sqrt{2}))^2.$$

This means that $\sigma(\sqrt{2}) = \pm\sqrt{2}$ and hence $\mathrm{Gal}(\mathbb{K}/\mathbb{F}) = \{\pm Id\}$.

2. Let $\mathbb{K} = \mathbb{Q}[\sqrt[3]{2}]$ and $\mathbb{F} = \mathbb{Q}$. Let σ be an element of $\mathrm{Gal}(\mathbb{K}/\mathbb{F})$. Then

$$2 = \sigma(2) = \sigma((\sqrt[3]{2})^3) = (\sigma(\sqrt[3]{2}))^3.$$

This means that $\sigma(\sqrt[3]{2})$ is also a root of the polynomial $X^3 - 2$ in $\mathbb{K}$. But the only root of $X^3 - 2$ that is in $\mathbb{K}$ is $\sqrt[3]{2}$. Therefore $\sigma(\sqrt[3]{2}) = \sqrt[3]{2}$ and this means that $\sigma = Id$. Hence $\mathrm{Gal}(\mathbb{K}/\mathbb{F}) = \{Id\}$.

3. Let $\mathbb{K}$ be the splitting field of the polynomial $f(X) = X^4 - 4$. Then the roots of the polynomial $f(X)$ are $\pm\sqrt{2}$ and $\pm i\sqrt{2}$. Therefore $\mathbb{K} = \mathbb{Q}[\sqrt{2}, i\sqrt{2}] = \mathbb{Q}[i, \sqrt{2}]$. Let σ be an element of $\mathrm{Gal}(\mathbb{K}/\mathbb{F})$. Then $2 = \sigma(2) = (\sigma(\sqrt{2}))^2$. Hence $\sigma(\sqrt{2}) = \pm\sqrt{2}$. similarly $\sigma(i) = \pm i$. Further we observe that, if $\sigma \neq Id$, then $\sigma^2(\sqrt{2}) = \pm\sigma(\sqrt{2}) = \sqrt{2}$ and $\sigma^2(i) = \sigma(\pm i) = \pm\sigma(i) = i$. This proves that an element $\sigma \neq Id$ is of order 2 and any σ in $\mathrm{Gal}(\mathbb{K}/\mathbb{F})$ must be one of the following four maps:

$$Id, \begin{pmatrix} i \mapsto i \\ \sqrt{2} \mapsto -\sqrt{2} \end{pmatrix}, \begin{pmatrix} i \mapsto -i \\ \sqrt{2} \mapsto \sqrt{2} \end{pmatrix}, \begin{pmatrix} i \mapsto -i \\ \sqrt{2} \mapsto -\sqrt{2} \end{pmatrix}$$

Thus $\mathrm{Gal}(\mathbb{K}/\mathbb{F})$ is a group of order 4 where each $\sigma \neq Id$ is of order 2. Hence $\mathrm{Gal}(f) = \mathrm{Gal}(\mathbb{K}/\mathbb{F})$ is isomorphic to $\mathbb{Z}_2 \times \mathbb{Z}_2$.

4. Let $f(X) = X^3 - 2$ in $\mathbb{Q}[X]$. Then the roots of $f(X)$ are $\alpha := \sqrt[3]{2}$, $\sqrt[3]{2}\,\zeta$ and $\sqrt[3]{2}\,\zeta^2$ where $\zeta \neq 1$ is a cube root of unity. Hence the splitting field of $f(X)$ over $\mathbb{Q}$ is $\mathbb{K} = \mathbb{Q}[\alpha, \alpha\,\zeta, \alpha\,\zeta^2]$.

Let us now observe that $\zeta = \frac{\alpha\zeta}{\alpha}$ is also in $\mathbb{K}$. Since $\zeta = \frac{-1 \pm i\sqrt{3}}{2}$, we can write $\mathbb{K}$ as $\mathbb{K} = \mathbb{Q}[\alpha, \zeta] = \mathbb{Q}[\alpha, i\sqrt{3}]$.

Let τ be the complex conjugation. Since α is real, $\tau(\alpha) = \alpha$ and $\tau(\zeta) = \overline{\zeta} = \zeta^2$. Hence τ is in $\mathrm{Gal}(f) = \mathrm{Gal}(\mathbb{K}/\mathbb{F})$ and it is of order 2.

Let $\sigma \neq \tau, Id$ be an element of $\mathrm{Gal}(f) = \mathrm{Gal}(\mathbb{K}/\mathbb{F})$. Since α is a root of $f(X)$ and $\sigma\,|_{\mathbb{Q}} = Id$, we see that $\sigma(\alpha)$ is also a root of $f(X)$. Since $\sigma \neq \tau$, we also conclude that $\sigma(\alpha)$ must be $\alpha\,\zeta$ or $\alpha\,\zeta^2$. Observe that $\sigma(\zeta)$ is also a cube root of unity. Let us assume $\sigma(\alpha) = \alpha\,\zeta$ and $\sigma(\zeta) = \zeta$. Then $\sigma^3 = Id$.

Let $x = a\alpha + b\alpha\zeta + c\alpha\zeta^2$ be an element of $\mathbb{K}$. Then

$$\begin{aligned} \tau\sigma\tau(x) &= \tau\sigma(a\alpha + b\alpha\overline{\zeta} + c\alpha\zeta) \\ &= \tau(a\alpha\zeta + b\alpha + c\alpha\zeta^2) \\ &= a\alpha\overline{\zeta} + b\alpha + c\alpha\zeta \\ &= \sigma^2(x). \end{aligned}$$

Hence the group Gal(f) is

$$\mathrm{Gal}(f) := \{Id, \tau, \sigma, \sigma^2, \sigma\tau, \sigma^2\tau : \tau^2 = Id = \sigma^3 \ \& \ \tau\sigma\tau = \sigma^{-1}\}$$

and we know this group is isomorphic to S_3. Thus Gal(f) is isomorphic to S_3.

5. Let $\mathbb{K} = \mathbb{Q}[\sqrt{2}, \sqrt{3}]$ and $\mathbb{F} = \mathbb{Q}$. Let $\sigma \neq Id$ be an element of Gal$(\mathbb{K}/\mathbb{F})$. Then $2 = \sigma(2) = \sigma((\sqrt{2})^2) = \sigma(\sqrt{2})^2$ and $3 = \sigma(\sqrt{3})^2$. Hence $\sigma(\sqrt{2}) = \pm\sqrt{2}$ and $\sigma(\sqrt{3}) = \pm\sqrt{3}$. As a consequence σ is of order 2 in Gal$(\mathbb{K}/\mathbb{F})$ and σ must be one of the following three maps:

$$\begin{pmatrix} \sqrt{2} \mapsto \sqrt{2} \\ \sqrt{3} \mapsto -\sqrt{3} \end{pmatrix}, \quad \begin{pmatrix} \sqrt{2} \mapsto -\sqrt{2} \\ \sqrt{3} \mapsto \sqrt{3} \end{pmatrix}, \quad \begin{pmatrix} \sqrt{2} \mapsto -\sqrt{2} \\ \sqrt{3} \mapsto -\sqrt{3} \end{pmatrix}.$$

This shows that Gal$(\mathbb{K}/\mathbb{F})$ is a group of order 4 and it is not cyclic. Hence Gal$(\mathbb{K}/\mathbb{F})$ is isomorphic to $\mathbb{Z}_2 \times \mathbb{Z}_2$.

Remark 4.1.73. In examples 1, 2, 4 and 5 of Examples 4.1.72, we notice that that the size of Gal$(\mathbb{K}/\mathbb{F})$ is same as the degree of the extension and in example 2 we also notice that Gal$(\mathbb{K}/\mathbb{F})$ is S_3. We will show that these are particular examples of more general results.

Exercise 4.1.74.

1. Let $f(X) = X^4 + 1$ in $\mathbb{Q}[X]$ and $\mathbb{K}$ be the splitting field of $f(X)$ over $\mathbb{Q}$. Show that Gal$(\mathbb{K}/\mathbb{F})$ is isomorphic to $\mathbb{Z}_2 \times \mathbb{Z}_2$.

2. Let $\mathbb{K} = \mathbb{Q}[\sqrt{2 + \sqrt{2}}]$. Show that Gal$(\mathbb{K}/\mathbb{F})$ is a cyclic group of order 4.

3. Let $f(X) = X^4 - 4X^2 + 3$ in $\mathbb{Q}[X]$. Find the splitting field $\mathbb{K}$ of $f(X)$ over $\mathbb{Q}$ and Gal(f).

4. Let $\mathbb{K} = \mathbb{Q}[\sqrt[8]{2}, i]$. Show that
   (a) Gal$(\mathbb{K}/\mathbb{F}) \simeq \mathbb{Z}_8$, if $\mathbb{F} = \mathbb{Q}[i]$,
   (b) Gal$(\mathbb{K}/\mathbb{F}) \simeq D_4$, if $\mathbb{F} = \mathbb{Q}[\sqrt{2}]$ and
   (c) Gal$(\mathbb{K}/\mathbb{F}) \simeq Q_8$, if $\mathbb{F} = \mathbb{Q}[i\sqrt{2}]$; here Q_8 denotes the group of quaternions of order 8. (See Example 2.1.23).

5. Let p be a prime number and $\mathbb{K} = \mathbb{Q}[\zeta_p]$ where ζ_p is the primitive p-th root of unity. Find Gal$(\mathbb{K}/\mathbb{F})$ if $\mathbb{F} = \mathbb{Q}$.

Now we study the structure of the group Gal$(\mathbb{K}/\mathbb{F})$ of an extension $\mathbb{K}/\mathbb{F}$.

Theorem 4.1.75. *Let $\mathbb{K}/\mathbb{F}$ be the splitting field of a polynomial $f(X)$ over $\mathbb{F}$. If the polynomial $f(X)$ has n-distinct roots, then Gal(f) is isomorphic to a subgroup of S_n.*

Proof. Let $S = \{\alpha_1, \alpha_2, \ldots, \alpha_n\}$ be the set of distinct roots of the polynomial $f(X)$. Let σ be an element of Gal(f). Then the elements $\sigma(\alpha_i)$ are all distinct and they are all roots of the polynomial $f(X)$. Hence σ maps the set S to itself, thus defining a permutation of S.

This induces a map $\theta : \text{Gal}(f) \to S_n$. It is very easy to show that this map is a homomorphism.

Since $\mathbb{K}$ is the splitting field of f and α_i for $1 \leq i \leq r$ are the distinct roots of f, we know that $\mathbb{K} = \mathbb{F}[\alpha_1, \alpha_2, \ldots, \alpha_r]$. Suppose there exists σ in Gal(f) such that $\sigma(\alpha_i) = \alpha_i$ for all i. Then $\sigma = Id$, the identity automorphism of $\mathbb{K}$. This proves that θ is a one-one map and hence Gal(f) is isomorphic to a subgroup of S_n. □

The size of the subgroup Gal$(\mathbb{K}/\mathbb{F})$ is related to the degree of the extension $\mathbb{K}/\mathbb{F}$ in the following theorem.

Theorem 4.1.76. *Let $\mathbb{K}$ be the splitting field of a polynomial $f(X)$ over a field $\mathbb{F}$. If all the roots of the polynomial $f(X)$ are distinct then*

$$|Gal(\mathbb{K}/\mathbb{F})| = [\mathbb{K} : \mathbb{F}].$$

Proof. This follows from Corollary 4.1.61 of Theorem 4.1.60. □

4.1.6 Radical Extensions

For the rest of this chapter we assume that all the fields in our discussion have characteristic zero.

Definition 4.1.77 (Simple Radical extension). An extension $\mathbb{K}/\mathbb{F}$ is called a simple radical extension if there exists an element $\alpha \in \mathbb{K}$ such that $\mathbb{K} = \mathbb{F}[\alpha]$ and α^n is in $\mathbb{F}$ for some $n \geq 2$.

Example 4.1.78.

1. Let $\mathbb{K} = \mathbb{Q}[\sqrt{2}]$. Then $\mathbb{K}$ is a simple radical extension of $\mathbb{Q}$. Further $\mathbb{K}$ is also the splitting field of the polynomial $f(X) = X^2 - 2$ over $\mathbb{Q}$.

2. Let $\mathbb{K} = \mathbb{Q}[\sqrt[3]{2}]$. Then $\mathbb{K}$ is a simple radical extension of $\mathbb{Q}$. The element $\sqrt[3]{2}$ is a root of the irreducible polynomial $f(X) = X^3 - 2$ over $\mathbb{Q}[X]$. However $\mathbb{Q}[\sqrt[3]{2}]$ is not the splitting field of the polynomial $f(X)$. Hence this extension is not normal. If we now adjoin a cube root of unity $\zeta \neq 1$ to $\mathbb{Q}[\sqrt[3]{2}]$, then the extension $\mathbb{Q}[\sqrt[3]{2}, \zeta]$ is the splitting field of the polynomial $f(X)$ over $\mathbb{Q}$.

   Further each of these extensions $\mathbb{Q}[\sqrt[3]{2}]/\mathbb{Q}$ and $\mathbb{Q}[\sqrt[3]{2}, \zeta]/\mathbb{Q}[\sqrt[3]{2}]$ are simple radical extensions.

The example above suggests the following proposition.

Proposition 4.1.79. *Let $\mathbb{K} = \mathbb{F}[\alpha]/\mathbb{F}$ be a simple radical extension such that $\alpha^n = \beta$ is in $\mathbb{F}$ for some $n \geq 2$. If $\mathbb{F}$ contains a primitive n-th root of unity $\zeta \neq 1$, then $\mathbb{K}$ is the splitting field of the polynomial $f(X) = X^n - \beta$ over $\mathbb{F}$.*

Proof. Let ζ be a primitive n-th root of unity in $\mathbb{F}$. Then $\alpha, \alpha\zeta, \alpha\zeta^2, \ldots, \alpha\zeta^{n-1}$ are all the roots of the polynomial $f(X)$ and all these roots are in $\mathbb{K} = \mathbb{F}[\alpha]$. Since $\mathbb{F}[\alpha]$ is the smallest field containing $\mathbb{F}$ and α, it follows that $\mathbb{K} = \mathbb{F}[\alpha]$ is the splitting field of the polynomial $f(X) = X^n - \beta$. □

We will now give an example of a splitting field which is not a simple radical extension.

Example 4.1.80. Let $\mathbb{K}$ be the splitting field of the polynomial $\varphi(X) = X^3 - 3X + 1$ over $\mathbb{Q}$. The polynomial $\varphi(X)$ is irreducible over $\mathbb{Q}$. This can be seen by using Eisenstein's irreducibility criterion or by direct computation.

Observe that $\varphi(0) = 1$, $\varphi(1) = -1$, $\varphi(-2) = -1$ and $\varphi(2) = 3$. Hence by intermediate value theorem, the polynomial has three real roots. Therefore $\mathbb{K} \subseteq \mathbb{R}$.

We show that the degree of the extension $[\mathbb{K} : \mathbb{Q}]$ is 3.

Let ζ be primitive 9-th root of unity. Then ζ^3 is a primitive cube root of unity and hence $\zeta^3 + \overline{\zeta}^3 + 1 = 0$.

Let $\alpha = \zeta + \zeta^8 = \zeta + \overline{\zeta}$. Then

$$\alpha^3 = \zeta^3 + \overline{\zeta}^3 + 3\zeta\overline{\zeta}(\zeta + \overline{\zeta}) = 3\alpha - 1.$$

This shows that α is a root of the polynomial $\varphi(X)$.

Let $\beta = \zeta^2 + \overline{\zeta}^2$ and $\gamma = \zeta^4 + \overline{\zeta}^4$. Then we can easily verify that β and γ the other two roots of the polynomial $\varphi(X)$.

Further one can check that $\alpha^2 = \beta + 2$. This shows that the root β is in the field $\mathbb{Q}[\alpha]$. Since $\gamma = -(\alpha + \beta)$, the root γ is also in $\mathbb{Q}[\alpha]$. Thus the splitting field $\mathbb{K}$ of $\varphi(X)$ over $\mathbb{Q}$ is $\mathbb{Q}[\alpha]$ and the degree of the extension is 3.

We will now show that the extension is not a simple radical extension.

Suppose the extension is a simple radical extension. Then there exists an element $a \in \mathbb{K}$ such that $\mathbb{K} = \mathbb{Q}(a)$ and $a^n \in \mathbb{Q}$ for some $n \geq 3$. Let $f(X) = X^n - a^n$ and $g(X) \in \mathbb{Q}[X]$ be the minimal polynomial of a. Let us observe that the degree of the polynomial g is 3.

Then the polynomial $g(X)$ divides $f(X)$ in $\mathbb{Q}[X]$ and we can write $f(X) = g(X)h(X)$ for some polynomial $h(X)$ in $\mathbb{Q}[X]$. Let r be another root of the polynomial $g(X)$ in $\mathbb{K}$. Then $r^n - a^n = g(r)h(r) = 0$ and hence $(\frac{r}{a})^n = 1$. This proves that $\frac{r}{a} \in \mathbb{K} \subseteq \mathbb{R}$ is a root of unity. Since the only real roots of unity are ± 1, we conclude that $r = \pm a$ and the cubic polynomial $g(X)$ has only two roots, a contradiction.

One can produce many more examples.

Definition 4.1.81 (Radical extension). An extension $\mathbb{K}/\mathbb{F}$ is called a radical extension if there exist finite number of subfields $\mathbb{K}_i$ of $\mathbb{K}$ such that

1. $\mathbb{F} = \mathbb{K}_0 \subseteq \mathbb{K}_1 \subseteq \mathbb{K}_2 \subseteq \ldots \subseteq \mathbb{K}_{r-1} \subseteq \mathbb{K}_r = \mathbb{K}$ and

2. $\mathbb{K}_{i+1} = \mathbb{K}_i(\alpha_i)$ such that $\alpha_i^{n_i}$ is in $\mathbb{K}_i$ for some α_i in $\mathbb{K}_{i+1}$ and $n_i \geq 2$.

We say that a radical extension is a prime radical extension if the natural numbers n_i are prime numbers.

Example 4.1.82. 1. The extension $\mathbb{Q} \subseteq \mathbb{Q}[\sqrt[3]{2}] \subseteq \mathbb{Q}[\sqrt[3]{2}, \zeta]$ is a prime radical extension.

2. The extensions $\mathbb{Q} \subseteq \mathbb{Q}[\sqrt{2}] \subseteq \mathbb{Q}[\sqrt{2}, \sqrt{3}]$ is a prime radical extension.

3. The extension $\mathbb{Q} \subseteq \mathbb{Q}[\sqrt{2}] \subseteq \mathbb{Q}[\sqrt{2}, \sqrt{3}] \subseteq \mathbb{Q}[\sqrt{3}, \sqrt[4]{2}]$is a radical extension.

Lemma 4.1.83. *Let $\mathbb{K}/\mathbb{F}$ be a simple radical extension. Then there exists prime radical extension*

$$\mathbb{F} = \mathbb{K}_0 \subseteq \mathbb{K}_1 \subseteq \mathbb{K}_2 \ldots \subseteq \mathbb{K}_r = \mathbb{K}.$$

Proof. Let $\mathbb{K}/\mathbb{F}$ be a simple radical extension and let $[\mathbb{K} : \mathbb{F}] = n$. Let $\alpha \in \mathbb{K}$ be such that $\mathbb{K} = \mathbb{F}[\alpha]$ and $\alpha^n \in \mathbb{F}$.

Let p_1 be a prime divisor of n and we write $n = p_1 n_1$ for some $m_1 \in \mathbb{N}$. If we let $\alpha_1 = \alpha^{n_1}$, then $\alpha_1^{p_1} = \alpha^n$ is in $\mathbb{F}$ and $\alpha^{n_1} = \alpha_1$ is in $\mathbb{F}(\alpha_1)$. Further $\mathbb{F} \subseteq \mathbb{F}[\alpha_1] \subseteq \mathbb{F}[\alpha] = \mathbb{K}$ and the extension $\mathbb{F}(\alpha_1)/\mathbb{F}$ is a simple prime radical extension. Now by induction on the degree of the extension $\mathbb{K}/\mathbb{F}[\alpha_1]$, we get a tower of subfields $\mathbb{K}_1 \subseteq \mathbb{K}_2 \subseteq \ldots \subseteq \mathbb{K}_r = \mathbb{K}$ such that each extension $\mathbb{K}_{i+1}/\mathbb{K}_i$ is a simple radical extension. Thus we get a tower of simple prime radical extensions, namely $\mathbb{F} = \mathbb{K}_0 \subseteq \mathbb{K}_1 \subseteq \mathbb{K}_2 \subseteq \ldots \subseteq \mathbb{K}_r = \mathbb{K}$. This completes the proof. □

Corollary 4.1.84. *Let $\mathbb{K}/\mathbb{F}$ be a radical extension. Then there exist subfields $\mathbb{K}_i$ for $0 \leq i \leq r$ such that $\mathbb{F} = \mathbb{K}_0 \subseteq \mathbb{K}_1 \subseteq \mathbb{K}_2 \subseteq \ldots \subseteq \mathbb{K}_r = \mathbb{K}$ is a prime radical extension.*

Proof. The proof follows easily from the lemma. □

Lemma 4.1.85. *Let $\mathbb{K}/\mathbb{L}$ and $\mathbb{L}/\mathbb{F}$ be two radical extensions. Then $\mathbb{K}/\mathbb{F}$ is also a radical extension.*

Proof. The proof is very easy and left to the reader. □

4.1.7 Solvable Radical Extensions

Definition 4.1.86 (Solvable Radical extension). A polynomial $f(X)$ over a field $\mathbb{F}$ is said to be solvable by radicals over $\mathbb{F}$ if there is a radical extension $\mathbb{L}/\mathbb{F}$ such that the splitting field $\mathbb{K}$ of the polynomial $f(X)$ is contained in $\mathbb{L}$.

Example 4.1.87. Let $f(X) = X^2 + bX + C \in \mathbb{Q}[X]$ and $\mathbb{K} = \mathbb{Q}(\sqrt{b^2 - 4c})$. Then $\mathbb{K}/\mathbb{Q}$ is a radical extension and therefore $f(X)$ is solvable by radicals over $\mathbb{Q}$.

Theorem 4.1.88. *Let $\mathbb{F}$ be a field of characteristic zero and $\mathbb{K}/\mathbb{F}$ be a radical extension. Then there exists an extension $\mathbb{L}/\mathbb{F}$ such that*

1. *$\mathbb{K} \subseteq \mathbb{L}$,*
2. *$\mathbb{L}$ is radical over $\mathbb{F}$ and*
3. *$\mathbb{L}$ is normal over $\mathbb{F}$.*

Proof. Since $\mathbb{K}/\mathbb{F}$ is a radical extension, there exists a tower $\mathbb{F} = \mathbb{K}_0 \subseteq \mathbb{K}_1 \subseteq \mathbb{K}_2 \subseteq \ldots \mathbb{K}_r = \mathbb{K}$ such that $\mathbb{K}_{i+1}/\mathbb{K}_i$ is a simple radical extension. Now we prove the result by induction on the length of the string r.

If $r = 0$, it is obvious. Now we assume that the result to be true for all extensions with length $k < r$.

Now consider the tower $\mathbb{F} = \mathbb{K}_0 \subseteq \mathbb{K}_1 \subseteq \mathbb{K}_2 \subseteq \ldots \mathbb{K}_{r-1}$. Then this is a radical tower. Hence by induction hypothesis there exists a field $\mathbb{L}$ such that $\mathbb{K}_{r-1} \subseteq \mathbb{L}$, $\mathbb{L}/\mathbb{F}$ is radical and $\mathbb{L}/\mathbb{F}$ is normal.

Since the extension $\mathbb{L}/\mathbb{F}$ is normal it is the splitting field of a polynomial $g(X)$ over $\mathbb{F}$.

Note that there exists an element $a \in \mathbb{K}$ such that $\mathbb{K} = \mathbb{K}_r = \mathbb{K}_{r-1}[a]$ and $a^n = b \in \mathbb{K}_{r-1}$. This means that a is a root of the polynomial $X^n - b$ in $\mathbb{L}$. Let $f(X) = \min_{a,\mathbb{L}}(X)$ be the minimal polynomial of a over $\mathbb{L}$ and let $\mathbb{E}$ be the splitting field of $f(X)$ over $\mathbb{L}$. Since a is in $\mathbb{K}$ and $\mathbb{K}_{r-1} \subseteq \mathbb{L} \subseteq \mathbb{E}$, we have $\mathbb{K}_r = \mathbb{K} \subseteq \mathbb{E}$.

Let $a = \alpha_1, \alpha_2, \ldots, \alpha_t$ be all the roots of the polynomial $f(X)$. We make use of the following observation again: If $\mathbb{E}$ is the splitting field of a polynomial $f(X)$ over $\mathbb{F}[X]$ and α_i for $1 \leq i \leq t$ are the roots of the polynomial $f(X)$, then $\mathbb{E} = \mathbb{F}[\alpha_1, \alpha_2, \cdots, \alpha_t]$.

Now the polynomial $f(X)$ divides the polynomial $X^n - b$ in $\mathbb{L}[X]$. Let $X^n - b = f(X)q(X)$ for some polynomial $q(X)$ in $\mathbb{L}[X]$. Then $\alpha_i^n - b = f(\alpha_i)q(\alpha_i) = 0$ and hence $\alpha_i^n = b$ is in the field $\mathbb{L}$ for all i. This means that the extension $\mathbb{E}/\mathbb{L}$ is a radical extension. Since the extension $\mathbb{L}/\mathbb{F}$ is a radical extension, we conclude from Lemma 4.1.85 that the extension $\mathbb{E}/\mathbb{F}$ is also a radical extension.

We will now show that the extension $\mathbb{E}/\mathbb{F}$ is normal.

The polynomial $g(X)$ splits in $\mathbb{L}$, the polynomial $f(X)$ splits in $\mathbb{E}$ and $\mathbb{L} \subseteq \mathbb{E}$. Hence the polynomial $f(X)g(X)$ splits in $\mathbb{E}$. Let $\mathbb{E}_1$ be the splitting field $f(X)g(X)$ over $\mathbb{L}$. Since $\mathbb{E}$ is the splitting field of $f(X)$ over the field $\mathbb{L}$, we see that $\mathbb{E} \subseteq \mathbb{E}_1$. The field $\mathbb{L}$ is also contained in $\mathbb{E}_1$. But the field $\mathbb{E}$ contains all the roots of the polynomials $f(X)$ and $g(X)$. Hence $\mathbb{E} = \mathbb{E}_1$. Thus we have shown that $\mathbb{E}$ is the splitting field of a polynomial over $\mathbb{L}$ and the extension $\mathbb{E}/\mathbb{L}$ is normal. □

Corollary 4.1.89. *Let $\mathbb{F}$ be a field of characteristic zero and let $\mathbb{K}/\mathbb{F}$ be a radical extension. Then there exists an extension $\mathbb{E}/\mathbb{F}$ such that*

1. $\mathbb{K} \subseteq \mathbb{E}$,
2. $\mathbb{E}$ *is prime radical over* $\mathbb{F}$ *and*
3. $\mathbb{E}$ *is normal over* $\mathbb{F}$.

Proof. The proof is easy: It follows from Lemma 4.1.83. □

Corollary 4.1.90. *Let $\mathbb{F}$ be a field of characteristic zero and let $\mathbb{K}/\mathbb{F}$ be a radical extension. Then there exists an extension $\mathbb{E}/\mathbb{F}$ such that*

1. $\mathbb{K} \subseteq \mathbb{E}$,
2. $\mathbb{E}$ *is normal over* $\mathbb{F}$ *and*
3. $\mathbb{E}/\mathbb{F}$ *is a prime radical extension such that* $\mathbb{F} = \mathbb{E}_0 \subseteq \mathbb{E}_1 \subseteq \mathbb{E}_2 \subseteq \ldots \mathbb{E}_r = \mathbb{E}$ *such that*
   (a) *for each j, there exists α_j in $\mathbb{E}_j$ and a prime number p_j such that* $\mathbb{E}_j = \mathbb{E}_{j-1}(\alpha_j)$ *and* $\alpha_j^{p_j}$ *is in* $\mathbb{E}_{j-1}$, *and*

(b) *if α_j is not a p_j-th root of unity, then $\mathbb{E}_{j-1}$ contains the p_j-th root of unity.*

Proof. The proof is easy and left as an exercise. □

Before we end this section we give an example due to Hungerford. (See [8]). This is an example of a radical extension that is not normal.

Example 4.1.91. Let $\mathbb{K} = \mathbb{Q}[i\sqrt{2}, (1-i)\sqrt[4]{2}]$. Then

1. $(\sqrt{2}i)^2 = -2 \in \mathbb{Q}$ and
2. $((1-i)\sqrt[4]{2})^2 = -2i\sqrt{2} \in \mathbb{Q}[i\sqrt{2}] = \mathbb{F}_1$.

Therefore $\mathbb{K}/\mathbb{Q}$ is a radical extension and $\mathbb{Q} := \mathbb{F}_0 \subseteq \mathbb{F}_1 \subseteq \mathbb{F}_2 = \mathbb{K}$ is a tower of simple radical extension. Hence by tower law

$$[\mathbb{K} : \mathbb{Q}] = [\mathbb{K} : \mathbb{F}_1][\mathbb{F}_1 : \mathbb{F}_0] \leq 4.$$

We will now show that the extension $\mathbb{K}/\mathbb{Q}$ is not normal.

Let us observe that $((1-i)\sqrt[4]{2})^4 = -8$. Hence $\alpha := (1-i)\sqrt[4]{2}$ is a root of the irreducible polynomial $f(X) = X^4 + 8$. If this extension were normal, the field $\mathbb{K}$ will have all the roots of the polynomial $f(X)$. The roots of the polynomial $f(X)$ are $\pm\alpha$ and $\pm(1+i)\sqrt[4]{2}$. Let $\beta = (1+i)\sqrt[4]{2}$.

We will show that the roots $\pm\beta$ are not in $\mathbb{K}$. Suppose β is in $\mathbb{K}$. Then $\alpha + \beta = 2\sqrt[4]{2}$ and $\beta - \alpha = 2i\sqrt[4]{2}$ are in $\mathbb{K}$. Therefore $\frac{\beta-\alpha}{\beta+\alpha} = i$ is in $\mathbb{K}$ and we have the tower $\mathbb{Q} \subseteq \mathbb{Q}[\sqrt[4]{2}] \subseteq \mathbb{Q}[\sqrt[4]{2}, i] = \mathbb{K}$.

The degree of the extension $[\mathbb{Q}[\sqrt[4]{2}] : \mathbb{Q}] = 4$ and $[\mathbb{Q}[\sqrt[4]{2}, i], \mathbb{Q}[\sqrt[4]{2}]] = 2$. As a consequence it follows that

$$4 \geq [\mathbb{K} : \mathbb{Q}] = [\mathbb{K} : \mathbb{Q}[\sqrt[4]{2}][\mathbb{Q}[\sqrt[4]{2}] : \mathbb{Q}] = 2 \cdot 4 = 8.$$

This is absurd. Hence the extension $\mathbb{K}/\mathbb{Q}$ is not normal.

Since the fourth root of unity, namely the complex number i is not there, the extension $\mathbb{K}/\mathbb{Q}$ fails to be normal.

We will see in our proof of Galois theorem that normality of the extension is very important.

4.2 Galois and Abel's Theorems

We are now going to prove an outstanding result due to Galois which relates the solvability of a polynomial by radicals to a problem in group theory.

We explain this briefly, more details are in Section 5.6 on Solvable Groups.

Given a group G, we denote by $[G, G]$ the subgroup generated by the subset $C := \{aba^{-1}b^{-1} : a, b \in G\}$ of G. This group is called the commutator subgroup of G and is also denoted by $G^{(1)}$. For $k \geq 2$, we define $G^{(k)} := [G^{(k-1)}, G^{(k-1)}]$, the commutator subgroup of $G^{(k-1)}$. We will now define Solvable groups.

Definition 4.2.1. A group G is called a solvable group if there exists an integer $k \geq 1$ such that $G^{(k)} = e$.

We have proved various properties of Solvable groups in Section 5.6 and we will be using them in our proof of Theorem 4.2.2, Theorem 4.2.3 and Theorem 4.2.4.

Theorem 4.2.2 (Galois). *If a polynomial $f(X) \in \mathbb{F}[X]$ is solvable by radicals then Gal(f) is solvable.*

We first explain the strategy of the proof.

Let $\mathbb{K}$ be the splitting field of the polynomial $f(X)$ in $\mathbb{F}[X]$. Then $\text{Gal}(f) = \text{Gal}(\mathbb{K}/\mathbb{F})$. We will embed $\mathbb{K}$ in an extension $\mathbb{L}$ such that the field $\mathbb{L}$ is prime radical and normal over the field $\mathbb{F}$.

We shall show that $\text{Gal}(\mathbb{L}/\mathbb{F})$ is solvable and $\text{Gal}(f)$ is the homomorphic image of the solvable group $\text{Gal}(\mathbb{L}/\mathbb{F})$. Hence $\text{Gal}(f)$ is solvable.

Let $f(X)$ be a polynomial in $\mathbb{F}[X]$ solvable by radicals over $\mathbb{F}$ and let $\mathbb{K}/\mathbb{F}$ be the splitting field of the polynomial $f(X)$. Then it follows from Corollary 4.1.89 there exists a prime radical extension $\mathbb{L}$ over $\mathbb{F}$ such that $\mathbb{L}$ contains the splitting field $\mathbb{K}$ of the polynomial $f(X)$ over $\mathbb{F}[X]$.

In this case we have the following theorem.

Theorem 4.2.3. *Let $f(X)$ be a polynomial in $\mathbb{F}[X]$. Assume $f(X)$ is solvable by radicals and $\mathbb{L}$ be an extension of $\mathbb{F}$ as above. Then Gal$(\mathbb{L}/\mathbb{F})$ is solvable.*

Proof. Without loss of generality, we may assume that $\mathbb{L}/\mathbb{F}$ is a prime radical extension such that $\mathbb{F} = \mathbb{L}_0 \subseteq \mathbb{L}_1 \subseteq \mathbb{L}_2 \subseteq \ldots \subseteq \mathbb{L}_r = \mathbb{L}$ such that

1. for each j, there exist α_j in $\mathbb{L}_j$ and a prime number p_j such that $\mathbb{L}_j = \mathbb{L}_{j-1}(\alpha_j)$, $\alpha_j^{p_j} \in \mathbb{L}_{j-1}$ and
2. if α_j is not a p_j-th root of unity, then $\mathbb{L}_{j-1}$ contains a p_j-th root of unity.

We denote by $G_j = \text{Gal}(\mathbb{L}/\mathbb{L}_j)$ for each j. Let σ be an element of G_j. Since $\sigma |_{\mathbb{L}_j} = Id$ and $\mathbb{L}_{j-1} \subseteq \mathbb{L}_j$, we see that $\sigma |_{\mathbb{L}_{j-1}} = Id$ too. Hence σ is also in G_{j-1} and we get inclusions $\{e\} = G_r \subseteq G_{r-1} \subseteq G_{r-2} \subseteq \ldots \subseteq G_0 = \text{Gal}(\mathbb{L}/\mathbb{F})$.

The field $\mathbb{L}_j = \mathbb{L}_{j-1}[\alpha_j]$ where α_j is a p_j-th root of unity or $\alpha_j^{p_j} = \beta_j \in \mathbb{L}_{j-1}$ and $\mathbb{L}_{j-1}$ contains a p_j-th roots of unity. Hence the extension $\mathbb{L}_j/\mathbb{L}_{j-1}$ is the splitting field of the polynomial $f_j(X) = X^{p_j} - \beta_j$.

Let σ be an element of $\text{Gal}(\mathbb{L}/\mathbb{L}_{j-1})$. Then $\sigma(\alpha_j^{p_j}) = \sigma(\alpha_j)^{p_j}$ and since $\beta_j = \alpha_j^{p_j}$ is in $\mathbb{L}_{j-1}$, it follows that $\sigma(\alpha_j)^{p_j} = \sigma(\alpha_j^{p_j}) = \alpha_j^{p_j} = \beta_j$. Thus $\sigma(\alpha_j)$ is also a root of the polynomial $f_j(X)$. Hence $\sigma(\alpha_j)$ is in $\mathbb{L}_j$.

We use this observation to get a map $\varphi_j : \text{Gal}(\mathbb{L}/\mathbb{L}_{j-1}) \to \text{Gal}(\mathbb{L}_j/\mathbb{L}_{j-1})$ defined by $\sigma \mapsto \sigma |_{\mathbb{L}_j}$. i.e, $\varphi_j(\sigma) = \sigma |_{\mathbb{L}_j}$. It can be seen that this map φ_j is a group homomorphism.

An element σ is in $\ker(\varphi)$, the kernel of the homomorphism φ iff $\sigma |_{\mathbb{L}_j} = Id$. Thus $\ker(\varphi_j) = \{\sigma \in G_{j-1} : \sigma |_{\mathbb{L}_j} = Id\}$. This is precisely the group $G_j = \text{Gal}(\mathbb{L}/\mathbb{L}_j)$. Hence by fundamental theorem of homomorphism of groups, it follows that G_{j-1}/G_j is isomorphic to the image of φ_j.

Let σ be an element of $\text{Gal}(\mathbb{L}_j/\mathbb{L}_{j-1})$. Since $\mathbb{L}_j$ is the splitting field of the polynomial $f_j(X)$ over $\mathbb{L}_{j-1}[X]$, the automorphism σ can be extended to an automorphism $\overline{\sigma} : \mathbb{L} \to \mathbb{L}$ such that $\overline{\sigma} |_{\mathbb{L}_j} = \sigma$. This proves that the map φ_j is on to. Therefore G_{j-1}/G_j is isomorphic to $\text{Gal}(\mathbb{L}_j/\mathbb{L}_{j-1})$. Now we observe that $|\text{Gal}(\mathbb{L}_j/\mathbb{L}_{j-1})| = [\mathbb{L}_j : \mathbb{L}_{j-1}] = p_j$ to conclude that $\text{Gal}(\mathbb{L}_j/\mathbb{L}_{j-1})$ is a cyclic group of order p_j. This proves that $\text{Gal}(\mathbb{L}/\mathbb{F})$ is a solvable group. □

We are now ready to prove Theorem 4.2.2 of Galois.

Proof. Let $f(X)$ be a polynomial in $\mathbb{F}[X]$ solvable by radicals. Then it follows from Theorem 4.2.3 that there is an extension $\mathbb{L}$ that is both normal and radical over $\mathbb{F}$ such that $\mathrm{Gal}(\mathbb{L}/\mathbb{F})$ is solvable.

Let $\mathbb{K}$ be the splitting field of $f(X)$ contained in $\mathbb{L}$. Let $\deg(f) = r$ and $\alpha_1, \alpha_2, \ldots, \alpha_r$ in $\mathbb{K}$ be the roots of $f(X)$. Then $\mathbb{K} = \mathbb{F}[\alpha_1, \alpha_2, \ldots, \alpha_r]$ as a vector space over $\mathbb{F}$.

Let $\sigma \in \mathrm{Gal}(\mathbb{L}/\mathbb{F})$. If α is a root of $f(X)$, then $\sigma(\alpha)$ is also a root of $f(X)$. This proves that σ permutes all the roots α_i of the polynomial $f(X)$. Hence $\{\sigma(\alpha_i) : 1 \leq i \leq r\} = \{\alpha_i : 1 \leq i \leq r\}$. As a consequence $\sigma(\mathbb{K}) \subseteq \mathbb{K}$. Thus every σ in $\mathrm{Gal}(\mathbb{L}/\mathbb{F})$ restricts to an element $\overline{\sigma}$ of $\mathrm{Gal}(f) = \mathrm{Gal}(\mathbb{K}/\mathbb{F})$. Since the extension $\mathbb{L}/\mathbb{K}$ is also normal (see Theorem 4.1.69), we can extend any homomorphism $\sigma \in \mathrm{Gal}(f)$ to an automorphism $\widetilde{\sigma} : \mathbb{L} \to \mathbb{L}$ such that $\widetilde{\sigma} \mid_{\mathbb{F}} = Id$. Thus we have proved that restriction map $\theta : \mathrm{Gal}(\mathbb{L}/\mathbb{F}) \to \mathrm{Gal}(f)$ is on-to and we can prove that this map is a homomorphism. The kernel of this map is $\ker(\theta) := \{\sigma \in \mathrm{Gal}(\mathbb{L}/\mathbb{F}) : \sigma \mid_{\mathbb{K}} = Id\} = \mathrm{Gal}(\mathbb{L}/\mathbb{K})$. Therefore the quotient group $\mathrm{Gal}(\mathbb{L}/\mathbb{F})/\ker(\theta)$ is isomorphic to the group $\mathrm{Gal}(f)$. Since quotient group of a solvable group is solvable, it follows that the group $\mathrm{Gal}(f)$ is also solvable. □

Now we come to the final result of Galois theory.

Theorem 4.2.4 (Abel). *There exists a polynomial $f(X)$ of degree 5 in $\mathbb{Q}[X]$ such that $f(X)$ is not solvable by radicals over $\mathbb{Q}$.*

Proof. Let $f(X) = X^5 - 6X + 3$. Using Eisenstein's irreducibility criterion, we conclude that the polynomial $f(X)$ is irreducible over $\mathbb{Q}$.

We will now show that the polynomial $f(X)$ has three real roots and two complex roots: Observe that $f(0) = 3 > 0$, $f(1) = -2 < 0$, $f(2) = 23 > 0$ and $f(-2) = -17 < 0$. Hence by intermediate value theorem, the polynomial $f(X)$ has at least three real roots. If we now show that the polynomial $f(X)$ remains positive in the interval $(2, \infty)$ and it remains negative in the interval $(-\infty, 2)$, then our claim will follow.

Let $\alpha < -2$. Then $\alpha^4 > 16$. We use this to conclude

$$f(\alpha) = \alpha^5 - 6\alpha + 3 = \alpha(\alpha^4 - 6) < 10\alpha + 3 < -20 + 3 = -17 < 0$$

for every $\alpha < -2$.

If $\alpha > 2$, then

$$f(\alpha) = \alpha(\alpha^4 - 6) + 3 > 23 > 0.$$

Let $\mathbb{K}$ be the splitting field of $f(X)$ over $\mathbb{Q}$. We may assume that $\mathbb{K}$ is a subfield of $\mathbb{C}$.

Let α be a root of $f(X)$. Since $f(X)$ is an irreducible polynomial of degree 5, the degree of the extension $\mathbb{Q}[\alpha]/\mathbb{Q}$ is $[\mathbb{Q}[\alpha] : \mathbb{Q}] = 5$. Hence by tower law $[\mathbb{K} : \mathbb{Q}] = [\mathbb{K} : \mathbb{Q}[\alpha]][\mathbb{Q}[\alpha] : \mathbb{Q}] = 5[\mathbb{K} : \mathbb{Q}[\alpha]]$. This shows that 5 divides $[\mathbb{K} : \mathbb{Q}] = |\mathrm{Gal}(\mathbb{K}/\mathbb{F})|$ and by Cauchy's Theorem 5.2.1, there exists an element σ in $\mathrm{Gal}(\mathbb{K}/\mathbb{F})$ of order 5.

The restriction of complex conjugation τ to the field $\mathbb{K}$ interchanges the two complex roots and keeps the real roots fixed. Hence τ is also in $\mathrm{Gal}(\mathbb{K}/\mathbb{F})$.

We have already seen in Theorem 4.1.75 that Gal($\mathbb{K}/\mathbb{F}$) is a subgroup of S_5. We have now shown that two elements, σ of order 5 and complex conjugation τ of order 2 are in Gal($\mathbb{K}/\mathbb{F}$). This shows that σ is a 5-cycle and τ is a transposition. Since the symmetric group S_n is generated by an n-cycle and a transposition, it follows that the group Gal($\mathbb{K}/\mathbb{F}$) is isomorphic to S_5. But the group S_5 is not solvable. Therefore the polynomial $f(X)$ is not solvable by radicals. □

Examples 4.2.5. We can show that the following polynomials are not solvable by radicals over $\mathbb{Q}$.

1. $f(X) = X^5 - 16X + 2$.

2. $f(X) = X^5 - 25X + 5$.

We will now show that cubic and quartic polynomials over $\mathbb{Q}$ are solvable by radicals.

4.3 Solvability of Cubic and Quartic polynomials

In this section we will show that polynomials with rational coefficients of degree less than or equal to four are solvable by radicals.

We know that a quadratic polynomial $f(X) = X^2 + bX + c$ is solvable by radicals and the solutions are $\frac{-b+\sqrt{b^2-4c}}{2}$ and $\frac{-b-\sqrt{b^2-4c}}{2}$. The solutions are in the field $\mathbb{K} = \mathbb{Q}[\sqrt{b^2 - 4c}]$ and the degree of the extension is 1 if $b^2 - 4c$ is a square and 2 is $b^2 - 4c$ is not a square.

4.3.1 Solvability of Cubic polynomials

Let $f(X) = X^3 + bX^2 + cX + d \in \mathbb{Q}[X]$ be a polynomial of degree 3. We will now show that $f(X)$ can be solved by radicals. The method is attributed to Cardano. For a history of this the reader can refer to any book on Galois Theory.

The first step in the proof is reducing the polynomial $f(X) = X^3 + bX^2 + cX + d$ to a polynomial of the form $X^3 + pX + q$. This is done by a substitution $X = Y + \alpha$ and making a suitable choice of α.

If $X = Y + \alpha$, then $X^3 = (Y + \alpha)^3 = Y^3 + \alpha^3 + 3\alpha Y(\alpha + Y)$, $X^2 = Y^2 + \alpha^2 + 2\alpha Y$. We substitute this in $f(X)$ to get

$$f(X = Y + \alpha) = Y^3 + (3\alpha + a)Y^2 + (3\alpha^2 + 2a\alpha + b)Y + (\alpha^3 + a\alpha^2 + b\alpha + c).$$

For the coefficient of Y^2 to vanish, we choose $\alpha = -\frac{a}{3}$. Then the polynomial in Y is $Y^3 + pY + q$, where $p = 3\alpha^2 + 2a\alpha + b$ and $q = \alpha^3 + a\alpha^2 + b\alpha + c$.

Therefore if suffices to solve a cubic polynomial $f(X) = X^3 + pX + q$ by radicals.

The method of solution starts with another substitution $X = U - V$. This substitution leads to the equation $f(X) = (U - V)^3 + p(U - V) + q = U^3 -$

$V^3 - (3UV - p)(U - V) + q$. If we can solve the two equations $3UV - p = 0$ and $U^3 - V^3 + q = 0$ by radicals, then we are through. If $3UV - p = 0$, then $V = \frac{p}{3U}$ and hence

$$U^3 - V^3 + q = U^3 - \left(\frac{p}{3U}\right)^3 + q = 0.$$

We write this equation as $3^3(U^3)^2 + 3^3qU^3 - p^3 = 0$ and make another substitution $z = U^3$ to write this equation as $3^3z^2 - 3^3qz - p^3 = 0$. This is a quadratic equation in z and the solutions are $z = -\frac{q}{2} \pm \sqrt{(\frac{q}{2})^2 + (\frac{p}{3})^3}$. If $z = -\frac{q}{2} + \sqrt{(\frac{q}{2})^2 + (\frac{p}{3})^3}$, then

$$U = \sqrt[3]{-\frac{q}{2} + \sqrt{\left(\frac{q}{2}\right)^2 + \left(\frac{p}{3}\right)^3}}.$$

Using the formula for U^3 in $V^3 = U^3 + q$, we get

$$V^3 = -\frac{q}{2} + \sqrt{\left(\frac{q}{2}\right)^2 + \left(\frac{p}{3}\right)^3} + q = \frac{q}{2} + \sqrt{\left(\frac{q}{2}\right)^2 + \left(\frac{p}{3}\right)^3}$$

and

$$V = \sqrt[3]{\frac{q}{2} + \sqrt{\left(\frac{q}{2}\right)^2 + \left(\frac{p}{3}\right)^3}}.$$

Now $Y = U - V$ is a solution of the polynomial $g(Y)$. This proves that a polynomial of degree 3 can be solved by radicals.

Given an irreducible cubic polynomial $f(X) = X^3 + aX^2 + bX + c$ in $\mathbb{Q}[X]$, we will now study the splitting field $\mathbb{K}$ of the polynomial $f(X)$. Let α_1, α_2 and α_3 be the roots of the polynomial $f(X)$.

Then

$$\alpha_1 + \alpha_2 + \alpha_3 = -a \quad (4.1)$$
$$\alpha_1\alpha_2 + \alpha_2\alpha_3 + \alpha_3\alpha_1 = b \quad (4.2)$$
$$\alpha_1\alpha_2\alpha_3 = -c \quad (4.3)$$

The first equation tells us that the third root is in the field $\mathbb{Q}(\alpha_1 + \alpha_2)$. Thus $\mathbb{K} = \mathbb{Q}(\alpha_1, \alpha_2)$ and we have the sequence $\mathbb{Q} \subseteq \mathbb{L} := \mathbb{Q}(\alpha_1) \subseteq \mathbb{K}$. Since $f(X)$ is irreducible we have $[\mathbb{L} : \mathbb{Q}] = 3$ and hence $[\mathbb{K} : \mathbb{Q}] = [\mathbb{K} : \mathbb{L}][\mathbb{L} : \mathbb{Q}] = 3[\mathbb{K} : \mathbb{L}]$.

Now the extension $\mathbb{K}/\mathbb{L}$ has degree 1 if α_2 can be expressed as a polynomial in α_1 with coefficients in $\mathbb{Q}$ or else it is 2.

Let us look at the following example.

Example 4.3.1. Let $f(X) = X^3 - 2$ in $\mathbb{Q}[X]$. Then $\alpha_1 = \sqrt[3]{2}$. If $\rho \neq 1$ is a cube root of unity, then $\alpha_2 = \rho\alpha_1$ and $\alpha_3 = \rho^2\alpha$ are the remaining roots of f. The root α_2 is not a polynomial in α_1. Hence $[\mathbb{K} : \mathbb{L}] = 2$ and $\mathbb{K} = \mathbb{Q}(\alpha_1, \alpha_2) = \mathbb{Q}(\alpha_1, \rho)$ and the degree of the extension $\mathbb{K}/\mathbb{Q}$ is 6. Hence $\mathrm{Gal}(\mathbb{K}/\mathbb{Q})$ is S_3.

In general we can write $f(X) = (X - \alpha_1)g(X)$ where $g(X)$ is a polynomial in $\mathbb{L}[X]$. The root α_2 is expressible as a polynomial α_1 iff the polynomial $g(X)$ is reducible in $\mathbb{L}[X]$. In this case $\mathbb{L} = \mathbb{K}$ and $\mathrm{Gal}(\mathbb{K}/\mathbb{Q})$ is A_3, the subgroup generated by a 3-cycle in S_3.

Let us look at the following example to illustrate this.

Example 4.3.2. Let $f(X) = X^3 - 3X + 1$. It has three real roots and its roots are $\alpha = \zeta + \overline{\zeta}$, $\beta = \zeta^2 + \overline{\zeta}^2$ and $\gamma = \zeta^4 + \overline{\zeta}^4$ where ζ is a primitive 9-th root of unity. (See Example 4.1.80).

we have shown in Example 4.1.80 that $\beta = \alpha^2 - 2$ and $\gamma = -(\alpha + \beta)$. Hence the splitting field of $f(X)$ over $\mathbb{Q}$ is $\mathbb{K} = \mathbb{Q}[\alpha]$ and the degree of the extension is 3.

Since the polynomial has three distinct roots, the Galois group $\mathrm{Gal}(\mathbb{K}/\mathbb{F})$ is a subgroup of S_3 and its order $|\mathrm{Gal}(\mathbb{K}/\mathbb{F})| = [\mathbb{K} : \mathbb{Q}] = 3$. Hence the Galois group $\mathrm{Gal}(\mathbb{K}/\mathbb{F})$ is A_3, the subgroup generated by a 3-cycle in S_3.

Example 4.3.3. Let $f(X) = X^3 + 3X + 1 = 0$. This polynomial has only one real root, say, α_1. Since the remaining roots are complex none of them can be expressed as a polynomial in α_1. Hence $[\mathbb{K} : \mathbb{Q}] = 6$ and $\mathrm{Gal}(\mathbb{K}/\mathbb{Q})$ is isomorphic to S_3.

4.3.2 Solvability of Quartic Polynomials

Let $f(X) = X^4 + aX^3 + bX^2 + cX + d$ be a quartic polynomial in $\mathbb{Q}[X]$. Just as we did while solving cubic equation, we make the substitution, $X = Y - \frac{a}{4}$ to get $f(X) = f(Y - \frac{a}{4}) = Y^4 + pY^2 + qY + r$ where $p = \frac{8b - 3a^2}{8}$, $q = \frac{a^3 - 4ab + 8c}{8}$ and $r = \frac{-3a^4 + 16a^2b - 64ac + 256d}{256}$.

Therefore it suffices to solve a quartic equation $f(X) = X^4 + pX^2 + qX + r$ by radicals.

Suppose $q = 0$. Then we can write $f(X) = (X^2 + \frac{p}{2})^2 - ((\frac{p}{2})^2 - r)$ and this can be factored as

$$f(X) = \left(X^2 + \frac{p}{2} + \sqrt{\left(\frac{p}{2}\right)^2 - r}\right)\left(X^2 + \frac{p}{2} - \sqrt{\left(\frac{p}{2}\right)^2 - r}\right).$$

This clearly shows that the equation can be solved by radicals.

Now we assume that $q \neq 0$. Now we introduce a parameter t to write this polynomial without X term. We write

$$\begin{aligned} f(X) &= X^4 + pX^2 + qX + r \\ &= X^4 + X^2 t + \left(\frac{t}{2}\right)^2 - X^2(t-p) + qX - \left(\left(\frac{t}{2}\right)^2 - r\right) \\ &= \left(X^2 + \frac{t}{2}\right)^2 - X^2(t-p) + qX - \left(\left(\frac{t}{2}\right)^2 - r\right). \end{aligned}$$

Observe that the expression $X^2(t-p) - qX + \left(\left(\frac{t}{2}\right)^2 - r\right)$ is a square iff $q^2 = 4(t-p)\left(\left(\frac{t}{2}\right)^2 - r\right)$. Simplifying this we get a cubic equation $t^3 - pt^2 - 4rt + 4rp - q^2 = 0$ in t with coefficients in $\mathbb{Q}$. This can be solved by radicals and let T be a solution to this equation. We can now write $f(X)$ as

$$f(X) = \left(X^2 + \frac{T}{2}\right)^2 - \left(\sqrt{T-p}\,X - \sqrt{\frac{q}{T-p}}\right)^2.$$

Notice that $T = p$ will force $q = 0$ which is not possible. Therefore

$$\begin{aligned} f(X) &= \left(X^2 + \sqrt{T-p}\,X - \left(\sqrt{\frac{q}{T-p}} - \frac{T}{2}\right)\right) \\ & \left(X^2 - \sqrt{T-p}\,X + \sqrt{\frac{q}{T-p}} + \frac{T}{2}\right). \end{aligned}$$

Each of these quadratic factors can be solved in terms of radicals. As a consequence the polynomial $f(X)$ can be solved by radicals.

The analysis of the splitting field of this polynomial gets harder. The readers can refer to advanced texts on Galois Theory.

4.4 Constructions with straightedge and compass

In high school geometry, we have learnt

1. to draw a straight line joining two points,
2. to draw a line perpendicular to a given line ℓ and passing through a given point,
3. to draw a line parallel to a given line ℓ and passing through a point p not on ℓ and
4. to bisect a given angle

using straightedge and compass. Here we are using the word straightedge to remind us that we did not use the lengths marked on the ruler to do these constructions.

There are three geometric problems with a long history starting with the Greek. Without going in to the history of the problems we state them below.

Using the straightedge and compass can we

1. **double the cube:** construct an edge of a cube whose volume is twice that of a cube with given segment as an edge?
2. **trisecting an angle:** divide an arbitrary angle in to three equal parts?
3. **Squaring the circle:** construct a square whose area is same as the area of a given circle?

We answer the first and third questions in negative and as for the second question we show that there exists an angle which can't be trisected.

In fact we will state a result without proof of a result due to Gauss which gives a necessary and sufficient condition on the constructibility of regular n-sided polygons.

We answer these questions by relating these problems to field extensions.

Let us first recall the rules for basic constructions with straightedge and compass.

1. Two points in the plane are given to start with. These points are considered to be constructed.
2. If two points can be constructed, we may draw a line through them or draw a circle with center at one point and passing through the other. The lines and circles constructed by this method are said to be constructed.
3. The point of intersections of constructed lines and constructed circles are also considered to be constructed.

We will now define constructible points in $\mathbb{R}^2$.

Definition 4.4.1 (Constructible Points). We say a point P in $\mathbb{R}^2$ is constructible, if P is a point of intersection of

1. two constructed lines or
2. a constructed line and a constructed circle or
3. two constructed circles.

With the rules described above we will describe few geometric constructions using straight edge and compass.

1. Drawing a line through a constructed point P and perpendicular to a constructed line ℓ.
   (a) Since a line is determined by two constructed points, we choose a constructed point Q on the line ℓ.
   (b) We draw a circular arc with center at P and passing through the point Q. This circular arc will meet the line at a point R.
   (c) We draw a circular arc with center Q and passing through R and a circular arc with center R and passing through Q. These two circular arcs meet at two points. One of the point is P and let us denote another point by S.
   (d) We construct the line joining P and S. This line is perpendicular to the line ℓ and passes through the point P.
2. Drawing a line parallel to a constructed line ℓ and passing through a constructed point P not on ℓ.
   (a) We draw a line ℓ_1 passing through the point P and perpendicular to the line ℓ.
   (b) We construct a line ℓ_2 perpendicular to the line ℓ_1 and passing through the point P. Then the constructed ℓ_2 is parallel to the line ℓ.
3. To mark off a given length (segment) defined by two points on a constructed line ℓ starting at a constructed point P on the line ℓ.

   We use construction of parallels to prove this.

   Let Q and R be the end points of the given length.

   (a) Draw the line joining the points P and Q.

(b) Draw the line parallel ℓ_Q to ℓ passing through the point Q.

(c) Draw a circular arc with radius as the given length and S be the point of intersection of the arc and the line ℓ_Q.

(d) Draw the line ℓ_S parallel to the line ℓ_{PQ} and T be the point of intersection of ℓ_S and ℓ_{PQ}. The length PT is then same as PQ.

These constructions allow us to introduce Cartesian coordinates: We assume that two given points are $(0,0)$ and $(1,0)$. With this convention we will now define constructible real numbers.

Definition 4.4.2 (Constructible Numbers)**.** We say that a real number a is constructible if $|a|$, the absolute value of the real number a, is the distance between two constructible points in $\mathbb{R}^2$.

The distance between the two original points is the unit length.

First we characterize the constructible points in the plane.

Proposition 4.4.3. *A point $P = (a,b)$ is constructible iff the Cartesian coordinates a and b are constructible.*

Proof. Let us denote the line joining the constructed points $(0,0)$ and $(1,0)$ by ℓ_X and the line perpendicular to the line ℓ_X and passing through $(0,0)$ by ℓ_Y for obvious reasons.

Let $P = (a,b)$ be a constructible point. From the constructed point P, we draw a line perpendicular to ℓ_X. The point of intersection of this line and the line ℓ_X is $(a,0)$ and the distance between the two constructed points $(0,0)$ and $(a,0)$ is $|a|$. Hence a is constructible.

To show that b is constructible, we draw a line perpendicular to ℓ_Y and passing through the point P. The point of intersection of these two lines is $(0,b)$. From this we conclude that the real number b is also constructible.

Conversely assume that the real numbers a and b are constructible. We mark of the lengths $|a|$ and $|b|$ on the lines ℓ_X and ℓ_Y starting at the point $(0,0)$. Then the coordinates of the two marked points are $(a,0)$ and $(0,b)$ respectively. If we now draw a line parallel to ℓ_Y and passing through the point $(a,0)$ and a line parallel to ℓ_X and passing through the point $(0,b)$, then the point of intersection of these two lines is the point with coordinates (a,b). Thus the point $P = (a,b)$ is constructible. □

We will now show that the set of constructible real numbers is a field. Let $\mathbb{K} = \{a \in \mathbb{R} : a \text{ is constructible}\}$.

Theorem 4.4.4. *The set $\mathbb{K}$ of all constructible real numbers is a field.*

Proof. Let a and b be two constructible numbers. Then, using the constructions illustrated earlier, we can mark off lengths $a \pm b$ on a constructed line. This proves that $a \pm b$ are constructible.

If we now show that the real numbers ab and $\frac{a}{b}$, if $b \neq 0$ are constructible, then we are through.

Now to show that the real numbers ab and $\frac{a}{b}$, if $b \neq 0$ are constructible, we use the similarity of right angled triangles.

We will first show that the real number ab is constructible.

1. From the origin $(0,0)$ we mark off the length 1 on the Y-axis ℓ_Y and mark of the length $|a|$ on the X-axis ℓ_X.
2. From the origin $(0,0)$ mark off length $|b|$ on the constructed line ℓ_Y joining $(0,0)$ and $(0,1)$.
3. Through the point $(0,|b|)$ draw a line parallel to the line joining the two constructed points $(0,1)$ and $(|a|,0)$.
4. This line meets the line ℓ_X at the point $(|ab|,0)$.

Thus ab is constructible.

We will now show that $\frac{a}{b}$ is constructible, if $b \neq 0$.

1. Starting at the origin $(0,0)$ we mark off the length $|a|$ on Y-axis ℓ_Y and unit length on X-axis ℓ_X.
2. On the line ℓ_Y, we mark off the length $|b|$ starting at the origin.
3. Through the constructed point $(0,|b|)$ draw the line parallel to the line joining the points $(0,a)$ and $(a,0)$.
4. The point of intersection of this line with the line ℓ_X is the point $(\frac{a}{b},0)$.

This proves that the number $\frac{a}{b}$ is also constructible and the proof is complete. □

Corollary 4.4.5. *The field $\mathbb{Q}$ is a subfield of $\mathbb{K}$, the field of constructible numbers.*

Proof. Let us first observe that every integer is constructible. Thus $\mathbb{Z} \subseteq \mathbb{K}$ and hence $\mathbb{Q} \subseteq \mathbb{K}$, since $\mathbb{K}$ is a field. □

We would like to understand the structure of the field $\mathbb{K}$.

Proposition 4.4.6. *Let $a > 0$ be a constructible number. Then $\sqrt{a}$ is constructible.*

Proof. Since $a > 0$ is constructible, the points $Q = (-a, 0)$ and $P = (1,0)$ are constructed points. Hence the point $(\frac{1-a}{2},0)$ is constructed.

1. Draw a circle C with center $(\frac{1-a}{2},0)$ and passing through the point $(1,0)$.
2. Let R be a point of intersection of the perpendicular line ℓ_Y with the circle C.
3. The triangle $\triangle(PRQ)$ is right angled at the point R.
4. The lengths OR, OP and OQ are constructed and $\frac{OQ}{OR} = \frac{OR}{OP}$. Hence $OR^2 = OQ = a$ and this proves that $\sqrt{a}$ is constructible.

□

Corollary 4.4.7. *If $a > 0$ is constructible, then all 2^n-th roots of a are constructible.*

Proof. The last proposition tells us that $\sqrt{a}$ is constructible. Hence $\sqrt[2^2]{a} = \sqrt{\sqrt{a}}$ is constructible. Now an induction argument finishes the proof. □

We will now relate the constructibility of a given real number to an extension of the field of rational numbers.

Let us recall that only ways of constructing

1. a point is getting it as a point of intersection of two constructed lines, a line and a circle or two circles,

2. a segment is constructing the end points of the segment

Therefore to know the set of all constructible numbers, we need to know all the points that can be got by these methods.

Lemma 4.4.8. *Let $\mathbb{F}$ be a subfield of $\mathbb{R}$. Let P, Q, R and S be four points in the plane $\mathbb{R}^2$ with coordinates in $\mathbb{F}$.Then the following holds.*

1. *If $\ell(P,Q)$ and $\ell(R,S)$ are two lines joining the pair (P,Q) and (R,S) and if they intersect, then the coordinates of the point of intersection of these two lines is again in $\mathbb{F}$.*

2. *If $\ell(P,Q)$ is a line joining the pair (P,Q), C is a constructible circle passing through the points R and S and if they intersect, then the coordinates of the points of intersection of the line $\ell(P,Q)$ and C lies in a quadratic extension of the field $\mathbb{F}$.*

3. *If C_1 is a constructible circle passing through the points P, Q, C_2 is a constructible circle passing through the points R and S and if they intersect, then the coordinates of the points of intersection of these two circles is in a quadratic extension of the field $\mathbb{F}$.*

Proof. The equation of the line $\ell(P,Q)$ can be written as $a_1x + b_1y + c_1 = 0$ and the equation of the line $\ell(R,Q)$ can be written as $a_2x + b_2y + c_2 = 0$ with a_i, b_i and c_i in the field $\mathbb{F}$. Subtracting these two equations, we get $(a_1 - a_2)x + (b_1 - b_2)y = c_1 - c_2$.

If $a_1 - a_2 \neq 0$, then $x = \frac{b_2-b_1}{a_1-a_2}y + \frac{c_1-c_2}{a_1-a_2}$ and we substitute this in $a_1x + b_1y + c = 0$ to get $a_1\left[\frac{b_2-b_1}{a-1-a_2}y + \frac{c_1-c_2}{a_1-a_2}\right] + b_1y + c = 0$. Hence $y\left(a_1\frac{b_2-b_1}{a_1-a_2} + b_1\right) + c_1 + \frac{c_1-c_2}{a_1-a_2} = 0$. This proves that y is in the field $\mathbb{F}$ and as a consequence x is also in the field $\mathbb{F}$. This proves the first part of the lemma.

We will now prove the second part.

Let $a_1x + b_1y + c_1 = 0$ be the equation of the line $\ell(P,Q)$ and $x^2 + y^2 + a_2x + b_2y + c_2 = 0$ be the equation of a circle C. As earlier the scalars a_i, b_i and c_i are in the field $\mathbb{F}$. To show that the scalars a_2, b_2 and c_2 are in $\mathbb{F}$, we need to show that the coordinates of the center of the circle C is also in the field $\mathbb{F}$.

If $b_1 \neq 0$, then $y = -(\frac{a_1}{b_1}x + \frac{c_1}{b_1})$. We substitute this in the equation of the circle and simplify the equation to get $\alpha x^2 + \beta x + \gamma = 0$, where $\alpha = \left(\frac{b_1^2+a_1^2}{b_1^2} + 1\right)$, $\beta = \frac{a_2b_1^2+2a_1c_1}{b_1^2}$ and $\gamma = -\frac{b_2}{b_1}(a_1 + c_1) + c_2$. Hence $x = \frac{-\beta+\sqrt{\beta^2-4\alpha\gamma}}{2\alpha}$ lies in the field $\mathbb{F}[\sqrt{\beta^2 - 4\alpha\gamma}]$.

Now let $x^2 + y^2 + a_1x + b_1y + c_1 = 0$ and $x^2 + y^2 + a_2x + b_2y + c_2 = 0$ be the equation of the circles C_1 and C_2 respectively. The points of intersection lie on the line $(a_1 - a_2)x + (b_1 - b_2)y + (c_1 - c_2) = 0$. If $b_1 - b_2 \neq 0$, then $y = \frac{(a_2-a_1)x+(c_2-c_1)}{b_1-b_2}$. We now substitute this in one of the equations of the circle and get a quadratic equation in x whose solution will lie in a quadratic extension of the field $\mathbb{F}$ as we saw in the previous case. This completes the proof. □

With this lemma we characterize the constructible numbers.

Theorem 4.4.9. *A real number a is constructible iff there exists a finite sequence of fields $\mathbb{Q} = \mathbb{F}_0 \subseteq \mathbb{F}_1 \subseteq \ldots \subseteq \mathbb{F}_n$ such that*

1. *the real number a is in $\mathbb{F}_n$ and*

2. *$\mathbb{F}_{j+1} = \mathbb{F}_j[a_j]$ where $a_j^2 \in \mathbb{F}_j$ for $j = 0, 1, \ldots, n-1$.*

Proof. Assume that a real number a is constructible.

we are given two points which we have taken them to be the origin $(0,0)$ and $(1,0)$. By hypothesis the length $|a|$ can be constructed. This means that the point $P := (a,0)$ is constructible.

The construction of the point P involves a finite number of fundamental constructions described earlier. From each of these fundamental constructions, we get new constructible points leading up to the construction of the point P.

Suppose the point P is constructed in k-steps by obtaining the points $P_0 = (0,0)$, $P_1 = (1,0)$, $P_2 \ldots, P_k = P$. The points P_0 and P_1 have coordinates in $\mathbb{Q}$. The constructed point $P_3 := (a_3, b_3)$ is obtained as a point of intersection of one of the three methods described above. Therefore the real numbers a_3 and b_3 either lie in $\mathbb{Q}$ or in a quadratic extension $\mathbb{Q}[\alpha]$ of $\mathbb{Q}$. Inductively we may assume that the coordinates of the point P_{k-1} lies in a field $\mathbb{E}$ constructed by this process. That is there exists a finite sequence of fields $\mathbb{Q} = \mathbb{F}_0 \subseteq \mathbb{F}_1 \subseteq \ldots \subseteq \mathbb{F}_r = \mathbb{E}$ such that

1. the coordinates of the point P_{k-1} lie in $\mathbb{E}$ and

2. $\mathbb{F}_{j+1} = \mathbb{F}_j[a_j]$ such that a_j^2 is in $\mathbb{F}_j$.

Finally the point $P = P_k$ is constructed by one of the fundamental constructions using the points $P_0, P_1, \ldots, P_{k-1}$. Therefore the co-ordinates a, 0 of the point P either lie in the field $\mathbb{E}$ or in a quadratic extension $\mathbb{E}[\beta]$ such that $\beta^2 \in \mathbb{E}$. Thus we have got a finite sequence of fields $\mathbb{Q} = \mathbb{F}_0 \subseteq \mathbb{F}_1 \subseteq \ldots \subseteq \mathbb{F}_r \subseteq \mathbb{F}_{r+1}$ such that

1. the real number a is in $\mathbb{F}_{r+1}$ and

2. $\mathbb{F}_{j+1} = \mathbb{F}_j[a_j]$ such that a_j^2 is in $\mathbb{F}_j$.

Conversely assume that there exists a finite sequence of fields $\mathbb{Q} = \mathbb{F}_0 \subseteq \mathbb{F}_1 \subseteq \ldots \subseteq \mathbb{F}_n$ satisfying the two properties.

We will prove the result by induction on n. Assume $n = 0$. Then $\mathbb{Q} = \mathbb{F}_0$ and $a \in \mathbb{Q}$. Hence a is constructible.

Now assume the result to be true for all $m < n$. That is, if there exists a sequence of fields $\mathbb{Q} \subseteq \mathbb{F}_1 \subseteq \mathbb{F}_1 \subseteq \ldots \subseteq \mathbb{F}_m$ satisfying the second property then any element a in $\mathbb{F}_m$ is constructible.

Let $\mathbb{Q} \subseteq \mathbb{F}_1 \subseteq \mathbb{F}_1 \subseteq \ldots \subseteq \mathbb{F}_n$ be a finite sequence of fields satisfying the two properties. By definition $\mathbb{F} = \mathbb{F}_{n-1}[\alpha]$ such that $\alpha^2 \in \mathbb{F}_{n-1}$.

By induction hypothesis all the elements in $\mathbb{F}_{n-1}$ are constructible. Therefore α^2 and hence α is constructible.

Since $a \in \mathbb{F}_n = \mathbb{F}_{n-1}[\alpha]$, there exists r and s in $\mathbb{Q}$ such that $a = r + s\alpha$. The real numbers r, s and α are constructible. Therefore a is constructible and the proof is complete. □

Corollary 4.4.10. *Assume a is a constructible. Then there exists a field extension $\mathbb{F}/\mathbb{Q}$ such that a is in $\mathbb{F}$ and $[\mathbb{F} : \mathbb{Q}] = 2^r$ for some $r \geq 1$.*

Proof. Since the number a is constructible, there is a finite sequence $\mathbb{Q} = \mathbb{F}_0 \subseteq \mathbb{F}_1 \subseteq \ldots \subseteq \mathbb{F}_n$ of fields with a in $\mathbb{F}_n$.

By tower law $[\mathbb{F}_n : \mathbb{Q}] = [\mathbb{F}_n : \mathbb{F}_{n-1}][\mathbb{F}_{n-1} : \mathbb{F}_{n-2}] \cdots [\mathbb{F}_1 : \mathbb{Q}]$. Since $[\mathbb{F}_i : \mathbb{F}_{i-1}]$ is either 1 or 2, it follows that $[\mathbb{F}_n : \mathbb{Q}] = 2^r$ for some $r \geq 1$. □

We will now answer the three questions.

4.4.1 Doubling the cube

Question 4.4.11. Given a cube C, can we construct a cube which is twice the area of the cube C?

Assume that an edge of the given cube C has unit length. Then the volume of C is 1. Suppose we can construct a cube C_1 such that the new cube C_1 is twice the volume of C. That is volume of C_1 is 2. Therefore length of each edge is $\sqrt[3]{2}$. We want to know if $\sqrt[3]{2}$ is constructible.

Theorem 4.4.12. *The real number $\sqrt[3]{2}$ is not constructible.*

Proof. Suppose $\sqrt[3]{2}$ is constructible. Then there exists finite sequence of fields $\mathbb{Q} \subseteq \mathbb{F}_0 \subseteq \mathbb{F}_1 \subseteq \ldots \subseteq \mathbb{F}_n$ such that $\sqrt[3]{2}$ is in $\mathbb{F}_n$ and $[\mathbb{F}_n : \mathbb{Q}] = 2^r$ for some $r \geq 1$. Since $\sqrt[3]{2}$ is in $\mathbb{F}_n$, it must have $\mathbb{Q}[\sqrt[3]{2}] \subseteq \mathbb{F}_n$.

On the other hand we have seen that $[\mathbb{Q}[\sqrt[3]{2}] : \mathbb{Q}] = 3$ and hence

$$[\mathbb{F}_n : \mathbb{Q}] = [\mathbb{F}_n : \mathbb{Q}[\sqrt[3]{2}][\mathbb{Q}[\sqrt[3]{2} : \mathbb{Q}] = 3[\mathbb{F}_n : \mathbb{Q}[\sqrt[3]{2}]].$$

This means that 3 divides $[\mathbb{F}_n : \mathbb{Q}] = 2^r$ a contradiction. □

Exercise 4.4.13. Can a cube be tripled and quadrupled?

4.4.2 Squaring the circle

The question here is

Question 4.4.14. Given a circle C is it possible to construct a square with the same area as the circle C?

The answer to this question is again negative.

Assume the radius of the circle is 1. Then the area of the circle is π. We would like to know if we can construct a square whose area is π. This is equivalent to constructing a real number a such that $a^2 = \pi$. This means that π is constructible. In particular π is algebraic over $\mathbb{Q}$.

However we know that π is transcendental over $\mathbb{Q}$. (See [1] or [5]). Therefore π is not constructible.

4.4.3 Trisecting an angle

Question 4.4.15. Given an angle θ, is it possible to construct the angle $\frac{\theta}{3}$?

The answer to this question is not straight forward yes or no!

We prove a negative result showing that the angle $\theta = 60°$can't be trisected. First we prove the following proposition.

Proposition 4.4.16. *Suppose an angle θ is constructible. Then the real numbers $\cos\theta$ and $\sin\theta$ are constructible.*

Proof. Suppose the angle θ is constructed as angle between two lines. We may assume one of the line to be X-axis ℓ_X. We denote by ℓ another line making an angle θ with ℓ_X at the origin. We draw a line perpendicular to ℓ_X and passing through the constructed point $P = (1, 0)$. This line meets the line at a point say Q. Hence the lengths OP, OQ and PQ are constructible. As a consequence the real numbers $\cos\theta = \frac{OP}{OQ}$ and $\sin\theta = \frac{PQ}{OQ}$ are constructible and the proof is complete. □

We will now prove that the angle $\theta = 60°$can't be trisected. Suppose we can trisect the angle $\theta = 60°$and let $\phi = \frac{\theta}{3}$. Now using cos 60 $= \frac{1}{2}$ we get

$$\frac{1}{2} = \cos 3\phi = 4\cos^3\phi - 3\cos\phi.$$

Hence $8\cos^3\phi - 6\cos\phi - 1 = 0$. Thus $\cos\phi$ is a solution of the polynomial $8X^3 - 6X - 1 = 0$. By Eisenstein criterion the polynomial $8X^3 - 6X - 1$ is irreducible. This polynomial has only one real root between 0 and 1 and it is $\cos\phi$. Therefore the degree of the extension field $\mathbb{F}$ which contains $\cos\phi$ must be a multiple of 3.

On the other hand if $\cos\phi$ were constructible, the degree of the extension must be a power of 2, a contradiction.

Exercise 4.4.17. The real numbers cos 1 and cos 2 are not constructible.

> If the real numbers cos 1 and cos 2 are constructible, then using trigonometric identities, we can show that cos 20 is constructible, a contradiction.

On the other hand we can show that cos 3 is constructible. We indicate an outline of proof below.

An equilateral triangle is constructible. A regular 5-gon is constructible. Therefore $\cos 12 = \cos(\frac{2\pi}{5} - 60) = \cos\frac{2\pi}{5}\cos 60 + \sin\frac{2\pi}{5}\sin 60$ is constructible. Now we use the identity $\cos\theta = \frac{\sqrt{1+\cos 2\theta}}{2}$ to show that $\cos 6 = \frac{\sqrt{1+\cos 12}}{2}$ is constructible. Once again we apply the trigonometric identity to the real number $\cos 6$ to conclude that $\cos 3$ is constructible.

4.4.4 Constructing a regular k-sided polygon

The idea of trisecting an angle can be used to prove result on constructing a regular k-sided polygon.

Suppose a regular k-sided polygon is constructible. This means that the angle $\theta = \frac{2\pi}{k}$ is constructible and the real numbers $\cos\theta$ and $\sin\theta$ are constructible by Proposition 4.4.16.

Now $1 = \cos 2\pi + i\sin 2\pi = \cos k\theta + i\sin k\theta = (\cos\theta + i\sin\theta)^k$.

If $\zeta = \cos\theta + i\sin\theta$, then ζ is a root of the polynomial $1+X+X^2+\cdots+X^{k-1}$.

We will now choose k to be a prime number, say, p. Then the polynomial $1+X+X^2+\cdots+X^{p-1}$ is irreducible and the degree of the extension $\mathbb{Q}[\zeta]$ over $\mathbb{Q}$ is $p-1$. Since $\cos\theta$ and $\sin\theta$ are constructible, the degree of the extension of the field $\mathbb{F}$ containing $\cos\theta$ and $\sin\theta$ will be 2^k for some k. If $\mathbb{K} = \mathbb{F}[i]$, then the field $\mathbb{Q}[\zeta]$ is contained in $\mathbb{K}$ and

$$2^{k+1} = [\mathbb{K} : \mathbb{F}][\mathbb{F} : \mathbb{Q}] = [\mathbb{K} : \mathbb{Q}] = [\mathbb{K} : \mathbb{Q}(\zeta)][\mathbb{Q}(\zeta) : \mathbb{Q}] = (p-1)[\mathbb{K} : \mathbb{Q}(\zeta)]$$

Hence $p-1 = 2^r$ for some $r \geq 1$. Thus we have proved that if a regular p-sided polygon is constructible, then $p = 2^r + 1$ for some $r \geq 1$.

More generally Gauss proved the following theorem which gives necessary and sufficient condition on the constructibility of a regular n-gon.

Theorem 4.4.18. *A regular n-sided polygon is constructible iff*

$$n = 2^a p_1 p_2 \cdots p_r$$

where a is a non-negative integer and p_is are prime numbers of the form $p_i = 2^{k_i} + 1$ for some some $k_i \in \mathbb{N}$.

4.5 Finite Fields

In this section we prove some important properties of Finite fields.

Let $\mathbb{F}$ be a field. Its characteristic is either zero or a prime number.

If the characteristic is zero, then $\mathbb{Z}$ is a subring of $\mathbb{F}$. As a consequence $\mathbb{Q}$ is a subfield of $\mathbb{F}$.

If the field $\mathbb{F}$ is finite, then the characteristic is a prime number. If the characteristic is p and if we denote by $\chi : \mathbb{Z} \to \mathbb{F}$ the characteristic homomorphism, then $\ker(\chi) = p\mathbb{Z}$. Hence by first fundamental theorem of homomorphism the field $\mathbb{Z}/p\mathbb{Z}$ becomes a subfield of $\mathbb{F}$. The field $\mathbb{Z}/p\mathbb{Z}$ is called the *prime subfield* of $\mathbb{F}$.

Our first result characterizes a finite field as the splitting field of a polynomial over $\mathbb{Z}_p$. More precisely we prove the following.

Theorem 4.5.1. *Let $\mathbb{F}$ be a finite field of characteristic p. Then there exists $k \geq 1$ such that $|\mathbb{F}| = p^k$ and $\mathbb{F}$ is the splitting field of the polynomial $f(X) = X^{p^k} - X$ over $\mathbb{Z}_p$.*

Conversely given a p-ime number p and $k \geq 1$, there exists a field $\mathbb{F}$ with p^k elements and it is the splitting field of the polynomial $f(X) = X^{p^k} - X$ over $\mathbb{Z}_p[X]$.

Proof. Let $\mathbb{F}$ be a finite field. Then $\mathbb{F}$ is a finite dimensional vector space over its prime subfield $\mathbb{Z}_p = \mathbb{Z}/p\mathbb{Z}$. Let $k = \dim_{\mathbb{Z}_p} \mathbb{F}$. Then $\mathbb{F}$ has p^k elements:

Let $e_1, e_2, \ldots, e_k$ be a $\mathbb{Z}_p$ basis of $\mathbb{F}$. Then the map

$$\varphi : \overbrace{\mathbb{Z}_p \times \mathbb{Z}_p \times \cdots \times \mathbb{Z}_p}^{k-\text{times}} \to \mathbb{F}$$

defined by

$$\varphi((a_1, a_2, \ldots, a_k)) = \sum_{i=1}^{k} a_i e_i$$

is a vector space isomorphism.

Since the field $\mathbb{F}$ has p^k elements, the multiplicative group $\mathbb{F}^* = \mathbb{F} \setminus \{0\}$ has $p^k - 1$ elements. Therefore $\alpha^{p^k-1} = 1$ for every element $\alpha \in \mathbb{F}^*$ and this shows that $\alpha^{p^k} = \alpha$ for all $\alpha \in \mathbb{F}$. Thus every element $\alpha \in \mathbb{F}$ is a root of the polynomial $f(X) = X^{p^k} - X$ in $\mathbb{Z}_p[X]$. Since $\deg(f(X)) = p^k$, it has at most p^k roots. However we just saw that all the p^k elements of $\mathbb{F}$ are roots of the polynomial $f(X)$. This proves that $\mathbb{F}$ is the splitting field of the polynomial $f(X) = X^{p^k} - X$ over $\mathbb{Z}_p$.

Conversely let p be a prime number and $k \geq 1$ be an integer. Let $\mathbb{K}$ be the splitting field of the polynomial $f(X) = X^{p^k} - X$ over $\mathbb{Z}_p[X]$.

We will now show that the roots of $f(X)$ are distinct and all the elements of $\mathbb{K}$ are roots of the polynomial $f(X)$.

The derivative of the polynomial $f(X) = X^{p^k} - X$ is $f'(X) = p^k X^{p^k-1} - 1 = -1$ in $\mathbb{Z}_p[X]$. Hence $f'(\alpha) = -1 \neq 0$ for every element $\alpha \in \mathbb{K}$. This proves that all the roots of the polynomial are distinct.

Let $\mathbb{F} := \{\alpha \in \mathbb{K} : f(\alpha) = 0\}$. Since the degree of the polynomial $f(X)$ is p^k and $\mathbb{K}$ contains all the roots of the polynomial $f(X)$ follows that $\mathbb{F}$ has p^k elements. If we now show that $\mathbb{F}$ is a field we will be through. (Why?)

Let α and β be two distinct elements of $\mathbb{F}$. That is $\alpha^{p^k} = \alpha$ and $\beta^{p^k} = \beta$. Then

$$(\alpha \pm \beta)^{p^k} = \alpha^{p^k} \pm \beta^{p^k} + \sum_{r=1}^{p^k-1} (\pm)^r \binom{p^k}{r} \alpha^{p^k-r} \beta^r.$$

Since p divides $\binom{p^k}{r}$ for $1 \leq r \leq p^k - 1$, it follows that $\binom{p^k}{r} \alpha^{p^k-r} \beta^r = 0$ in $\mathbb{K}$. Therefore

$$(\alpha \pm \beta)^{p^k} = \alpha^{p^k} \pm \beta^{p^k} = \alpha \pm \beta.$$

Hence $\alpha \pm \beta$ is also in $\mathbb{F}$.

It is easy to show that $\alpha\beta$ is also in $\mathbb{F}$: $(\alpha\beta)^{p^k} = \alpha^{p^k} \beta^{p^k} = \alpha\beta$.

This in particular shows that, if α is in $\mathbb{F}$, then all the powers of α are also roots of the polynomial $f(X)$. Hence they are also in $\mathbb{F}$.

Let $\alpha \neq 0, 1$ be an element in $\mathbb{F}$. Then $\alpha^{p^k} = \alpha$ shows that $\alpha^{p^k-1} = 1$. We write this as $1 = \alpha(\alpha)^{p^k-2}$ to conclude that α^{p^k-2} in $\mathbb{F}$ is the inverse of the element α.

This proves that $\mathbb{F}$ is a field with p^k elements. □

Another important result in finite fields is the following theorem.

Theorem 4.5.2. *Let* $\mathbb{F}$ *be a finite field. Then* $\mathbb{F}^* = \mathbb{F} \setminus \{0\}$, *the multiplicative group of nonzero elements in* $\mathbb{F}$, *is cyclic.*

This theorem is a particular case of the following more general result.

Theorem 4.5.3. *Let* $\mathbb{K}$ *be a field and* $\mathbb{K}^*$ *be the multiplicative group of nonzero elements of* $\mathbb{K}$. *If* G *is a finite subgroup of* $\mathbb{K}^*$, *then* G *is cyclic.*

Proof. Let G be a nontrivial finite subgroup of $\mathbb{K}^*$.

Let $k = \max\{|y| : y \in G\}$ and x in G be an element of order k. That is $x^k = 1$ and $x^t \neq 1$ for $1 \leq t < k$. By Lagrange's Theorem 2.3.1, k divides $|G|$. In particular $k \leq |G|$.

We will prove the theorem by showing that $k = |G|$.

Let $y(\neq x)$ be an element of G. We claim that $r := |y|$ divides k.

Suppose not. Let $d = \gcd(r, k)$ and $l := \operatorname{lcm}(r, k)$. Then d is less than r and $l := \operatorname{lcm}(r, k) = \frac{rk}{\gcd(r,k)} = \frac{rk}{d} = \frac{r}{d}k > k$. Since G is abelian, by Proposition 2.2.27, there exists an element z in G of order $l > k$, a contradiction to our assumption that x is of maximal order in G. This proves our claim.

Since r divides k, we can write $k = rq$ for some $q \in \mathbb{N}$. Let us now observe that $y^k = y^{rq} = (y^r)^q = 1$. This proves that every element y in G is a root of the polynomial $f(X) = X^k - 1$ in $\mathbb{K}$. However the polynomial $f(X) = X^k - 1$ has at most k roots in the field $\mathbb{K}$. Hence $|G| \leq k$. On the other hand we already know that $k \leq |G|$. This proves that x is an element of order $k = |G|$ in G. Thus G is a cyclic group. □

Chapter 5

Group Theory II

In this chapter we study some advanced topics in group theory.

5.1 Group Action

A symmetry of a geometric figure P is a permutation of the points of P that preserves the geometric structure of P.

For an equilateral triangle the symmetries are the rotation of the vertices and reflections along the medians.

As another example consider the map $z := x + iy \mapsto \overline{z} := x - iy$ of $\mathbb{C}$, the field of complex numbers. Geometrically this map is the reflection about x-axis of the plane $\mathbb{R}^2$. However this complex conjugation preserves the algebraic properties of $\mathbb{C}$, namely $\overline{z+w} = \overline{z} + \overline{w}$ and $\overline{zw} = \overline{z}\overline{w}$ for all z and w in $\mathbb{C}$. Thus it is an automorphism of the field $\mathbb{C}$.

Therefore we may define symmetry of a set X to be a permutation of its points that is compatible with the structure of the set X.

In the first example, symmetry refers to a permutation which preserves the geometric structure of the equilateral triangle, while in the second symmetry refers to an automorphism of the field of complex numbers $\mathbb{C}$. These examples are special cases of more general concept, that of a group acting on a set.

Definition 5.1.1. Let X be a set and G a group. An action of G on X is a map

$$\varphi : G \times X \to X$$

satisfying

1. $\varphi(e, x) = x$ for all $x \in X$ and

2. $\varphi(g, \varphi(h, x)) = \varphi(gh, x)$ for all x in X and g, h in G.

We write $\varphi(g, x)$ as $g.x$. In this notation, the condition 2 can be written as $g.(h.x) = (gh).x$ for all x in X and g, h in G, where gh is the product of the elements g and h in G.

Let G be a group acting on a set X. Give an element g in G, define a map $\varphi_g : X \to X$ by $\varphi_g(x) := \varphi(g, x)$. Then the map φ_g is a bijection from X to X.

Proposition 5.1.2. *If a group G acts on a set X, then $\varphi_g : X \to X$ defined by $\varphi_g(x) := \varphi(g,x)$ is a permutation of X, for every element g in G.*

Proof. If x and y be two elements in X such that $\varphi_g(x) = \varphi_g(y)$, then $\varphi(g,x) = \varphi(g,y)$. Since φ is a group action, it follows that $x = \varphi(g^{-1}, \varphi(g,x)) = \varphi(g^{-1}, \varphi(g,y)) = \varphi(g^{-1}g, y) = y$. This proves that φ_g is one-one.

Let x be in X. Then $\varphi(g^{-1}, x)$ is such that $\varphi_g(\varphi(g^{-1}, x)) = x$. Hence φ_g is onto and φ_g is a permutation of X for each $g \in G$. □

Thus, if a group G acts on a set X, given by a map $\varphi : G \times X \to X$, we get a map $\psi : G \to S(X)$ defined by $g \mapsto \varphi_g$. This map is a homomorphism as proved below.

Theorem 5.1.3. *Let X, G and $\psi : g \mapsto \varphi_g$ be as above. Then the map $\psi : G \to S(X)$ is a homomorphism.*

Conversely, if there exists a homomorphism $\psi : G \to S(X)$, then there is a unique action $\varphi : G \times X \to X$ such that $\varphi_g = \psi(g)$ for all $g \in G$.

Proof. Let $\varphi : G \times X \to X$ be an action of the group G on X.

Let g and h be two elements in G, and x be a pint in X. Then

$$\begin{aligned}
\psi(gh)(x) &= \varphi_{gh}(x) \\
&= \varphi(gh, x) \\
&= \varphi(g, \varphi(h, x)) \\
&= \varphi_g(\varphi(h, x)) \\
&= \varphi_g(\varphi_h(x)) \\
&= \varphi_g \circ \varphi_h(x) \\
&= \psi(g)(\psi(h)(x)) \\
&= (\psi(g)\psi(h))(x).
\end{aligned}$$

This proves that $\psi : G \to S(X)$ defined by $\psi(g) = \varphi_g$ is a homomorphism.

Conversely let $\psi : G \to S(X)$ be a homomorphism. Then we define $\varphi : G \times X \to X$ by $\varphi(g,x) := \psi(g)(x)$. We can easily show that this is an action of the group G on X.

Uniqueness is easy and left as an exercise. □

Examples 5.1.4. 1. Let G be a group and X be a set. Then the map $\varphi : G \times X \to X$ defined by $\varphi(g,x) := x$ is an action of G on X and this action is called *trivial action* of G on X.

2. Let $X = \mathbb{Z}$ and $G = \mathbb{Z}$. The map $\varphi : G \times X \to X$ by $\varphi(k,x) := k + x$ is an action of G on X.

3. Let $G = \mathbb{R}^*$ and $X = \mathbb{R}^n$ for $n \geq 1$. Then G acts on X via the map

   (a) $(t, (x_1, x_2, \cdots, x_n)) \mapsto (tx_1, tx_2, \cdots, tx_n)$.

   (b) $(t, (x_1, x_2, \cdots, x_n)) \mapsto (e^t x_1, e^t x_2, \cdots, e^t x_n)$.

   Similarly if $G = \mathbb{C}^*$ and $X = \mathbb{C}^n$, then G acts on X.

   Observe that same group G has different actions on the same set X.

4. Let $X = \mathbb{R}^n$ and $G = \mathrm{GL}(n, \mathbb{R})$. Then $\varphi : G \times X \to X$ defined by $\varphi(A, v) := Av$ is an action of G on X.

5. Let $X = \mathbb{R}^2$ and $G = \mathbb{R}$. Then

   (a) $\varphi : G \times X \to X$ by $\varphi(t, (x, y)) := (e^t x, e^t y)$ and

   (b) $\psi : G \times X \to X$ by $\psi(t, (x, y)) := (e^t x, e^{-t} y)$

   are two different actions of G on X.

6. Let $X = \mathbb{R}^2$ and $G = \mathbb{Z}_2$. Define $\varphi : G \times \mathbb{R}^2 \to \mathbb{R}^2$ by $\varphi(0, (x, y)) := (x, y)$ and $\varphi(1, (x, y)) := (y, x)$. Then φ is an action of G on $\mathbb{R}^2$.

   There is another action of $G = \mathbb{Z}_2$ on $\mathbb{R}^2$: Let $\psi : G \times \mathbb{R}^2 \to \mathbb{R}^2$ be the map $\psi(0, (x, y)) := (x, y)$ and $\psi(1, (x, y)) := (-x, -y)$.

7. Let $X = \mathbb{R}^n$ and $G = S_n$. Then $(\sigma, (x_1, x_2, \ldots, x_n)) \mapsto (x_{\sigma(1)}, x_{\sigma(2)}, \ldots, x_{\sigma(n)})$ is an action of G on X.

8. Let $X = \mathbb{R}^n$ and $G = \mathrm{GL}(n, \mathbb{R}) \ltimes \mathbb{R}^n$. The map $\varphi : G \times X \to X$ defined by $\varphi((A, v), x) := Ax + v$ is an action.

9. Let G be a group and $X = G$. Then

   (a) the map $(g, x) \mapsto gx$ is an action of G on G and it is called the *left action* of G on G.

   (b) the maps $(g, x) \mapsto xg^{-1}$ is an action of G on G and it is called the *right action* of G on G.

   (c) the map $(g, x) \mapsto gxg^{-1}$ is also an action of G on G and it is called the *conjugate action* of G on G.

10. Let H be a subgroup of a group G. Then H acts on G via the map $(h, g) \mapsto hg$ and also via the map $(h, g) \mapsto gh^{-1}$. The first one is called the *left action* of H on G and the second one the *right action.* This action is called the *induced action* of H.

11. Let G be a group and H a subgroup of G. Then G acts on the set of left cosets G/H of G via $(x, gH) \mapsto xgH$.

12. Let m and $n \geq 2$. Let $G = S_m \times S_n$, the product of the permutation groups S_m and S_n and $X = \{k \in \mathbb{N} : 1 \leq k \leq m + n\}$. Then $\varphi : G \times X \to X$ defined by $\varphi((\sigma, \tau))(x) = \begin{cases} \sigma(x) & \text{if } x \leq m \\ \tau(x - m) + m & \text{if } m + 1 \leq x \leq m + n \end{cases}$ is an action of G on X.

Remark 5.1.5. Observe that in the above examples, a group G acts in more than one way on a set X.

Exercise 5.1.6. 1. Let $X := \{z \in \mathbb{C} : \mathrm{Im}(z) > 0\}$ and $G = \mathrm{SL}(2, \mathbb{R})$. Show that, if $A = \begin{pmatrix} a & b \\ c & d \end{pmatrix}$ is in $\mathrm{SL}(2, \mathbb{R})$ and $z \in X$, then $A \cdot z := \frac{az+b}{cz+d}$ is also in X. Hence prove that the map $(A, z) \mapsto A \cdot z$ is an action of $\mathrm{SL}(2, \mathbb{R})$ on X.

We write

$$\begin{aligned} A \cdot z &= \frac{az+b}{cz+d} \\ &= \frac{[(ax+b)iay][(cx+d)-icy]}{|cz+d|^2} \\ &= \frac{[(ax+b)(cx+d)+acy^2]+iy(ad-bc)}{|cz+d|^2}. \end{aligned}$$

Since A is in $SL(2,\mathbb{R})$ and $y>0$, it follows that Az is in X. It is routine to check that $A(B.z)=(AB).z$ for z in X.

2. Let $G=S_n$ and $S^{n-1}=\{(x_1,x_2,\ldots,x_n)\in\mathbb{R}^n : \sum_{i=1}^n x_i^2=1\}$. Show that for every point $x=(x_1,x_2,\ldots,x_n)$ in S^{n-1} and every element σ in S_n, the point $\sigma(x):=(x_{\sigma(1)},x_{\sigma(2)},\ldots,x_{\sigma(n)}$ is also in S^{n-1} and conclude that the map $(\sigma,x)\mapsto\sigma(x)$ is an action of S_n on S^{n-1}.

3. Let $X=S^{n-1}$ and $G=\mathrm{O}(n,\mathbb{R})$. Show that, for every A in $\mathrm{O}(n,\mathbb{R})$ and $x\in S^{n-1}$, the vector Ax is again in S^{n-1} and $(A,x)\mapsto Ax$ is an action of G on S^{n-1}.

4. Let $G=\mathrm{GL}(2,\mathbb{R})$ and X is the set of all lines passing through origin in $\mathbb{R}^2$. Define an action of G of X. Can you generalize this to n-dimensions?

5. Let $G=\mathrm{GL}(n,\mathbb{R})$ and $X=\mathrm{SL}(n,\mathbb{R})$. Show that $(A,B)\mapsto ABA^{-1}$ is an action of $\mathrm{GL}(n,\mathbb{R})$ on $\mathrm{SL}(n,\mathbb{R})$.

   More generally if H is a normal subgroup of a group G, then show that G acts on H, via the map $(g,h)\mapsto ghg^{-1}$.

Definition 5.1.7. Let G be a group acting on a set X and H be a subgroup of G. Then H also acts on X: If $\varphi: G\times X\to X$ is an action of G on X, then $\varphi_H: H\times X\to X$ defined by $\varphi_H(h,x):=\varphi(h,x)$ is an action of H on X and it is called *the induced action* of H on X.

Example 5.1.8. 1. Let $G=\mathbb{C}^*$ and $\mathbb{T}:=\{z\in\mathbb{C}^* : |z|=1\}$. Then $(z,(z_1,z_2,\cdots,z_n)\mapsto(zz_1,zz_2,\cdots,zz_n)$ is an induced action of $\mathbb{T}$ on $\mathbb{C}^n$.

2. Let $G=\mathbb{C}^*$ and $H_k:=\{z\in\mathbb{C}^* : z^k=1\}$ for $k\geq 2$. Then the map $(z,(z_1,z_2,\cdots,z_n))\mapsto(zz_1,zz_2,\cdots,zz_n)$ is an induced action of H_k on $\mathbb{C}^n$.

3. Let $\mathbb{T}=\{z\in\mathbb{C}^* : |z|=1\}\subseteq G=\mathbb{C}^*$ and $X=\mathbb{C}^*$. The group G acts on $X=G$ through the left multiplication. Hence $\mathbb{T}$ also acts on X.

4. Let $n\geq 2$ and $G_n:=\{z\in\mathbb{C}^* : z^n=1\}$ the set of all n-th roots of unity. Then G_n is a subgroup of $\mathbb{T}$ and $\mathbb{C}^*$. Hence the group G_n acts on $\mathbb{T}$ and also on $\mathbb{C}^*$.

5. Let $X=\mathbb{R}^n$ and $G=\mathrm{GL}(n,\mathbb{R})$ and $H=\mathrm{SL}(n,\mathbb{R})$. Then the map $\varphi_H: H\times X\to X$ defined by $\varphi(A,v):=Av$ is an induced action of $\mathrm{SL}(n,\mathbb{R})$ on X.

6. Let $X = \mathbb{R}^n$ and $H = \mathrm{O}(n, \mathbb{R})$. The group H is a subgroup of $\mathrm{GL}(n, \mathbb{R})$ and the map $\varphi_H : H \times X \to X$ defined by $\varphi(A, v) := Av$ is an induced action.

7. Let $H := \langle (1, 2, 3, \cdots, n) \rangle$ be the cyclic subgroup of S_n generated by the n-cycle $(1, 2, 3, \cdots, n)$. Then H acts on S^{n-1} through the action of S_n on S^{n-1}.

Exercise 5.1.9. Find the homomorphism of G in to $S(X)$ of the examples and exercises of the group action given in Examples 5.1.4 and Exercises 5.1.6. Find also the image of these homomorphisms.

Exercise 5.1.10. Show that the action of $S_m \times S_n$ on $\{1, 2, 3, \ldots, m + n\}$ as defined in Example 5.1.4(12) gives us a one-one homomorphism from $S_m \times S_n$ to S_{m+n}.

Definition 5.1.11. Let G be a group acting on a set X. If x is a point in X, the trajectory $\{\varphi(g, x) : g \in G\}$ of the point x is called the orbit of the point x and it is denoted it by $\mathcal{O}_x$.

Example 5.1.12. 1. Let G act on $X = G$ via the left action $(g, x) \mapsto gx$. If x is in G, the orbit $\mathcal{O}_x$ of x is G.

If z is a point in G and $g = zx^{-1}$, then $gx = (zx^{-1})x = z$ is in $\mathcal{O}_x$. Thus $G = \mathcal{O}_x$.
Similarly for the right action too the orbit $\mathcal{O}_x = X$ for every point x in $X = G$.

2. Let $X = \mathbb{R}^n$ and $G = \mathrm{GL}(n, \mathbb{R})$. Let $e_1 = (1, 0, \ldots, 0)$ in $\mathbb{R}^n$. Then $\mathcal{O}_{e_1} = \mathbb{R}^n \setminus \{0\}$.

Let v be a non zero vector in $\mathbb{R}^n$. Let $B := \{v = v_1, v_2, v_3, \ldots, v_{n-1}\}$ be a basis of $\mathbb{R}^n$ extending $\{v\}$. We write $v_i = \sum_{j=1}^n v_{ji} e_j$ where $\{e_j : 1 \leq j \leq n\}$ is the standard basis of $\mathbb{R}^n$. If

$$A = \begin{pmatrix} v_{11} & v_{12} & \cdots & v_{1n} \\ v_{21} & v_{22} & \cdots & v_{2n} \\ \vdots & \vdots & \ddots & \vdots \\ v_{n1} & v_{n2} & \cdots & v_{nn} \end{pmatrix}$$

then $Ae_1 = \sum_j v_{j1} e_j = v$. This shows that $\mathcal{O}_{e_1} = \mathbb{R}^n \setminus \{0\}$.
Using same arguments, we can show that $\mathcal{O}_x = \mathbb{R}^n \setminus \{0\}$, for every point $x = (x_1, x_2, \ldots, x_n) \neq (0, 0, \ldots, 0)$ in $\mathbb{R}^n$.

Exercise 5.1.13. 1. Let $G = \mathbb{R}$ and $X = \mathbb{R}^2$. Then G acts on $\mathbb{R}^2$ given by the map $(t, (x, y)) \mapsto (e^t x, e^t y)$. If (x, y) is fixed find $\{(e^t x, e^t y) : t \in \mathbb{R}\}$.

Observe that $\{e^t(0, 0) : t \in \mathbb{R}\} = \{(0, 0)\}$.
Let $(x, y) = (1, 0)$. Then, since $e^t > 0$ for all $t \in \mathbb{R}$, it follows that $\{e^t(1, 0) : t \in \mathbb{R}\}$ is the positive x-axis. If $(x, y) = -(1, 0)$, then $\{e^t(-1, 0) : t \in \mathbb{R}\}$ is the negative real axis.
In general, $\{e^t(x, y) : t \in \mathbb{R}\}$ is the positive ray passing through the point (x, y), if $(x, y) \neq (0, 0)$.

2. Let $G = \mathbb{R}$ act on $X = \mathbb{R}^2$ via the map $(t, (x, y)) \mapsto (e^t x, e^{-t} y)$. If (x, y) is fixed find $\{(e^t x, e^{-t} y) : t \in \mathbb{R}\}$.

   Let $(x, y) = (1, 0)$. Then, since $e^t > 0$ for all $t \in \mathbb{R}$, it follows that $\{e^t(1, 0) : t \in \mathbb{R}\}$ is the positive x-axis. If $(x, y) = -(1, 0)$, then $\{e^t(-1, 0) : t \in \mathbb{R}\}$ is the negative real axis.

   Let $(x, y) = (0, 1)$. Then, since $e^t > 0$ for all $t \in \mathbb{R}$, it follows that $\{(0, e^{-t}) : t \in \mathbb{R}\}$ is the positive y-axis. If $(x, y) = -(0, 1)$, then $\{(0, -e^{-t}) : t \in \mathbb{R}\}$ is the negative y-axis.

   Let (x, y) be point in $\mathbb{R}^2$ such that $xy \neq 0$. Then the orbit of the point (x, y) is $\{(e^t x, e^{-t} y) : t \in \mathbb{R}\}$. If we take a point (u, v) in this orbit then $(u, v) = (e^t x, e^{-t} y)$ for some t in $\mathbb{R}$ and hence $uv = e^t x e^{-t} y = xy$. Thus the orbit of the point (x, y) is the rectangular hyperbola $\{(u, v) \in \mathbb{R}^2 : uv = xy\}$ lying in the same quadrant as the point (x, y).

3. Let G_n be the group of n-th roots of unity and $\mathbb{T}$ be the group of unit complex numbers. If z is in $\mathbb{T}$, find $\{wz : w \in G_n\}$.

   The group G_n is a cyclic group of order n. If ζ is a primitive n-th root of unity, then ζ is a generator of G_n. Therefore the trajectory $\{wz : w \in G_n\} = \{\zeta^k z : 1 \leq k \leq n\}$ is the set of vertices of a regular n-sided polygon on the unit circle with z as one of the vertices.

   Thus for every point z in H, the trajectory of the point z under the action of G_n is the regular n-sided polygon with z as one of the vertices.

   Notice that in this example all the orbits have same size.

Exercise 5.1.14. Find the orbit of a point of the group actions given below.

1. Let $G = \mathbb{R}$ act on $X = \mathbb{R}^2$ via $(t, (x, y)) \mapsto (e^t x, y)$. What is the orbit of a point (x_0, y_0)?

2. Let $G = \mathrm{GL}(2, \mathbb{R})$ act on $X = \mathbb{R}^2$ through $(A, v) \mapsto A.v$. What is the orbit of

   (a) $e_1 = (1, 0)$?

   (b) $e_2 = (0, 1)$?

   (c) any non-zero vector (x, y)?

3. Let $G = O(2, \mathbb{R})$ and let it act on $X = \mathbb{R}^2$ as $(A, v) \mapsto Av$. What is the orbit of

   (a) $e_1 = (1, 0)$?

   (b) $e_2 = (0, 1)$?

   (c) any non-zero vector (x, y)?

4. Let $G = SL(2, \mathbb{R})$ and $X = \{z \in \mathbb{C} : \mathrm{Im}(z) > 0\}$. Find the orbit of the point i under the action of G described in Exercise 5.1.6(1).

Let y be a positive real number. If $A = \begin{pmatrix} \sqrt{y} & 0 \\ 0 & \frac{1}{\sqrt{y}} \end{pmatrix}$, then $A.i = iy$ and $\begin{pmatrix} 1 & x \\ 0 & 1 \end{pmatrix}(iy) = x + iy$. Thus every element $z = x + iy$ is in the orbit of i and $\begin{pmatrix} \sqrt{y} & \frac{x}{\sqrt{y}} \\ 0 & \frac{1}{\sqrt{y}} \end{pmatrix}$ is matrix sending i to z. Hence $\mathcal{O}_i = X$.

5 Let $H := \left\{ \begin{pmatrix} t & 0 \\ 0 & \frac{1}{t} \end{pmatrix} : t \subset \mathbb{R}^* \right\}$. Then H as a subgroup of $SL(2, \mathbb{R})$ acts on $\mathbb{R}^2$. Let $(x, y) \neq (0, 0)$. Find the orbit $\mathcal{O}_{(x,y)}$ under the action of H.

Let H be a subgroup of a group G. For the left action $(h, g) \mapsto hg$ of the group H on G, the orbits are the right cosets of H. All the orbits are of same "size".

Let G be a group acting on G via conjugation $(g, x) \mapsto gxg^{-1}$. If x is in G, then the orbit of x is the set $\{gxg^{-1} : g \in G\}$ and it is called the conjugacy class of x. It is also denoted by C_x.

If x is in $Z(G)$, the center of G, then $C_x = \{x\}$. That is the orbit is singleton and every element g of G fixes the point x.

If $x \notin Z(G)$, then only the points in the centralizer of x fix the point x and the orbit can be of different size for different points.

We will see few more examples where the orbits are of different size.

Examples 5.1.15. Let $X = \mathbb{R}^n$ and $G := S_n$.

1. Let x be a real number and $p = (x, x, x, \cdots, x)$. Then the orbit of the point p is $(x, x, x, \cdots, x)$ and every element of the group G fixes the point p.

2. Let $x \neq 1$ be a real number and $p = (1, x, x, x, \cdots, x)$. Then the orbit is of size n. (Why?) An element σ is S_n fixes the point p, iff it keeps 1 fixed and permutes the x in any order. Thus an element σ in S_n fixes the point p iff σ is a permutation of the set $\{2, 3, 4, \cdots, n\}$. They are totally $(n-1)!$ in numbers.

3. Let $x \neq 1$ be a real number and $p = (1, 1, x, x, \cdots, x)$. Then the orbit is of size $n(n-1)$. (Why?) An element σ is S_n fixes the point p, iff it permutes 1 in the first position to the second position and permutes the x or it keeps 1 fixed and permutes the x in any order. Thus an element σ in S_n fixes the point p iff $\sigma = (1, 2)\tau$ where τ is a permutation of the set $\{3, 4, \cdots, n\}$. They are totally $2(n-2)!$ in numbers.

4. Let $p = (1, 1, 1, x, x, \cdots, x)$. Then we can see that the size of the orbit is $\frac{n(n-1)(n-2)}{3!}$ and an element σ fixes the point p iff $\sigma = \tau_1\tau_2$ where τ_1 is a permutation of $\{1, 2, 3\}$ and τ_2 is a permutation of $\{4, 5, 6, \cdots, n\}$.

5. Let $p = (x_1, x_2, x_3, \cdots, x_n)$ where all the x_i's are distinct. Then the size of the orbit is $n!$ and an element σ in S_n fixes the point p iff $\sigma = Id$.

We strongly recommend the readers to work out the details when $n = 6$ to understand what is happening.

In these example, we notice that when the size of an orbit $\mathcal{O}_x$ is small, the set of group elements fixing the point x is big. In geometric terms this says that more symmetry means the set of points fixing an element in the orbit is big!

Now we want to study this relation more.

Definition 5.1.16. If G is a group acting on a set X and x is a point in X then the set $G_x := \{g \in G : gx = x\}$ is called the stabilizer of x.

We first show that the stabilizer of point is a subgroup of G.

Proposition 5.1.17. *Let G be a group acting on a set X and x in X. Then G_x is a subgroup of G.*

Proof. By definition of group action the identity element e in G is in the stabilizer of every point x.

Let x be a point in X and $a, b \in G_x$. Then $a^{-1}x = x$ and $(ab)x = a(bx) = ax = x$. Hence G_x is a subgroup of G. □

Examples 5.1.18. 1. Let $X = \mathbb{R}^n$ and $G = \mathrm{GL}(n, \mathbb{R})$. Let $0 \in \mathbb{R}^n$. Then $\mathcal{O}_0 = \{0\}$ and $G_0 = \mathrm{GL}(n, \mathbb{R})$.

2. Let $X = \mathbb{R}^n$ and $G = \mathrm{GL}(n, \mathbb{R})$. Let $e_1 = (1, 0, \ldots, 0)$ in $\mathbb{R}^n$. Then $\mathcal{O}_{e_1} = \mathbb{R}^n \setminus \{0\}$ and $G_{e_1} := \{A \in \mathrm{GL}(n, \mathbb{R}) : Ae_1 = e_1\}$.

   Let v be a non zero vector in $\mathbb{R}^n$. Let $B := \{v = v_1, v_2, v_3, \ldots, v_{n-1}\}$ be a basis of $\mathbb{R}^n$ extending $\{v\}$. We write $v_i = \sum_{j=1}^n v_{ji}e_j$ where $\{e_j : 1 \leq j \leq, n\}$ is the standard basis of $\mathbb{R}^n$. If

$$A = \begin{pmatrix} v_{11} & v_{12} & \cdots & v_{1n} \\ v_{21} & v_{22} & \cdots & v_{2n} \\ \vdots & \vdots & \ddots & \vdots \\ v_{n1} & v_{n2} & \cdots & v_{nn} \end{pmatrix}$$

   then $Ae_1 = \sum_j v_{j1}e_j = v$. This shows that $\mathcal{O}_{e_1} = \mathbb{R}^n \setminus \{0\}$. We will now describe the stabilizer G_{e_1}. Let

$$A = \begin{pmatrix} a_{11} & a_{12} & \cdots & a_{1n} \\ a_{21} & v_{22} & \cdots & a_{2n} \\ \vdots & \vdots & \ddots & \vdots \\ a_{n1} & a_{n2} & \cdots & a_{nn} \end{pmatrix}$$

   be in $\mathrm{GL}(n, \mathbb{R})$ such that $Ae_1 = e_1$. But $Ae_1 = \sum_{j=1}^n a_{j1}e_j$. Therefore $Ae_1 = e_1$ iff $a_{j1} = 0$ for $j \geq 2$ and $a_{11} = 1$. Hence we can write the matrix A as

$$A = \begin{pmatrix} 1 & w \\ 0 & A_1 \end{pmatrix}$$

   where $0 = (0, 0, \ldots, 0)^t$ in $\mathbb{R}^{n-1}$ is written as a column vector, and $w = (a_{12}, a_{13}, \ldots, a_{1n})$ in $\mathbb{R}^{n-1}$ is written as a row vector and $A_1 = \begin{pmatrix} a_{22} & a_{23} & \cdots & a_{2n} \\ a_{32} & a_{33} & \cdots & a_{3n} \\ \vdots & \vdots & \ddots & \vdots \\ a_{n2} & a_{n3} & \cdots & a_{nn} \end{pmatrix}$. Let us now observe that $\det(A) = \det(A_1)$. Since $\det(A) \neq 0$, we see $\det(A_1) \neq 0$. Hence A_1 is in $\mathrm{GL}(n-1, \mathbb{R})$.

If we now start with an element A_1 in $\mathrm{GL}(n-1,\mathbb{R})$ and a vector w in $\mathbb{R}^{n-1}$ written as a row vector, then the matrix

$$A = \begin{pmatrix} 1 & w \\ 0 & A_1 \end{pmatrix}$$

fixes the vector e_1. Thus

$$G_{e_1} = \left\{ \begin{pmatrix} 1 & w \\ 0 & A_1 \end{pmatrix} \in \mathrm{GL}(n,\mathbb{R}) : \det(A_1) \neq 0 \text{ and } w \in \mathbb{R}^{n-1} \right\}$$

is the stabilizer of e_1.

3. Let $X = \mathbb{R}^2$ and $G = O(2,\mathbb{R})$. Let p be a point in $\mathbb{R}^2$. Then the orbit $\mathcal{O}_p = \{q \in \mathbb{R}^2 : \|q\| = \|p\|\}$. If $p = (0,0)$, the orbit $\mathcal{O}_p = \{(0,0)\}$ and if $p \neq (0,0)$, the orbit $\mathcal{O}_p$ is a circle of radius $\|p\|$, centered at the origin.

Exercise 5.1.19. 1. Let $X = \mathbb{R}^n$ and $G = \mathrm{GL}(n,\mathbb{R}) \ltimes \mathbb{R}^n$. Let $x = 0 \in \mathbb{R}^n$. Find $\mathcal{O}_x$ and G_x. Show that G_{e_1} in Example 5.1.18(2) is isomorphic to $\mathrm{GL}(n-1,\mathbb{R}) \ltimes \mathbb{R}^{n-1}$.

Let $A = \begin{pmatrix} 1 & w \\ 0 & A_1 \end{pmatrix}$ and $B = \begin{pmatrix} 1 & x \\ 0 & B_1 \end{pmatrix}$ be two elements in G_{e_1}. Then

$$AB = \begin{pmatrix} 1 & x + wB_1 \\ 0 & A_1B_1 \end{pmatrix}$$

where wB_1 is the product of the $1 \times (n-1)$ matrix w and the matrix B_1.

Now it is easy to define an isomorphism.

2. Let $X = \mathbb{R}^n$ and $G = \mathrm{O}(n,\mathbb{R}) \ltimes \mathbb{R}^n$. Let $x = 0 \in \mathbb{R}^n$. Find $\mathcal{O}_x$ and G_x.

3. Let $X = \mathbb{H}^2$ and $G = \mathrm{SL}(2,\mathbb{R})$. Let $z = i$. Find $\mathcal{O}_i$ and G_i for the action defined in Exercise 5.1.6(1). Let $z = x + iy$ be a point in $\mathbb{H}^2$. Find $\mathcal{O}_z$ and G_z.

We have already shown that $\mathcal{O}_i = \mathbb{H}^2$.

Let $A = \begin{pmatrix} a & b \\ c & d \end{pmatrix}$ be an element of $\mathrm{SL}(2,\mathbb{R})$ such that $Ai = i$. This means that $ai + b = i(ci + d) = -c + id$. Comparing the real and imaginary part, we see that A is of the form $A = \begin{pmatrix} a & -b \\ b & a \end{pmatrix}$. Since $\det(A) = a^2 + b^2 = 1$, there exists $\theta \in [0, 2\pi)$ such that $A = \begin{pmatrix} \cos\theta & -\sin\theta \\ \sin\theta & \cos\theta \end{pmatrix}$. Conversely we can show that every element in $\mathrm{SO}(2,\mathbb{R})$ fixes the point i. Thus the subgroup $G_i = \mathrm{SO}(2,\mathbb{R})$.

Let z be a point in $\mathbb{H}^2$. Let A in $\mathrm{SL}(2,\mathbb{R})$ be such that $A.i = z$. Therefore $\mathcal{O}_z = \{Bz : B \in \mathrm{SL}(2,\mathbb{R})\} = \{BA.i : B \in \mathrm{SL}(2,\mathbb{R})\} = \mathcal{O}_i$.

Let B be an element of $\mathrm{SL}(2,\mathbb{R})$ such that $Bz = z$. Then, since $z = A.i$, it follows that $A.i = z = Bz = BA.i$. Thus $(A^{-1}BA).i = i$. This proves that an element B in $\mathrm{SL}(2,\mathbb{R})$ is in G_z iff $A^{-1}BA$ is in G_i. Therefore $G_z = \{A^{-1}BA : B \in G_i\}$.

4. Let $G = S_n$ and $X = \mathbb{R}^n$. Find $\mathcal{O}_x$ and G_x if

(a) $x = e_2 + e_4 + \cdots + e_{2k}$, where $n = 2k$,

(b) $x = e_1 + e_3 + \cdots + e_{2k+1}$, where $n = 2k + 1$,

(c) $x = e_1$, $x = (1, 1, 1, \ldots, 1)$, $x = e_1 + e_j$ for $j \geq 2$.

5. Let $X = S^{n-1}$ and $G = S_n$. Find the orbit and stabilizer of a point $x = (x_1, x_2, \ldots, x_n)$ under the action of G on S^{n-1}.

6. Let $n \geq 2$ and $H_n := \{z \in \mathbb{C}^* : z^n = 1\}$. Let $X = S^1$. Given a point $z \in S^1$, find the orbit $\mathcal{O}_z$ and the stabilizer G_z.

Definition 5.1.20. Let X be a set and G be a group acting on the set X. We say that the group G acts transitively, if given two points x and y in X, there exists $g \in G$ such that $gx = y$.

Note that if G acts transitively on a set, then there is only one orbit and conversely:

Let x be a point in X and $\mathcal{O}_x$ be the orbit of x. Let y be a point in X. Since the group G acts transitively on X, there exists an element g in G such that $y = gx$. This means that y is in $\mathcal{O}_x$ and this is true for all points in X. Thus $X = \mathcal{O}_x$.

Converse is easy.

Examples 5.1.21. 1. Let G be a group and $X = G$. Then the left action of G on X is transitive.

2. Let $X = \mathbb{R}^n \setminus \{0\}$ and $G = \mathrm{GL}(n, \mathbb{R})$. Then the action of G on X in Example 5.1.4(4) is transitive.

3. Let $G = \mathrm{SL}(2, \mathbb{R})$ act on $X = \{z \in \mathbb{C} : \mathrm{Im}(z) > 0\}$ as in Exercise 5.1.6(1). This action is transitive.

4. Let $X = \mathbb{R}^n$ and $G = \mathrm{GL}(n, \mathbb{R}) \ltimes \mathbb{R}^n$. Then given any point x in $\mathbb{R}^n$, the element $(I, x) \in \mathrm{GL}(n, \mathbb{R}) \ltimes \mathbb{R}^n$ is such that $(I, x)(0) = x$. This shows that every point $x \in \mathbb{R}^n$ is in the orbit of 0.

   Let x and y be two points in $\mathbb{R}^n$. Then we know from linear algebra that there exists a non-singular matrix $A \in \mathrm{GL}(n, \mathbb{R})$ such that $Ax = y$. If now take $v = 0$, then $(A, 0)(x) = y$. This shows that for any point $x \in \mathbb{R}^n$, the orbit $\mathcal{O}_x = \mathbb{R}^n$.

5. Let $X = \mathbb{R}^n$ and $G = \mathrm{GL}(n, \mathbb{R})$. Then the action of G on X in Example 5.1.4(4) is not transitive. What are the orbits of this action?

6. Let G be a group and H a proper subgroup of G. Then H acts on G via the left action of G on G. This action is not transitive.

7. Let $X = \mathbb{R}^n \setminus \{0\}$ and $G = \mathrm{O}(n, \mathbb{R})$. The action of G on X in Example 5.1.8(6) is not transitive.

Exercise 5.1.22. 1. Let $X = \mathbb{R}^n$ and $G = S_n$. Is the action of G on X as in Example 5.1.4(7)transitive?

2. Let X be a set and G be a group acting on X. The group G acts on $\mathcal{O}_x$ defined by $(h, gx) \mapsto (hg)x$. Show that this is a transitive action of G.

3. Let G be a group and H a normal subgroup of G. Then the group G acts on H via the conjugate action $(g, h) \mapsto ghg^{-1}$. Show that this action of G on H is not transitive.

4. Let the group $G = \mathrm{GL}(n, \mathbb{R}) \times \mathrm{GL}(n, \mathbb{R})$ act on $X = M(n, \mathbb{R})$ by $((A, B), X) \mapsto AXB^{-1}$. Is this action transitive? Find the orbit and the stabilizer of the point I.

5. Let $X = \mathbb{R}^n \setminus \{0\}$ and $G = \mathrm{SL}(n, \mathbb{R})$. Is the action of G on X transitive? First do this for $n = 2$.

6. Let G be a group. The group $\mathbb{Z}_2$ acts on G by $(0, g) \mapsto g$ and $(1, g) \mapsto g^{-1}$. Find the orbit and the stabilizer of a point x in G. When is the orbit singleton?

7. Show that a finite group G can not act transitively on an infinite set X. That is if a finite group G acts transitively on a set X, then X is finite.

Let G be a group acting on a set X. We say that two points x_1 and x_2 are related and write $x_1 \sim x_2$ if there exists an element g in G such that $x_2 = gx_1$. That is an element g in G moves x_1 to x_2. Now it is easy to see that this relation is an equivalence relation.

Let $x \sim y$. Then $y = gx$ for some g in G. This means that y is in the orbit $\mathcal{O}_x$ of x. Therefore $hy = hgx$ is also in $\mathcal{O}_x$. Thus $\mathcal{O}_y \subseteq \mathcal{O}_x$. The point $x = g^{-1}y$ is in $\mathcal{O}_y$. Hence we conclude from what we have shown above that $\mathcal{O}_x \subseteq \mathcal{O}_y$ and this proves that $= \mathcal{O}_x = \mathcal{O}_y$ for any point $y \in \mathcal{O}_x$.

Let x and y be two points in same orbit. Then $y = ax$ for some a in G. Let h be a point in G_y. Then $ax = y = hy = hax$. This means that $a^{-1}hax = x$ and hence $a^{-1}ha$ is in G_x. This proves that the stabilizers G_x and G_y are conjugate in G.

We summarize this in the following theorem.

Theorem 5.1.23. *Let G be a group acting on the set X. Then*

1. *the orbits of X are either disjoint or same.*

2. *the set X is a disjoint union of its orbits*

3. *the stabilizers G_x and G_y are conjugate if x and y are in the same orbit.*

Proof. Since the equivalence classes of an equivalence relation are pairwise disjoint, the first assertion follows from the equivalence relation defined above.

Since every point x is in the orbit $\mathcal{O}_x$, the second assertion also follows.

The third part is already done above. □

Example 5.1.24. 1. Let H be a subgroup of a group G. Then the subgroup H acts on the left on G and the orbits are the right cosets of H. We know that the cosets are disjoint and their union is the group G.

As a specific example let $X = \mathbb{Z}$ and $G = k\mathbb{Z}$ for some $k \neq 0, \pm 1$. Then the orbits of the action of H on X are the left cosets of the group H.

2. The group H in Exercise 5.1.14(5) acts on $\mathbb{R}^2$ and the orbits are rectangular hyperbolas, positive x-axis, negative x-axis, positive y-axis, negative y-axis and the origin.

Exercise 5.1.25. 1. Let H be a subgroup of a group G and $X = G/H$. Find the orbit and stabilizer of a point xH of the G action on $X = G/H$, given by $(g, xH) \mapsto gxH$.

The action is transitive and the stabilizer of an element xH is xHx^{-1}.

2. Let G, H and X be as in Exercise 5.1.25 above. The group H acts on X via the action of G on X. Given a point $\overline{x}$ in G/H, we denote the H-orbit $\{h\overline{x} : h \in H\}$ by $\mathcal{O}^H_{\overline{x}}$.
   (a) If $\overline{x}$ is in X, when is $\mathcal{O}^H_{\overline{x}}$ a singleton? Find the stabilizer in this case.
   (b) When is $\mathcal{O}^H_{\overline{x}}$ a singleton for all points $\overline{x}$ in X?
3. Let G be a group and H a subgroup of G. Show that the group G acts transitively on G/H and this action induces a homomorphism $\varphi : G \to S(G/H)$, the group of bijections of G/H. Further show that $\ker(\varphi) \subseteq \bigcap_{x \in G} xHx^{-1}$.

Definition 5.1.26. Let G be a group acting on two sets X and Y. We say that a map $f : X \to Y$ is a G-equivariant, if $f(gx) = gf(x)$ for all x in X and g in G.

If $f : X \to Y$ is a bijection and G-equivariant, then we say that f is a G-isomorphism and the sets X and Y are said to be G-isomorphic.

Let G be a group acting transitively on a set X. Let x in be a point X and $H = G_x$ be the stabilizer of x in G. Let y be a point in X. Then $y = gx$ for some g in G.

If h is in H, then $y = g.x = (gh).x$. Thus $\varphi(\cdot, x)$ is constant on each of the left cosets gH. Observe that this is true for every point in X. We use this to prove the following theorem.

Theorem 5.1.27. *Let G be a group acting transitively on a set X. Let x_0 be a point in X and $H = G_{x_0}$ be its stabilizer in G. Then the sets X and G/H are G-isomorphic.*

Proof. Fix a point x_0 in X and $H = G_{x_0}$ be the stabilizer of x_0 in G. There is a natural map

$$\theta : G/H \to X \text{ defined by } \theta(gH) = g.x_0.$$

Since the action of G on X is transitive, for every point x in X, there exists an element g in G such that $gx_0 = x$. Therefore the map $\theta(gH) = gx_0 = x$ and the map θ is on-to.

Suppose $g_1x_0 = g_2x_0$ for g_1 and g_2 in G. Then $g_2^{-1}g_1$ is in $H = G_{x_0}$ and hence $g_1H = g_2H$. This proves that the map is one-one and the map $\theta : G/H \to X$ is a bijection. Further $\theta(g_1(gH)) = g_1(gx_0) = g_1\theta(gH)$. This proves that θ is a G-isomorphism. □

Let G be a group acting on a set X. If x is in X and $\mathcal{O}_x$ denotes the orbit of x, then the group G acts transitively on $\mathcal{O}_x$. Hence there is a bijection between G/G_x and $\mathcal{O}_x$ for every point x in X.

We record this as a theorem.

Theorem 5.1.28 (The Orbit-Stabilizer Theorem). *Let G be a group acting on a set X. Then for each x in X, the correspondence $gx \mapsto gG_x$ is a bijection between the orbit $\mathcal{O}_x$ and the set of left cosets of G_x in G.*

In case of finite groups we have the following very useful corollary of Theorem 5.1.23.

Corollary 5.1.29. *Let G be a finite group acting on a set X. Then $|\mathcal{O}_x|$ divides the order of G for every x in X.*

Proof. Let x be a point in X and G_x be its stabilizer. Since G is a finite group we know that $|G| = |G/G_x||G_x|$.

We have shown that there is a bijection between G/G_x and $\mathcal{O}_x$. Hence $|G/G_x| = |\mathcal{O}_x|$ and $|G| = |\mathcal{O}_x||G_x|$. This completes the proof. □

This is a very useful result. For example, let G be a finite group of prime order acting on a set X. Since $|\mathcal{O}_x|$ divides p, a prime number, it follows that, either $\mathcal{O}_x = \{x\}$ and $G_x = G$ or $|\mathcal{O}_x| = p$ and $G_x = \{e\}$.

We will now see some applications of these results.

Exercise 5.1.30. Let G be a group and H a subgroup of G. Show that H is normal in G iff for each element xH in G/H, the H orbit $\mathcal{O}^H_{\overline{x}} = \{xH\}$.

Assume H is normal in G. Then $x^{-1}hx$ is in H for all x in G and for all $h \in H$. Therefore $hx \in xH$ for all $h \in H$. This means that $hxH = xH$ for all $h \in H$. Thus $\mathcal{O}^H_{\overline{x}} = \{xH\}$ for all $xH \in G/H$.

Conversely assume that $\mathcal{O}^H_{\overline{x}} = \{xH\}$ for all $xH \in G/H$. This means that $hxH = xH$ for all $h \in H$. Hence $hx = xh'$ for some $h' \in H$ and $x^{-1}hx = h'$ is in H for all $x \in G$.

Lemma 5.1.31. *Let G be a finite group acting on a set X. Let H be a subgroup of G acting on X via the induced action. Then $|\mathcal{O}^H_x|$ divides the order of G.*

Proof. Let G be a finite group and H a subgroup acting on a set X.

In view of Proposition 5.1.29, we see that $|\mathcal{O}^H_x|$ divides the order of H. By Lagrange's Theorem 2.3.1 the order of H divides the order of G. Hence $|\mathcal{O}^H_x|$ divides the order of G. □

The importance of this lemma is in the following theorem.

Theorem 5.1.32. *Let G be a group and H a subgroup of index p in G. If p is the smallest prime number dividing the order of G, then H is normal in G.*

Proof. Let $X := G/H$. Then the group G acts on X and hence the group H acts on X through the induced action.

If $\mathcal{O}^H_{gH} = \{gH\}$ for all $gH \in X$, then using Exercise 5.1.30, we can conclude that H is a normal subgroup of G.

Thus it suffices to show that $\mathcal{O}^H_{gH} = \{gH\}$ for all $gH \in X$.

First observe that $\mathcal{O}_{eH}^{H} = \{eH\} = \{H\}$. Let $gH \neq H$ be an element in X. Then $|\mathcal{O}_{gH}^{H}|$ is strictly less than p (Why?) and it divides the order of G. Since p is the smallest prime number dividing the order of G and $|\mathcal{O}_{gH}^{H}| < p$ divides $|G|$, it follows that $|\mathcal{O}_{gH}^{H}| = 1$ for all elements $gH \in G/H$. □

Exercise 5.1.33. 1. Let G be a group and H be a normal subgroup of order 2 in G. Then H is contained in $Z(G)$, the center of G.

2. Let G be a finite group and H be a normal subgroup of order p, where p is the smallest prime dividing the order of G. Then H is contained in the center of G.

Since the group H is normal in G, the group G acts on H via the map $(g, h) \mapsto ghg^{-1}$.
Let $h(\neq e) \in H$ and $\mathcal{O}_h = \{ghg^{-1} : g \in G\}$. The element h is in $\mathcal{O}_h$ and the identity element e is not in $\mathcal{O}_h$. Therefore $\mathcal{O}_h$ is a proper subset of H. Now $|\mathcal{O}_h|$ divides the order of G and $|\mathcal{O}_h| < p$. Hence $|\mathcal{O}_h| = 1$ and this means that $ghg^{-1} = h$ for all $g \in G$. This proves that $H \subseteq Z(G)$, the center of G.

Let $X^G := \{x \in X : gx = x \text{ for all } g \text{ in } G\}$ for a a group G acting on a set X. The set X^G is also equal to the set $\{x \in X : \mathcal{O}_x = \{x\}\}$. Since the set X is a disjoint union of the orbits of the action, we write $X = X^G \cup (\cup_{x \in S} \mathcal{O}_x)$, where S is a subset of X and for $x \in S$ the orbit $\mathcal{O}_x$ is not a singleton.

If X is a finite set, then we have $X = X^G \cup (\cup_{x \in F} \mathcal{O}_x)$ where F is a finite subset of X. Hence

$$|X| = |X^G| + \sum_{x \in F} |\mathcal{O}_x|. \tag{5.1}$$

This equation is called the *class equation* of the group action.

Example 5.1.34. Let G be a finite group and $X = G$. Let G act on X via conjugate action. Then

$$\begin{aligned} X^G &= \{x \in G : gxg^{-1} = x \text{ for all } g \text{ in } G\} \\ &= \{x \in G : gx = xg \text{ for all } g \text{ in } G\} \\ &= Z(G) \end{aligned}$$

the center of G and $\mathcal{O}_x = \{gxg^{-1} : g \in G\}$ is the conjugacy class C_x. Therefore

$$|G| = |Z(G)| + \sum_{x \in F} |C_x| \tag{5.2}$$

where F is a finite set. Classically this equation was called the class equation.

Lemma 5.1.35. *Let $n \geq 1$ and p be a prime number. If G is a group of order p^n acting on a finite set X, then $|X| \equiv |X^G|$ (mod p).*

Proof. Let G be a group of order p^n acting on a finite set X. Since $|\mathcal{O}_x|$ divides the order of G, it follows that $|\mathcal{O}_x| = p^{n_x}$ for some $n_x \geq 1$. Hence, using Equation 5.1, we get that

$$|X| \equiv |X^G| \pmod{p}.$$

This completes the proof. □

Corollary 5.1.36. *Let G and X be as in Lemma 5.1.35. Then the following holds.*

1. *If p does not divide $|X|$, then $|X^G| \geq 1$.*
2. *If p divides $|X|$, then $|X^G| \equiv 0 \pmod p$. In particular $X^G \neq \emptyset$ and X^G has at least p elements.*

Proof. Easy. □

An application of Corollary 5.1.36 is the following proposition.

Proposition 5.1.37. *Let G be a group of even order, say $2k$ for some $k \geq 1$. Then G has an element of order 2.*

Proof. Let G be a group of even order. We know that there is a natural action of $\mathbb{Z}_2$ on any group: $(0, g) \mapsto g$ and $(1, g) \mapsto g^{-1}$.

By Corollary 5.1.36, the set $G^{\mathbb{Z}_2}$ has at least two elements. This shows that there exists an element $x \neq e$ in $G^{\mathbb{Z}_2}$. This element x is of order 2.

We can also conclude that the number of elements of order two is odd! □

Another application of Lemma 5.1.35 and its Corollary 5.1.36 is the following theorem.

Theorem 5.1.38. *Let G be a group of order p^n for some prime number p and $n \geq 1$. Then the center of G is non-trivial.*

Proof. Let G be a group of order p^n. The group G acts on the set $X = G$ by conjugate action. Now we use Equation 5.2 to get

$$p^n = |G| \equiv |X^G| = |Z(G)| \pmod p$$

and use Corollary 5.1.36 to conclude that $Z(G)$ has at least p-elements. This proves that $Z(G)$ is non-trivial. □

We use Theorem 5.1.38 to classify all groups of order p^2 for any prime number p.

Theorem 5.1.39. *Let G be a group of order p^2 for some prime number $p \geq 2$. Then*

1. *either G is a cyclic group of order p^2 or*
2. *isomorphic to the group $\mathbb{Z}_p \times \mathbb{Z}_p$.*

Proof. If there exists an element $x \in G$ of order p^2, then G is cyclic and we are through.

Let us assume that every element in G is of order p. (why?) Since the center $Z(G)$ is non-trivial, we choose an element $x (\neq e)$ in $Z(G)$ and denote by $H := \langle x \rangle$ the cyclic group generated by x. Let $y \notin H$ and $K = \langle y \rangle$ be the cyclic group generated by y. Since x is in the center of G, it follows that $xy = yx$.

Let $S := \{x^r y^s : 1 \leq r, s \leq p\}$. We claim that $S = G$. If we show that $x^k y^l = x^r y^s$ imply that $k = r$ and $l = s$, then S will have p^2 elements proving that $S = G$.

If $x^k y^l = x^r y^s$ for two pairs (k, l) and (r, s), then $x^{k-r} = y^{s-l}$ is in $H \cap K$. But $H \cap K = \{e\}$. This proves that $k = r$ and $s = l$.

Further since $xy = yx$, it follows that G is abelian and hence the subgroups H and K are normal in G. Further $H \cap K = \emptyset$. We can now appeal to Theorem 2.6.8 to conclude that $H \times K$ is isomorphic to $HK = G$. Thus the group G is isomorphic to $H \times K$ and the proof is complete.

Another proof: Let G_y be the stabilizer of y under the conjugate action of G on itself. The subgroup G_y contains the center of G and also the element y. Since y is not in H and H is contained in $Z(G)$, it follows that the group G_y has at least $p + 1$ elements. By Lagrange's theorem $|G_y|$ divides $|G| = p^2$. Hence the order of G_y is p^2. This proves that y is in $Z(G)$.

Let $K := \langle y \rangle$, the cyclic subgroup generated by the element y. Then, as in Theorem 2.6.8 the map

$$\varphi : H \times K \to G$$

defined by

$$\varphi(h, k) := hk$$

is an isomorphism. This completes the proof. □

Another application of Theorem 5.1.38 is the following theorem.

Theorem 5.1.40. *Let p be a prime number and G be a group of order p^n for $n \geq 1$. Then, for every $r \leq n$, there exists a subgroup H of G of order p^r.*

Proof. We will prove this result by induction on n.

Let $n = 1$. Then the result is true.

Assume the result to be true up to a natural number $m \geq 2$. That is, if $t \leq m$ and G is a group of order p^t, then for every $r \leq t$, there is a subgroup H of G of order p^r.

Let G be a group of order p^{m+1}. By Theorem 5.1.38, the center $Z(G)$ of G is non-trivial. Let $x \neq e$ be an element in $Z(G)$ and let $K = \langle x \rangle$. Let $|x| = |K| = p^k$ for some $k \geq 1$. Observe that K is a subgroup of $Z(G)$.

Let $r \leq m + 1$. If $r \leq k$, by induction hypothesis the group K has a subgroup H of $Z(G)$ of order p^r.

Assume $r > k$. Then G/K is a subgroup of order p^{m+1-k} which is strictly less than p^m. Since $r \leq m + 1$ and $r - 1 \leq m$ and by induction hypothesis, the group G/K has a subgroup H_1 of order p^{r-k} in G/K. Let $H := \pi_K^{-1}(H_1)$. Then H is a subgroup of G of order p^r. □

Exercise 5.1.41. Let p and q be two distinct primes and G be a group of order pq. Then

1. either the $Z(G)$ is trivial or

2. G is a cyclic group isomorphic to $\mathbb{Z}_{pq}$.

Let us assume that the center is non-trivial and let $x(\neq e)$ be an element in the center $Z(G)$. Let $H := \langle x \rangle$ be the cyclic subgroup generated by x. If H is a proper subgroup of G, then there exists an element $y \notin H$. Let

$K := \langle y \rangle$. Since x is in the center $xy = yx$. Further $H \cap K = \{e\}$. In view of this $x^r y^s = y^s x^r$ for all integers r and s and the map

$$(x^r, y^s) \mapsto x^r y^s$$

is an isomorphism between $H \times K$ and G.

If the center is trivial, we need to do some more work to understand the structure of a group of order pq. This will be done later.

Exercise 5.1.42. Use exercise 2.7.8 to show that a group of order p^2 is abelian.

As another application of Theorem 5.1.38, we prove the following theorem.

Theorem 5.1.43. *Let G be a group of order p^n for some prime $p \geq 2$. Then G is non-abelian iff the order of $Z(G)$ is less than p^{n-1}.*

Proof. Since the order of the group G is p^n, it follows from Theorem 5.1.38 that the center $Z(G)$ is non-trivial. Therefore the order of $Z(G)$ is p^k for some $k \leq n$.

Suppose G is non-abelian. Let $x \in G$ be an element such that $x \notin Z(G)$. Then the stabilizer G_x of x is a proper subgroup of G. Let the order of G_x be p^r for some $r < n$. Since the element x is not in $Z(G)$, it follows that $Z(G)$ as a proper subgroup of G_x. Hence $p^k = |Z(G)| < |G_x| < p^r < |G| = p^n$. Thus we have shown that k is less than $n - 1$.

Conversely if the order of $Z(G)$ is less than p^{n-1}, then G is non-abelian. □

Corollary 5.1.44. *Let p be a prime number and G be a non-abelian group of order p^3. Then $G/Z(G)$ is isomorphic to the group $\mathbb{Z}_p \times \mathbb{Z}_p$, where $Z(G)$ is the center of G.*

Proof. By previous theorem the order of $Z(G)$ is less than or equal to p. We also know that $Z(G)$ is non-trivial. Therefore $|Z(G)| = p$ and $G/Z(G)$ is a group of order p^2. We have shown in Theorem 5.1.39 that a group of order p^2 is either cyclic or isomorphic to $\mathbb{Z}_p \times \mathbb{Z}_p$. If $G/Z(G)$ is cyclic, then G is abelian and this is a contradiction. This proves that $G/Z(G)$ is isomorphic to $\mathbb{Z}_p \times \mathbb{Z}_p$. □

Remark 5.1.45. This corollary tells us that, if G is a non-abelian group of order p^3 for a prime number p, then every element $x \neq e$ in $G/Z(G)$ is of order p.

Exercise 5.1.46. Classify all groups of order 8.

The possible order of elements of G are 1, 2, 4 or 8.
If G is abelian, then G is either isomorphic to $\mathbb{Z}_8$, $\mathbb{Z}_4 \times \mathbb{Z}_2$ or $\mathbb{Z}_2 \times \mathbb{Z}_2 \times \mathbb{Z}_2$. (Why?)
Suppose G is non-abelian. Then there is an element a of order 4. (Why?). Let $H := \langle a \rangle$ and b be an element not in H.

Since the index of H in G is 2, the group H is normal in G and $G = H \cup aH = \{e, a, a^2, a^3, b, ba, ba^2, ba^3\}$. Since b is not in H, the element b^2 can't be in the coset bH. (Why?)

If $b^2 = a$ or a^3, then b is an element of order 8, a contradiction to our assumption that G is non-abelian. Therefore $b^2 = e$ or a^2.
If $b^2 = e$, then

$$G := \{e, a, a^2, a^3, b, ba, ba^2, ba^3 : a^4 = e = b^2 \text{ and } bab^{-1} = a^3\}.$$

If $b^2 = a^2$, then

$$G := \{e, a, a^2, a^3, b, ba, ba^2, ba^3 : a^4 = e = b^4 \text{ and } bab^{-1} = a^3\}.$$

Exercise 5.1.47. Let G be a group of even order. Then there exists an element of order 2.

Let G be a group of order $2k$ for some $k \geq 1$. The group $\mathbb{Z}_2 := \{\pm 1\}$, subgroup of $\mathbb{R}^*$, acts on G as $(1, x) \mapsto x$ and $(-1, x) \mapsto x^{-1}$. (why is this an action?) Then using class Equation 5.1, we can write

$$|G^{\mathbb{Z}_2}| \equiv |G| \equiv 0 \mod 2.$$

The set set $\{x \in G : x^2 = e\}$ is non-empty as the identity element e is in it. Hence it has at least two elements. Thus there are odd number of elements of order 2, more than what we set out to prove!

We will finish this section with an important theorem called The Counting Theorem. This theorem is also called *Burnside's theorem* and it has important applications in combinatorics!

Theorem 5.1.48 (Counting Theorem). *Let G be a finite group acting on a set X and for each g in G, let*

$$X^g := \{x \in X : gx = x\}.$$

Then the number of distinct orbits is

$$\frac{1}{|G|}\sum_{g \in G} |X^g|.$$

Proof. For the group action $G \times X \to X$, we start counting the pair (g, x) such that $gx = x$. By definition of X^g, this number is the sum $\sum_{g \in G} |X^g|$. This number can also be alternatively written as

$$|\{(g, x) \in G \times X : gx = x\}|.$$

Since the set G_x is the set of all elements in G such that $gx = x$, we can write $\sum_{g \in G} |X^g| = |\{(g, x) \in G \times X : gx = x\}| = \sum_{x \in X} |G_x|$.

Let N be the number of orbits of this action and let us label them as $X_1, X_2, \ldots, X_N$. Then we can write the sum $\sum_{x \in X} |G_x|$ as

$$\sum_{x \in X} |G_x| = \sum_{i=1}^{N} \sum_{x \in X_i} |G_x|.$$

We have seen that for two points in the same orbit, the stabilizers are conjugate in G. So, if we fix an element x_i in X_i, then

$$\sum_{x\in X_i} |G_x| = \sum_{x\in X_i} |G_{x_i}| = |\mathcal{O}_{x_i}||G_{x_i}| = |G|.$$

Therefore

$$\sum_{g\in G} |X^g| = \sum_{x\in X} |G_x| = \sum_{i=1}^{N}\sum_{x\in X_i} |G_x| = N|G|$$

proving that $N = \frac{1}{|G|}\sum_{g\in G}|X^g|$. □

5.2 Cauchy's Theorem

Lagrange's theorem (Theorem 2.3.1) tells us that, if H is a subgroup of a finite group G, then the order of H divides the order of the group G. However the converse is not true as we have seen from Exercise 2.8.24 and Proposition 2.8.25.

On the other hand if G is a finite abelian group of order n and if k divides n, then we have shown in Theorem 2.7.10 that G has a subgroup of order k.

Naturally, we would like to know to what extent the converse of Lagrange's theorem is true. A partial converse to this is Cauchy's Theorem 5.2.1 proved in this section. Though this is a partial converse, it is a very important result as will be seen in the applications of Cauchy's Theorem 5.2.1.

Theorem 5.2.1 (Cauchy's Theorem). *Let G be a finite group and $p \geq 2$ be a prime number such that p divides the order of G. Then there exists an element of order p in G.*

Proof. Let $X := \{(g_1, g_2, \cdots, g_p) \in \overbrace{G\times G\times\cdots\times G}^{p-\text{ times}} : g_1g_2\ldots g_p = e\}$. The set X is non-empty as the element $(e, e, \ldots, e)$ is in X. We will now show that there exists $x \neq e$ in G such that $(x, x, \ldots, x)$ is in X.

Let $\sigma = (1\ 2\ \ldots\ p) \in S_p$ be the p-cycle and $H := \langle\sigma\rangle$ be the cyclic subgroup generated by $\sigma \in S_p$. The group H acts on X as follows: The map

$$\varphi : H \times X \to X$$

is defined by

$$\varphi(\sigma^k, (g_1, g_2, \cdots, g_p)) := (g_{k+1}, g_{k+2}, \cdots, g_{k+p}).$$

For example $\varphi(\sigma, (g_1, g_2, \ldots, g_p)) := (g_2, g_3, \ldots, g_p, g_1)$.

Since $g_1g_2\ldots g_p = e$, it follows that $g_2g_3\ldots g_p = g_1^{-1}$. Therefore $g_2g_3\ldots g_pg_1 = g_1^{-1}g_1 = e$ and $\varphi(\sigma, (g_1, g_2, \ldots, g_p)) := (g_2, g_3, \ldots, g_p, g_1)$ is in S, if $(g_1, g_2, \ldots, g_p)$ is in S. Consequently

$$\varphi(\sigma^2, (g_1, g_2, \ldots, g_p)) = \varphi(\sigma, (g_2, g_3, \ldots, g_p, g_1)) = (g_3, g_4, \cdots, g_p, g_1, g_2)$$

is also in S.

Inductively we can show that

$$\varphi(\sigma^k, (g_1, g_2, \ldots, g_p)) := (g_{k+1}, g_{k+2}, \ldots, g_p, g_1, g_2, \ldots, g_k)$$

is also in S for every k.

First observe that $|X| = |G|^{p-1}$.

Let $(g_1, g_2, \ldots, g_p) \in X$. Then $g_1 g_2 \ldots g_p = e$ and hence we can write $g_p = g_{p-1}^{-1} g_{p-2}^{-1} \cdots g_1^{-1}$. This proves that the element is uniquely determined by the choice of the elements $g_1, g_2, \ldots, g_{p-1}$ and

$$X = \left\{(g_1, g_2, \ldots, g_{p-1}, g_{p-1}^{-1} g_{p-2}^{-1} \cdots g_1^{-1}) : g_i \in G \text{ for } 1 \leq i \leq p-1\right\}.$$

Hence $|X| = |G|^{p-1}$.

Since p divides the order of G, we conclude that p divides $|X|$. By Class Equation 5.1, we can write $|X| = |X^H| + \sum_{x \in F} |\mathcal{O}_x|$ for a finite subset F of X. Now we observe that, for $x \in F$, $|\mathcal{O}_x| \geq 2$ and $|\mathcal{O}_x|$ divides the order of H, a prime number. Hence an orbit $\mathcal{O}_x$ for $x \in F$ has exactly p-elements. Therefore $|X^H| \equiv 0 \pmod{p}$. Since $(e, e, \ldots, e)$ is in X^H, the set X^H has at least p-elements. Therefore there exists an element $(g_1, g_2, \cdots, g_p) \neq (e, e, \ldots, e)$ in X^H. This shows that $\varphi(\sigma, (g_1, g_2, \cdots, g_p)) = (g_1, g_2, \cdots, g_p)$. As a consequence $g_1 = g_2 = g_3 = \cdots = g_p$ and $g_1^p = e$. This completes the proof of the theorem. □

5.2.1 Applications of Cauchy's Theorem

Exercise 5.2.2. Let G be a finite group such that every element $a \neq e$ is of same order, say q. Then q is a prime number and G is a group of order q^n for some $n \geq 1$.

Suppose the order of the group G is not a power of any prime number p. Then there exist two prime numbers $q \neq p$ such that both p and q divide the order of G. By Cauchy's theorem there exists an element x of order p and an element y of order q, a contradiction.

Exercise 5.2.3. Let G be a finite group. Then the order of G is a power of p iff order of every element is a power of p.

Let G be a finite group and the order of G be p^n for some $n \geq 1$. Let $a(\neq e)$ be in G. Since the order of a divides the order of G, the order of a is also a power of p.

Let the order of every element of G be a power of p. Suppose the order of G is not a power of p. Then there exists a prime $q \neq p$ such that q divides the order of G. By Cauchy's theorem, there exists an element of order q, a contradiction.

Exercise 5.2.4. Let G be a finite group of order n and let $f : G \to G$ be defined by $f(x) = x^k$. Then f is a bijection iff the integers k and n are relatively prime. If G is abelian, then map is an isomorphism.

We will show the map f is on to if k and n are relatively prime. Since k and n are relatively prime there exists an integer l such that $k\,l \equiv 1 \pmod{p}$. Let x be an element of G. Then $f(x^l) = x^{lk} = x$ and f is on to. Since G is finite every onto map to itself is one-one.

Assume the $\gcd(k, n) = d > 1$. Let p be a prime factor of d. Then p divides n and k. By Cauchy's theorem there exists an element g of order p. Then $g^k = (g^p)^{\frac{k}{p}} = e$. Therefore $f(g) = f(e) = e$ and $g \neq e$, a contradiction.

Theorem 5.2.5. *Let G be a finite abelian group of order $m\,n$, where* $\gcd(m, n) = 1$. *Then G is isomorphic to a direct product of two subgroups H and K of G of order m and n respectively.*

Proof. First let us observe that we have already shown this for cyclic groups.

If either m or n is 1, the result is clear: We take H or K to be the trivial group.

Assume $m, n \geq 2$. Since the order of G is $m\,n$, every $g \in G$ satisfies $g^{mn} = e$.

Since G is an abelian group, the set $\{g \in G : g^k = e\}$ is a subgroup of G for every $k \geq 1$. (Why?)

Let $H = \{g \in G : g^m = e\}$ and $K = \{g \in G : g^n = e\}$. Then H and K are subgroups of G.

We will first show that $G = HK$, i.e., every element of G is the product of an element of H and an element of K. Since m and n are relatively prime, using Bezout identity, we write $1 = m\,r + n\,s$ for some r and s in $\mathbb{Z}$. Therefore any $g \in G$ can be written as $g = g^{m\,r+n\,s} = (g^r)^m (g^s)^n$. Notice that $(g^r)^m$ is in K: $((g^r)^m)^n = (g^r)^{m\,n} = e$. Similarly $(g^s)^n \in H$. Therefore, re-ordering the terms as $g = (g^s)^n (g^r)^m$ shows that every element $g \in G$ is the product of an element in H and an element K, so $G = HK$.

Now we show that G is isomorphic to the direct product $H \times K$. Let

$$f : H \times K \to G$$

be defined by $f(h, k) := hk$. Then, since G is abelian,

$$\begin{aligned} f((h,k)(h_0,k_0)) &= f(hh_0, kk_0) \\ &= hh_0kk_0 \\ &= (hk)(h_0k_0) \\ &= f(h,k)f(h_0,k_0). \end{aligned}$$

This proves that the map f is a homomorphism. We have already shown above that $G = HK$, so f is onto. To show f is one-to-one, we show that the kernel is trivial. Suppose $f(h, k) = e$. Then $hk = e$, so $h = k^{-1}$. This equation shows h and k are in $H \cap K$. Since the order of H and K are relatively prime, the intersection $H \cap K$ is trivial. Thus h and k are identity. This proves f has trivial kernel.

Now that we know f is an isomorphism of $H \times K$ with G, we can show that the order of H is m and the order of K is n. Since f is a bijection, $|H \times K| = |G|$, so $|H||K| = mn$.

We will now show that $\gcd(|H|, n) = 1$. To prove this we bring in (at last!) Cauchy's theorem. If $\gcd(|H|, n) > 1$, let p be a prime dividing both $|H|$ and n. Cauchy's Theorem 5.2.1 tells us that the subgroup H has an element x of order p. But the order of every element in H divides m. Hence $p = o(x)$, the order of x, divides m. This contradicts our assumption that $\gcd(m, n) = 1$. Thus $\gcd(|H|, n) = 1$. Similarly, $\gcd(|K|, m) = 1$. Since $|G| = |H||K = mn$, we see that $|H|$ divides m. Arguing similarly, we show that $|K|$ divides n. Therefore, $|H| = m$ and $|K| = n$. □

Corollary 5.2.6. *Any finite abelian group is isomorphic to a direct product of finite abelian groups of prime-power order.*

Proof. Let G be a finite abelian group of order $n \geq 2$. Let $n = p_1^{r_1} p_2^{r_2} \cdots p_k^{r_k}$ be the prime factorization of n. It follows from Theorem 5.2.5 that there exists subgroups H_i of G order $p_i^{r_i}$ such that G is isomorphic to the direct product $H_1 \times H_2 \times \cdots \times H_k$. □

Corollary 5.2.7 (Converse of Lagrange's Theorem). *Let G be a finite abelian group of order n. Let k be a divisor of n. Then G has a subgroup H of order k.*

Proof. The proof is easy and left as an exercise. □

Lemma 5.2.8. *Let G be a group of order pq, where $p > q$ are two distinct primes. Then there is unique normal subgroup of order p.*

Proof. By Cauchy's Theorem 5.2.1, there exists an element x of order p. Let $H := \langle x \rangle$ be the cyclic subgroup generated by x.

Then the index of the subgroup H is q and q is the smallest prime dividing the order of G. Hence it follows from Theorem 5.1.32, that the subgroup H is normal in G. □

Theorem 5.2.9. *Let G be as in Lemma 5.2.8. If q does not divide $p - 1$, then the group G is cyclic.*

Proof. Let x and y be two elements in G of order p and q. We denote by $H = \langle x \rangle$ and $K = \langle y \rangle$ be the cyclic subgroups of G generated by x and y respectively. It follows from Lemma 5.2.8 that the subgroup H is normal in G. Therefore $yxy^{-1} = x^i$ for some $i \leq (p-1)$ and $y^r x y^{-r} = x^{i^r}$ for any $r \leq q$. Since $y^q = 1$, it follows that $x = y^q x y^{-q} = x^{i^q}$. Hence $x^{i^q - 1} = e$ and by Lagrange's theorem p divides $i^q - 1$. This means that $i^q \equiv 1 \pmod{p}$. We claim that $i = 1$.

Suppose not. Since p is a prime number the integer i and p are relatively prime and hence i is in the group $U_p = \{k < p : \gcd(k, p) = 1\}$. Therefore q, the order of the element i, divides $|U_p| = p-1$. Thus q divides $p-1$ a contradiction to our assumption that q does not divide $p - 1$. Hence $i = 1$ and $yx = xy$. □

Exercise 5.2.10. Let G be a group of order $2p$ for some prime number $p \geq 2$. Then

1. either G is a cyclic group isomorphic to $\mathbb{Z}_{2p}$ or

2. G is isomorphic to D_p the dihedral group of order p.

Exercise 5.2.11. Let $p > q$ be two prime numbers. Find all groups of order

1. $p^2 q$ if q does not divide $p-1$.

2. $p^2 q^2$ if q does not divide $p-1$.

Remark 5.2.12. Some of the applications of Cauchy's Theorem 5.2.1 proved in this section are usually done after Sylow's Theorems proved in next section.

We use Cauchy's Theorem to prove another important result.

Theorem 5.2.13. *The group A_5 is simple.*

Proof. Let H be a normal subgroup of $G = A_5$. Then the possible prime divisors of $|H|$ are 2, 3 or 5.

Suppose 2 divides $|H|$. Then by Cauchy's theorem there exists and element σ in H such that $\sigma^2 = Id$. Since σ is an even permutation, it must be a product of two transpositions, say $\sigma = (1\ 2)(3\ 4)$. Since all permutations with same cycle structure are conjugate, all the permutations of the form $(i\ j)(k\ l)$ for $1 \leq i, j, k, l \leq 5$ are conjugate to σ and this coupled with the normality of H in G, shows that the subgroup H contains all permutations of the form $(i\ j)(k\ l)$.

By writing the 3-cycle $\tau = (1\ 2\ 3)$ as $\tau = (1\ 2)(2\ 3)$, we conclude that the 3-cycle $(1\ 2\ 3)$ is in H. Again we use normality of H in A_5 to conclude that all the 3-cycles are in H. Since A_n is generated by 3-cycles, we conclude that $H = A_5$.

Suppose 3 divides the order of H. Then H contains a 3-cycle and by the normality it contains all 3-cycles forcing H to be A_5.

Suppose 5 divides the order of H. Then H must contain all the 5-cycles. In all there are 24 of them. Since the order of H divides $|A_5| = 60$, it follows that $|H| \geq 30$. Therefore 3 divides the order of H and we appeal to the earlier argument to conclude that $H = A_5$. □

5.3 Sylow Theorems and applications

Cauchy's Theorem 5.2.1 gave us a partial converse to Lagrange's theorem. Using this we were able to show that converse of Lagrange's theorem is true for finite abelian groups.

This section is devoted to proving one of the most important results in finite Group Theory. These results extend Cauchy's Theorem and helps us understand the structure of finite non-abelian groups.

Theorem 5.3.1 (Sylow's first theorem). *Let G be a finite group of order $p^n m$ where $p \geq 2$ is a prime number and n, m are natural numbers such that* $\gcd(p, m) = 1$. *Then there exists a subgroup H of G of order p^n.*

Corollary 5.3.2. *Let G be a finite group of order $p^n m$ where p, n and m are as in Theorem 5.3.1. Then there exists a subgroup H_r of order p^r for every $r \leq n$.*

Theorem 5.3.3 (Sylow's second theorem). *Let G be a finite group of order $p^n m$ where p, n and m are as in Theorem 5.3.1. Let H and K be two Sylow-p subgroups of G. Then they are conjugate in G. i.e., there exists $g \in G$ such that $K = gHg^{-1}$.*

Theorem 5.3.4 (Sylow's third theorem). *Let G be a finite group of order $p^n m$ where p, n and m are as in Theorem 5.3.1. Let t be the number of Sylow-p subgroups of G. Then*

1. *t divides the order of G and*
2. *$t \equiv 1 \pmod p$.*

5.3.1 Proofs of Sylow Theorems

Proof of first Sylow Theorem 5.3.1

Let G be a finite group and let $|G| = p^n m$ where p is a prime number, n and m are natural numbers such that $(p, m) = 1$.

Let $X = \{S \subseteq G : |S| = p^n\}$. If the result is true, then such a subgroup must be in X. One can now think of an action of the group G on X, namely $(g, S) \mapsto gS$.(Why is this an action?) If H is a Sylow p-subgroup of G, then the orbit $\mathcal{O}_H$ of H is the set of all left cosets of H. Hence $|\mathcal{O}_H| = m$. Since p and m are relatively prime, p does not divide $|\mathcal{O}_H|$. This means that Sylow subgroups must be found from the orbits whose size is not divisible by p. We will now show that there are orbits whose size is not divisible by p and the stabilizer of any element of such orbits is a Sylow p-subgroup of G.

First let us observe that $|X| = \binom{p^n m}{p^n}$. Then we have the following lemma.

Lemma 5.3.5. *Let p be a prime number and $n, m \in \mathbb{N}$ be such that $(p, m) = 1$. Then the prime number p does not divide $\binom{p^n m}{p^n}$.*

We will assume the lemma and complete the proof of Sylow's first theorem.

Let $\mathcal{O}_j$ for $1 \le j \le q$ be the number of distinct orbits of the action of G on X. Then $X = \cup_{j=1}^q \mathcal{O}_j$ is a disjoint union. Hence $|X| = \sum_{j=1}^q |\mathcal{O}_j|$. Since p does not divide $|X|$, p does not divide $\sum_{j=1}^q |\mathcal{O}_j|$. Hence there exists a j such that p does not divide $|\mathcal{O}_j|$.

Let $S \in \mathcal{O}_j$ and G_S be the stabilizer of S. Then we know that $|G| = |G/G_S||G_S| = |G_S||\mathcal{O}_j|$. Since p^n divides $|G|$ and p does not divide $|\mathcal{O}_j|$, it follows that p^n divides G_S. Hence $|G_S| \ge p^n$.

We will now show that $|G_S| \le p^n$. Fix $s_0 \in S$. Then $g \mapsto gs_0$ is a one-one map from G_S to S. (Why?) Therefore $|G_S| \le |S| = p^n$. This completes the proof of the theorem. □

Remark 5.3.6. Let us fix an orbit $\mathcal{O}$ such that p does not divide $|\mathcal{O}|$. If S is in $\mathcal{O}$ then the stabilizer $H := G_S$ is a Sylow p-subgroup of G and if s_0 is a point in S then $S = Hs_0$.

Since the group G acts transitively on $\mathcal{O}$, every element in $\mathcal{O}$ is of the form $gS = gHs_0$ for g in G. If we choose $g = s_0^{-1}$, then $s_0^{-1}Hs_0$ is in $\mathcal{O}$ and it is a subgroup of G order p^n. Let us denote this subgroup by K.

We don't stop here. For any g in G, we can write $gS = gs_0(s_0^{-1}Hs_0) = gs_0K$. This shows that the elements of the orbit $\mathcal{O}$ are the set of left cosets of a Sylow p-subgroup of G.

Proof of the lemma 5.3.5

First we expand $\binom{p^n m}{p^n} = \frac{(p^n\, m)!}{p^n!(p^n\, m - p^n)!}$ as

$$\frac{p^n\, m \cdot (p^n m - 1)(p^n\, m - 2) \cdots (p^n\, m - (p^n - 1))}{p^n(p^n - 1) \cdots (p^n - (p-1))(p^n - p) \cdots p(p-1)(p-2) \cdots 2 \cdot 1}.$$

For $1 \leq i \leq n-1$, a prime power p^i divides a factor $(p^n\, m - r)$ of the numerator in the expansion above iff $r = kp^i$ for $1 \leq k \leq (p-1)$ and each such $r = kp^i$ occurs as a factor in the expansion of $p^n!$. Therefore $(p^n m - r)$ and $(p^n - r)$ are divisible by p^i but not by p^{i+1}.

After the cancellation of the powers of p^i in the numerator and denominator, each of the remaining terms in the numerator and denominator are not divisible by p. Hence $\binom{p^n m}{p^n}$ is not divisible by p. □

Proof of Corollary 5.3.2

Let H be a Sylow p-subgroup of order p^n and $r \leq n$. We appeal to Theorem 5.1.40 to conclude that H has a subgroup H_r of order p^r. □

Proof of second Sylow Theorem 5.3.3

Let G be a group and H be a Sylow p-subgroup of G. The group G acts transitively on $Y = G/H$, the set of left cosets of H. The stabilizer of the point $eH = H$ is the subgroup H. If gH is a point in G/H, then we have seen in Theorem 5.1.23 that stabilizer of gH and H are conjugate in G.

Therefore to prove this theorem it is enough to show that, given a Sylow p-subgroup K of G, there exists a left coset aH of H such that the subgroup K is the stabilizer of the left coset aH.

The Sylow p-subgroup subgroup K acts on Y under the induced action of G on Y. Let us denote by Y^K, the K-fixed points of Y. If we now show that Y^K is non-empty, we are done. But we know from Class Equation 5.1 that $|Y| \equiv |Y^K| \pmod p$. Observe that $|Y| = m$. Since $\gcd(p, m) = 1$, it follows that Y^K is non-empty. This means that there exists a coset aH of H such that $kaH = aH$ for all $k \in K$. That is $k \in aHa^{-1}$ for every $k \in K$. Thus $K \subseteq aHa^{-1}$. Since the orders of H and K are same it follows that $K = aHa^{-1}$. □

Proof of Third Sylow theorem 5.3.4

Let $X = \{S \subseteq G : |S| = p^n\}$. Let $X_1 := \{S \in X : p \text{ divides } |\mathcal{O}_S|\}$ and $X_2 := \{S \in X : p \text{ does not divide } |\mathcal{O}_S|\}$. First let us observe that the sets X_1 and X_2 are disjoint. Therefore $|X| = |X_1| + |X_2|$.

We saw in Remark 5.3.6 that, if H is a Sylow subgroup of G, then the set of left cosets is an orbit of G and it is in X_2. Further every element in X_2 is in an orbit of a left coset of a Sylow p-subgroup. Thus the number of orbits in X_2 is same as the number of Sylow p-subgroups of G.

Each orbit in X_2 is of size m. Therefore if t is the number of Sylow p-subgroups of G, then $|X_2| = tm$. Since p divides $|\mathcal{O}_S|$ for each $S \in X_1$, it follows that $|X_1|$ is a multiple of p. Hence $|X| = |X_1|+|X_2| = k\,p+t\,m$ for some $k \in \mathbb{N}$. This proves that $t\,m \equiv |X| \equiv m \pmod{p}$. That is $(t-1)m \equiv 0 \pmod{p}$. Since $p \nmid m$, it follows from Euclid's Lemma 1.1.28 that $t \equiv 1 \pmod{p}$.

Let T be the set of all Sylow subgroups of G. We have shown in second Sylow Theorem 5.3.3 that any two Sylow p-subgroups of G are conjugate in G. Hence the group G acts on T via conjugation and the action is transitive. Since the size of an orbit divides the order of the group, it follows that $t = |T|$ divides the order of G and with this the proof is complete. □

5.3.2 Applications of Sylow's theorems

In this section we discuss some applications of Sylow theorems.

Example 5.3.7. 1. Let G be a group of order 10. Then G has unique Sylow 5-subgroup.

Let n_5 denote the number of Sylow 5-subgroups of G. Then $n_5 = 5r+1$ for some non-negative integer r and further $5r+1$ divides 10. Hence $r = 0$ and $n_5 = 1$. Therefore G has a unique Sylow 5-subgroup.

2. Let G be a group of order 35. Then the Sylow 7-subgroup and Sylow 5-subgroup are normal in G.

As in the previous example, we can show that G has unique Sylow 5-subgroup and unique Sylow 7-subgroup.

Exercise 5.3.8. 1. Show that a group of order 20 has unique Sylow 5-subgroup.

2. Let G be a group of order 87. Show that it has a normal subgroup of order 29 and a normal subgroup of order 3.

Example 5.3.9. 1. Let G be a group of order 15. Then G is cyclic.

Let G be a group of order $15 = 3 \times 5$. Let n_5 denote the number of Sylow 5-subgroups of G. Then by Sylow Theorem 5.3.4, $n_5 = 5r+1$ and $5r+1$ divides 15. Thus $n_5 = 1$ and using Sylow Theorem 5.3.3, we conclude that G has a unique normal subgroup H of order 5.

Similarly we can show that G has unique normal subgroup K of order 3.

Since $H \cap K$ is empty and both are normal in G, it follows from Theorem 2.6.8 that the map $\varphi : H \times K \to G$ defined by $\varphi(h,k) := hk$ is a one-one homomorphism.

If $H = \langle x \rangle$ and $K = \langle y \rangle$ then we can show that the order of the element xy is 15. Thus G is cyclic.

2. Let G be a group of order 119. Then G is cyclic.

   Let G be a group of order $119 = 7 \times 17$. Since 7 does not divide $16 = 17 - 1$, it follows from the previous examples that G is cyclic.

   Alternatively, we can also do it as follows: Since $n_{17} = 17t + 1$, the number of 17 Sylow subgroups divides 7, it follows that the 17 Sylow subgroup U is unique and hence normal in G.

   Similarly we can show that G has a unique 7 Sylow subgroup. Let us call it W. Both the subgroups are cyclic, normal and their intersection is identity. Therefore, if we write $U = \langle u \rangle$ and $W := \langle w \rangle$, then the element uw is of order 119 and hence the group U is cyclic.

3. Let G be a group of order 21. Then G is isomorphic to the group $\mathbb{Z}_{21}$ or isomorphic to $\{y^i x^j : 1 \leq i \leq 3, 1 \leq j \leq 7, y^3 = x^7 = e \text{ and } yxy^{-1} = x^2\}$.

   - First we conclude that Sylow 7-subgroup of G is unique and hence normal in G. Let $H = \langle x \rangle$ denote the unique Sylow 7-subgroup of G.
   - The number of Sylow 3-subgroups $n_3 = 3r + 1$ must divide 21. Therefore the number of Sylow 3-subgroups is either 1 or 7. Let $K = \langle y \rangle$ be a Sylow 3-subgroup of G.
   - If K is unique then both H and K are normal in G and $H \cap K = \{e\}$. Hence we conclude as earlier that the group G is isomorphic to $H \times K$ and it is a cyclic group of order 21. Thus G is isomorphic to $\mathbb{Z}_{21}$.
   - Let us now assume that $n_3 = 7$. Since H is normal in G, the element yxy^{-1} is in H and it is not equal to x. Let $yxy^{-1} = x^r$ for some r such that $2 \leq r \leq 6$. Now $x = y^3 x y^{-3} = x^{r^3}$. Therefore $r^3 \equiv 1 (\text{mod } 7)$ and $r = 1$, 2 and 4 are the solutions of the equation $r^3 \equiv 1(\text{mod } 7)$.
     The case $r = 1$ is already done.
     Let $r = 2$. Then $G = \{y^i x^j : 1 \leq i \leq 3, 1 \leq j \leq 7, y^3 = x^7 = e \,\&\, yxy^{-1} = x^2\}$.
     Let us now assume that $r = 4$. Then $yx^4y^{-1} = (yxy^{-1})^4 = x^{16} = x^2$. This shows that $y^{-1}x^2 = x^2$. Since H is a cyclic group of prime order, every element $z \neq e$ is a generator of H. Therefore if we now take $z = x^2$, then we have shown that $y^{-1}zy = z^2$. Thus in this case too the group G is isomorphic to $G = \{y^i x^j : 1 \leq i \leq 3, 1 \leq j \leq 7, y^3 = x^7 = e \,\&\, yxy^{-1} = x^2\}$.

Exercise 5.3.10. 1. Let G be a group of order 33. Show that G is cyclic.

2. Find all groups of order 48 and 56.

   The Sylow 2-subgroups are normal in both the cases.

3. Find all groups of order 111.,

4. Let G be a group of order 30. Find all groups of order 30 by showing that

   (a) either Sylow 5-subgroup H or Sylow 3-subgroup K is normal in G.

   (b) the set HK is a cyclic subgroup of G.

5. Let G be a finite group such that all its Sylow subgroups are normal in G. Show that G is a direct product of its Sylow subgroups.

   Hint: Appeal to internal direct product.

Exercise 5.3.11. 1. Let G be a group of order pq where p and q are distinct prime numbers. Show that G has a normal subgroup.

   Let G be a group of order pq where p and q are distinct prime numbers. Let $p > q$ and H be a Sylow p-subgroup. Then index of H in G is q and q is the smallest prime dividing the order of G. Hence H is normal in G.

2. Let G be a group of order pq where $p > q$ are two primes. Then G is a cyclic group if q does not divide $p - 1$.

   This has already been done as an application of Cauchy theorem (See Corollary 5.2.9.)

   We will do it again as application of Sylow theorems.

   Let t be the number of Sylow p-subgroups of G. Then from Theorem 5.3.4, it follows that $t = kp + 1$ for some $k \geq 0$ and t divides $|G| = pq$. From this we conclude that $k = 0$. Hence there is a unique Sylow p-subgroup in G. Let us denote this subgroup by H. From Theorem 5.3.3 it follows that H is normal in G.

   Let K be a Sylow q-subgroup of G, generated by an element y. That is $K = \langle y \rangle$. If $H = \langle x \rangle$, then yxy^{-1} is also in H, say x^r for $1 \leq r \leq p - 1$. From here on the proof is same as in Corollary 5.2.9.

3. Let G be a group of order pqr where p, q and r are distinct prime numbers. Show that one of the Sylow subgroup of G is normal.

   We may assume that $p < q < r$. Let n_s denote the Sylow s-subgroups of G for $s. = p, q$ or r. Suppose none of the Sylow subgroups of G are normal in G. Then $n_s \geq 1 + s$.

   Since $p < q < r$ and n_p divides the order of G, it follows that $n_p = q, r$ or qr. That is $q \leq n_p \leq qr$.

   Similarly n_q must be either p, r or pr. Since $p < r < pr$, we see that $n_q \geq r$.

   The number of Sylow r-subgroups n_r must be pq.

   Now we count the number of elements in each of these Sylow s-subgroups and get

$$n_p(p-1) + 1 + n_q(q-1) + 1 + n_r(r-1) + 1 = pqr.$$

   This shows that

$$\begin{aligned} pqr &= n_p(p-1) + 1 + n_q(q-1) + 1 + n_r(r-1) + 1 \\ &\geq n_p(p-1) + n_q(q-1) + n_r(r-1) \\ &\geq q(p-1) + r(q-1) + pq(r-1) \\ &= pq - q + qr - r + pqr - pq \\ &= pqr + qr - q - r. \end{aligned}$$

From this we get

$$qr \leq q + r < 2r.$$

This proves that $q < 2$, a contradiction. Hence one of the Sylow subgroups of G must be normal in G.

Example 5.3.12. Let $n \geq 2$ and $G = D_n$. Then every Sylow subgroup of G is cyclic. But G is not cyclic. Can you construct another example?

Exercise 5.3.13.

1. Let G be a group of order p^2q^2 where p and q are two distinct primes such that $p \nmid q^2 - 1$ and $q \nmid p^2 - 1$. Show that G is abelian.

2. Show that any group of order $20,449$ is abelian. ($20449 = 11^2.13^2$.)

Exercise 5.3.14. Let G be a finite group and H be a normal subgroup of prime order p. Assume that p is the smallest prime dividing the order of G. Show that H is in the center of G.

Why do we need normality assumption on H? Consider $G = S_3$ and H subgroup generated by a two cycle. Is H normal in G? What do you observe?

Exercise 5.3.15.

1. Let G be a group of order p^2q. Show that G is not simple.

2. Let G be a group of order $p^n m$ such that $\gcd(p, m) = 1$. If $p \nmid (m-1)!$, then G has unique Sylow p-subgroup H of G. In particular G is not simple.

   Let H be a Sylow p-subgroup of G and $X = G/H$. Then G acts on X.

   Let $\varphi : G \to \text{Perm}(X) = S_m$, the symmetric group S_m, be the homomorphism induced by this action.

   Let g be an element in the kernel of φ. Then $\varphi_g(xH) = xH$ for all cosets xH of H. In particular $\varphi_g(H) = H$. This shows that g is in H. As a consequence $\ker(\varphi)$ is contained in H and $|\ker(\varphi)| = p^t$ for some $t \leq n$. Therefore the image of φ, denoted as $\text{Im}(\varphi)$ is a subgroup of order $p^{n-t}m$ and it must divide $m!$, the order of S_m. Since p does not divide m, we see that $p^{n-1} \mid (m-1)!$. Therefore, if $p \nmid (m-1)!$, it follows that $n = t$. This proves that $\ker(\varphi) = H$ and hence H is normal in G.

   As a consequence the group G is not simple.

3. Let p and q be distinct primes and G be a group of order p^2q^2. Show that G is not simple.

4. Show that any abelian group of order pq^2, where p and q are two distinct primes such that $q^2 > p$, is cyclic.

5. Show that any group of order 45 is abelian.

Example 5.3.16. Find all groups of order 595

- Let us first write $595 = 5 \times 7 \times 17$.

- Let $n_5 = 5t + 1$ be the number of 5 Sylow subgroups of G. Since $5t + 1$ divides 119, it follows that $n_5 = 1$. Let H be the 5 Sylow subgroup of G. Since H is unique it is normal in G.
- Since H is normal in G, the map $\varphi_g : h \mapsto ghg^{-1}$ is an automorphism of H and the map $\varphi : g \mapsto \varphi_g$ is a homomorphism from G in to $\mathrm{Aut}(H)$. Since the order of $\mathrm{Aut}(H)$ is $4 = 2^2$, the image of this homomorphism φ must be of order 2^r for some $r \leq 2$. By Lagrange's theorem 2^r divides 595, the order of the group G. Hence $r = 0$ and this means that the image is trivial- that is $\varphi_g(h) = ghg^{-1} = h$ for all $h \in H$ and for all $g \in G$. Therefore the subgroup H is contained in the center of G.
- Let $U := G/H$. The group U is of order 119 and hence U is cyclic. Since U is cyclic, it follows that G is abelian. Hence all its Sylow subgroups are normal. Let K be the 17 Sylow subgroup of G and L the 7 Sylow subgroup of G. Since the orders of H, K and L are prime, they are all cyclic. Hence the group G being isomorphic to the cyclic group $H \times K \times L$ is also cyclic. Thus we have proved that any group of order 595 is cyclic and hence isomorphic to $\mathbb{Z}_{595}$.

Another Proof:

- We have seen above that there is a unique Sylow 5-subgroup H in G. Then the quotient group G/H is of order 119.
- Let $\overline{K}$ be a Sylow 17-subgroup of G/H. It is easy to see that $\overline{K}$ is a normal subgroup of $/H$. If $\pi_H : G \to G/H$ is the quotient map, then $\pi_H^{-1}(\overline{K})$ is normal in G. Observe that the subgroup H is contained in $\pi_H^{-1}(\overline{K})$.

  Let K be a Sylow 17-subgroup of $\pi_H^{-1}(\overline{K})$. Then K is normal in $\pi_H^{-1}(\overline{K})$. Since H and K are normal in G and $H \cap K = \{e\}$, the product $H \times K$ is isomorphic to $HK = \pi_H^{-1}(\overline{K})$. Since H and K are cyclic groups, it follows that $\pi_H^{-1}(\overline{K})$ is also cyclic. Thus $\pi_H^{-1}(\overline{K})$ is a cyclic normal subgroup of G. Hence every subgroup of $\pi_H^{-1}(\overline{K})$ is normal in G. In particular the subgroup K is normal in G.
- Similarly, we can start with a Sylow 7-subgroup $\overline{L}$ of G/H and show that G has unique Sylow 7-subgroup L.
- We have now shown that all Sylow subgroups H, K and L of G are normal in G. Hence G is isomorphic to $H \times K \times L$ and it is also cyclic, isomorphic to $\mathbb{Z}_{595}$.

Exercise 5.3.17. 1. Classify groups of order 175, 255 and 275.

2. Show that there are only two groups of order 105 up to isomorphism.

Proposition 5.3.18. *Let G be a finite group of order less than* 60*. If the order of G is not prime, then G is not simple.*

Proof. Let G be a group of order $n < 60$.

1. If $n = p^k$ for a prime number p and $k \geq 2$, then the center $Z(G)$ is non-trivial. Hence G is not simple. Therefore, if $n = 4, 8, 9, 16, 25, 27, 32$ or 49, the group G is not simple.

2. If $n = pq$ where p and q are distinct primes such that $p < q$, then the Sylow q-subgroup is normal in G. (Why?)

   Hence G is not simple, if $n = 6, 10, 14, 15, 21, 22, 26, 33, 34, 35, 38, 39, 46, 51, 55, 57$ or 58.

3. If $n = p^2q$ where p and q are distinct primes, then G is not simple. (See Exercise 5.3.15).

   As a consequence the group G is not simple, if $n = 12, 20, 28, 44, 45, 50$ or 52.

4. If $n = pqr$ for distinct primes p, q and r then one of the Sylow subgroups of G is normal in G. Consequently, if $n = 30$ or 42 then G is not simple.

5. The remaining cases are $n = 24, 54$ and 56.

   If $n = 24$ and $n_3 = 3t + 1$ denotes the number of Sylow 3-subgroups of G, then n_3 must divide 8, an impossibility. Therefore Sylow 3-subgroup is normal in G and G is not simple. Similar reasoning proves that Sylow 3-subgroup is normal in G, if $n = 54$.

   Let $n = 56 = 2^3 \times 7$. Then a counting argument shows that either $n_7 = 1$ or $n_2 = 1$. Thus G is not simple, if $n = 56$.

Thus we have shown that if G is a group of order less than 60 and not prime, then G is simple. □

5.3.3 Groups of order 12

Let G be group of order 12. Let H be a Sylow 2-subgroup and K be a Sylow 3-subgroup of G.

Let n_2 be the number of Sylow 2-subgroups and n_3 be the number of Sylow 3-subgroups of G.

Since $n_2 \equiv 1 \pmod 2$ there exists $l \in \mathbb{N}$ such that $n_2 = 2l + 1$. Further n_2 divides 12, the order of the group G. Hence $n_2 = 1$ or 3.

Similarly if we denote by n_3, the number of Sylow 3-subgroups of G, then $n_3 = 1$ or 4. First we claim that $n_2 = 1$ or $n_3 = 1$. We can prove this result by appealing to Exercise 5.3.15. However for the sake of completeness, we give a proof here.

Assume the contrary. Let $H = H_1$, H_2 and H_3 be the Sylow 2-subgroups and $K = K_1$, K_2, K_3 and K_4 be the Sylow 3-subgroups of G. Observe that $K_i \cap K_j = \{e\}$.

The number of elements in $H_1 \cup H_2$ is at least 5 and and the number of elements in $\cup_{j=1}^{4} K_j$ is 9. Since $H_i \cap K_j = \{e\}$, the union $\cup_{i=1}^{3} H_i \cup (\cup_{j=1}^{4} K_j)$ will have at least 14 elements, a contradiction. Therefore either $n_2 = 1$ or $n_3 = 1$.

- $n_2 = 1$ and $n_3 = 1$.

  In this case H and K are the unique Sylow subgroups of G. Hence they are normal in G. Since $H \cap K = \{e\}$, the map

$$\varphi : H \times K \to G$$

defined by

$$\varphi(h,k) := hk$$

is an isomorphism. Further the subgroup K being a group of order 3 is cyclic and H being a group of order 4 is abelian and it is isomorphic to $\mathbb{Z}_4$ or $\mathbb{Z}_2 \times \mathbb{Z}_2$. Therefore G must be isomorphic to $\mathbb{Z}_4 \times \mathbb{Z}_3$ or $\mathbb{Z}_2 \times \mathbb{Z}_2 \times \mathbb{Z}_3$. We have already seen that $\mathbb{Z}_4 \times \mathbb{Z}_3$ is isomorphic to $\mathbb{Z}_{12}$ and $\mathbb{Z}_2 \times \mathbb{Z}_3$ is isomorphic to $\mathbb{Z}_6$. Thus G is either isomorphic to $\mathbb{Z}_{12}$ or $\mathbb{Z}_2 \times \mathbb{Z}_6$.

- $n_3 = 4$ and $n_2 = 1$.

  We shall prove that G is isomorphic to A_4. To prove this, we will first show that G is a subgroup of S_4 and then show that the image of G in S_4 contains all the three cycles.

  Let $X = \{K_1, K_2, K_3, K_4\}$ be the set of Sylow 3-subgroups of G. Since all the Sylow 3-subgroups of G are conjugates, there is an action of G on X, namely, $(g, K_i) \mapsto gK_ig^{-1}$. Let $\phi : G \to S_4 :=$ Aut(X) be the homomorphism induced by this action of G on X. We will now show that ϕ is one-one.

  Fix a subgroup K_i. Since the group G acts transitively on X, it follows that $\mathcal{O}_{K_j} = X$. Let G_{K_j} be the stabilizer of K_j. Then $G_{K_j} = N(K_j)$, the normalizer of K_j.(Why?). We know that $|X| = |G/N(K_j)|$. Since $|X| = 4$ and G is order 12, it follows that $N(K_j)$ of order 3. Hence $N(K_j) = K_j$.

  Let $g \in \ker(\phi) := \{g \in G : \phi(g) = \mathrm{Id}\}$. Then $\phi(g)(K_j) = gK_jg^{-1} = K_j$ for all j. This proves that an element $g \in G$ is in the kernel of ϕ iff $g \in \cap_{j=1}^{4} K_j = \{e\}$. i.e., $g = e$. Thus we have shown that the map $\phi : G \to S_4$ is one-one.

  Let $e \neq x$ be an element in K_j for some j. Since x is of order 3 and ϕ is one-one, the element $\phi(x)$ is also of order 3. Thus for every element x in K_j, the element $\phi(x)$ is an even permutation. This proves that the image of ϕ has at least $9 = |\cup_{j=1}^{4} K_j|$ even permutations. Since the order of Im(ϕ) divides the order of G and it has at least 9 even permutations, the image is contained in A_4, all even permutations in S_4. Further the order of Im(ϕ) is 12. Hence Im$(\phi) = A_4$, the group of even permutations and G is isomorphic to A_4.

- $n_2 = 3$ and $n_3 = 1$.

  Here we have two cases, namely, H is either a cyclic group of order 4 or $H \simeq \mathbb{Z}_2 \times \mathbb{Z}_2$.

  Let K be the unique Sylow 3-subgroup of G. Since K is normal in G and $H \cap K = \{e\}$, it follows that $HK = G$.

  First we assume that $H = H_1$ is cyclic and let $H := \langle x \rangle$. Let $e \neq y \in K$. If $xy = yx$, then we get a homomorphism from $H \times K \to HK = G$ which is also an isomorphism. This will prove that G is cyclic, a contradiction. Therefore $xy \neq yx$. On the other hand, since K is normal in G, the element $xyx^{-1} (\neq e)$ must be in K. Therefore $xyx^{-1} = y^2 = y^{-1}$ and

$$G := \{x^iy^j : 1 \leq i \leq 4, 1 \leq j \leq 3\ ,\ x^4 = y^3 = e \text{ and } xy = y^{-1}x\}.$$

Now we assume that $H = \mathbb{Z}_2 \times \mathbb{Z}_2$. Since K is normal in G, we get a non-trivial homomorphism $\varphi : H \to \mathrm{Aut}(K)$ defined by $\varphi_h(k) := hkh^{-1}$. Arguing exactly as in the previous paragraph, we can show that the homomorphism is non-trivial.

Since the order of Aut(K) is 2 and the order of H is 4 the kernel is of order 2 and the image is of order 2. Let $e \neq u$ be in the kernel of φ. Then $\varphi(u)(k) = uku^{-1} = k$. Let $w \in H$ be such that $\varphi(w) \neq Id$, the identity automorphism. Then $\varphi(w)(k) = wkw^{-1} = k^2 = k^{-1}$. (Why?).

Let $x = uk = ku$. Since H is abelian, $uw = wu$. We use this to conclude that $wx = wuk = uwk = uk^{-1}w = k^{-1}uw = k^{-1}u^{-1}w = x^{-1}w$ and x is of order 6. Hence

$$G := \{x^i w^j : x^6 = e = w^2 \,\&\, wx = x^{-1}w\}$$

is isomorphic to D_6.

Exercise 5.3.19. Find all groups of order 28 and 44.

5.4 Finitely generated Abelian groups

In this section we describe the structure of finitely generated abelian groups.
We shall write the binary operation as addition throughout this section.

Exercise 5.4.1. Let G be an abelian group and S be a finite subset of G. Show that

1. $\langle S\rangle := \left\{\sum_{a\in S} n_a.a : n_a \in \mathbb{Z}\right\}$ is a subgroup of G and
2. $\langle S\rangle = \cap_{S\subseteq H} H$ where H is a subgroup containing S.

Definition 5.4.2. Let G be an abelian group.

1. If S is a finite subset of G, then $\langle S\rangle$ is called the subgroup generated by S.
2. We say that G is finitely generated if there exists a finite subset $S \subseteq G$ such that $\langle S\rangle = G$.

Set of generators for a group need not be unique. We have few examples below.

Example 5.4.3. 1. Let $G = (\mathbb{Z}, +)$. Then G is a finitely generated abelian group. We can take $S = \{1\}$. The set $T = \{-1\}$ is also a set of generator for G. Let k be a non-zero integer and $S = \{k\}$. Does S a generate G?

2. Let $G = (\mathbb{Z}, +)$ and $S = \{2, 3\}$. Then S generates G.

   Observe that $1 = 3 - 2$. Since 1 is written as integral linear combination of elements of S, every integer $k = 3\,k - 2\,k$.

3. Let $k \in \mathbb{N}$ and $S = \{2, n = 2k + 1\}$. Then S generates the group $\mathbb{Z}$.

It is enough to show that 1 can be written as an integral linear combination of 2 and n, then we are through. We can write $1 = an + b2$, where $a = 1$ and $b = -k$.

4. Let $G = \mathbb{Z}$ and $S = \{p, q\}$ where p and q are two prime numbers. Then S generates $\mathbb{Z}$.

5. Let $S = \{a_1, a_2, \ldots, a_k\}$ be such that the $\gcd(a_1, a_2, \ldots, a_k) = 1$. Then S generates $\mathbb{Z}$.

   Since $\gcd(a_1, a_2, \ldots, a_k) = 1$, by Bezout's identity, there exist integers n_i, for $1 \le i \le k$, such that $1 = \sum_{i=1}^{k} n_i a_i$.

These examples tell us that a group can have generating sets of different sizes.

Exercise 5.4.4. 1. Let $G = k\mathbb{Z}$. Show that G is finitely generated and find a set of generator of G.

2. Let $n \geq 1$ and $G = \overbrace{\mathbb{Z} \times \mathbb{Z} \times \cdots \times \mathbb{Z}}^{n-\text{times}}$. Show that G is finitely generated.

Exercise 5.4.5. 1. Let $f : G_1 \to G_2$ be an isomorphism between two groups G_1 and G_2. If S is a set of generators of G_1, show that $f(S)$ is a set of generators of the group G_2.

2. Let G be a finitely generated subgroup of $(\mathbb{Q}, +)$. Show that G is cyclic.

   Suppose G is generated by two elements, say $\frac{m_1}{n_1}$ and $\frac{m}{n_2}$. Observe that $\frac{m_1}{n_1} = m_1 n_2 \frac{1}{n_1 n_2}$ and $\frac{m_2}{n_2} = m_2 n_1 \frac{1}{n_1 n_2}$. Therefore, we can write

   $$k_1 \frac{m_1}{n_1} + k_2 \frac{m_2}{n_2} = (k_1 m_1 n_2 + k_2 m_2 n_1) \frac{1}{n_1 n_2}$$

   for all k_1 and k_2 in $\mathbb{Z}$. This proves that every element of G is an integral multiple of $\frac{1}{n_1 n_2}$. Thus the subgroup G is contained in $\mathbb{Z}[\frac{1}{n_1 n_2}]$. Since $\mathbb{Z}[\frac{1}{n_1 n_2}]$ is cyclic, it follows that G is also cyclic.
   If G is generated by k elements, say $\frac{m_i}{n_i}$ for $1 \le i \le k$ then, following the arguments above, we can show that G is a subgroup of $\mathbb{Z}[\frac{1}{n_1 n_2 \cdots n_k}]$. Hence G is cyclic.

Exercise 5.4.6. 1. Let $G = \mathbb{Q}$. Is G finitely generated? How about $G = \mathbb{R}$?

2. Let G be a finite abelian group. Show that G is finitely generated.

We will now prove the main theorem in this section.

Theorem 5.4.7. *Let G be a finitely generated abelian group. Then G is a direct product of cyclic groups.*

An important step in proving this theorem is the following lemma.

Lemma 5.4.8. *Let $n \geq 2$ and $a_1, a_2, \ldots, a_n \in \mathbb{Z}$ be such that* $\gcd(a_1, a_2, \ldots, a_n) = 1$. *Then there exists an $n \times n$ matrix A with entries in $\mathbb{Z}$ such that*

1. $\det(A) = 1$ *and*

2. *the integer a_i is the i-th entry in the first row for $1 \leq i \leq n$.*

Let us do some numerical examples before starting with the proof.

Let $n = 2$ and a_1, a_2 be two relatively prime integers. Then there exists integers x and y such that $a_1x + a_2y = 1$ and the matrix A is given by $\begin{pmatrix} a_1 & a_2 \\ -y & x \end{pmatrix}$.

Let $n = 2$ and $a_1 = 2$ and $a_2 = 3$. Then a_1 and a_2 are relatively prime and the integers $x = -1$ and $y = 1$ such that $a_1x + a_2y = 1$. Therefore the matrix A is $\begin{pmatrix} 2 & 3 \\ -1 & -1 \end{pmatrix}$ then $\det(A) = 1$ and $(2, 3)$ occurs as the first row of the matrix A.

Let $a_1 = 7$ and $a_2 = 17$. Then $x = 5$ and $y = -2$. Therefore the matrix A is $\begin{pmatrix} 7 & 17 \\ 2 & 5 \end{pmatrix}$.

Exercise 5.4.9. Let $a_1 = 17$ and $a_2 = 19$. Find the matrix A.

We will work out an example when $n = 3$. Let $a_1 = 4$, $a_2 = 6$ and $a_3 = 7$. The integers 4, 6 and 7 are relatively prime. The gcd of 4 and 6 is 2. We write $4 = 2.2$ and $6 = 2.3$. Since 2 and 3 are relatively prime there exists a 2×2 matrix with determinant 1 and $(2, 3)$ appears as the first row, for example $\begin{pmatrix} 2 & 3 \\ -1 & -1 \end{pmatrix}$. Since $2 = (4, 6)$ and 7 are relatively prime there exists integers s and t such that $7s + 2t = 1$. The pair $(s = 1, t = -3)$ is a solution for the equation. Now consider the matrix $\begin{pmatrix} 2.2 & 2.3 & 7 \\ -1 & -1 & 0 \\ 1.2.e & 1.3.e & -3 \end{pmatrix}$ where $e = \pm 1$. The determinant of this matrix is $-6 - 7e$. If we choose $e = -1$, then the matrix is $A = \begin{pmatrix} 2.2 & 2.3 & 7 \\ -1 & -1 & 0 \\ 1.2.(-1) & 1.3.(-1) & -3 \end{pmatrix}$ and the determinant is 1.

Example 5.4.10. Let $(a_1, a_2, a_3) = (15, 33, 35)$. Then the gcd of 15 and 33 is 3. We write $15 = 3 \times 5$ and $33 = 3 \times 11$. Now $1 = (-2) \times 5 + 1 \times 11$. Therefore $\det \begin{pmatrix} 5 & 11 \\ -1 & -2 \end{pmatrix} = 1$. The gcd of 3 and 35 is 1 and we write $1 = 35 \times (-1) + 3 \times 12$. Now consider the determinant $\det \begin{pmatrix} 3 \times 5 & 3 \times 11 & 35 \\ -1 & -2 & 0 \\ e \times (-1) \times 5 & e \times (-1) \times 11 & 12 \end{pmatrix}$ where $e = \pm 1$ will be chosen to make this determinant 1. When we compute this determinant we get $36 + 35e$. Hence $e = -1$ and the matrix is $\begin{pmatrix} 15 & 33 & 35 \\ -1 & -2 & 0 \\ 5 & 11 & 12 \end{pmatrix}$. Let us now consider the triple $(15, 35, 33)$. Then we will get a different matrix A as shown below.

We write $15 = 3 \times 5$ and $35 = 5 \times 7$. Then 3 and 7 are relatively prime and $1 = (-2) \times 3 + 1 \times 7$. Therefore $\det \begin{pmatrix} 3 & 7 \\ -1 & -2 \end{pmatrix}$. The integers 5 and 33 are relatively prime and we write $1 = (-13) \times 5 + 2 \times 33$. We want to choose

e so that $\det \begin{pmatrix} 5\times 3 & 5\times 7 & 33 \\ -1 & -2 & 0 \\ e\times 2\times 3 & e\times 2\times 7 & -13 \end{pmatrix}$ is 1. In this case $e = -1$ and the matrix is $\begin{pmatrix} 5\times 3 & 5\times 7 & 33 \\ -1 & -2 & 0 \\ -6 & -147 & -13 \end{pmatrix} = \begin{pmatrix} 15 & 35 & 33 \\ -1 & -2 & 0 \\ -6 & -147 & -13 \end{pmatrix}$.

Proof of Lemma 5.4.8

Proof. Let $n \geq 2$ and a_i for $1 \leq i \leq n$ be mutually co-prime integers. If $n = 2$, then, by Bezout identity, there exists integers r and s such that $ar + bs = 1$. Hence the matrix $\begin{pmatrix} a & b \\ -s & r \end{pmatrix}$ has determinant 1 with first (a_1, a_2) and this completes the proof.

Now assume $n \geq 3$. We prove it by induction on n. Assume that the result is true for $k < n$. Let a_i for $1 \leq i \leq n$ be mutually co-prime integers. Let $d := (a_1; a_2; \ldots; a_{n-1})$, the gcd of $a_1, a_2, \ldots, a_{n-1}$ and we write $a_i := db_i$ for $1 \leq i \leq n-1$. Then the integers $b_1, b_2, \ldots, b_{n-1}$ are mutually co-prime and by induction there exists a matrix B such that $\det(B) = 1$ and the first row of B is $B_1 = (b_1, b_2, \ldots, b_{n-1})$.

By assumption d and a_n are relatively prime. Hence there exists integers rb and s such that $dr + a_n s = 1$.

Now the matrix $\begin{pmatrix} d.b_1 & d.b_2 & \cdots & d.b_{n-1} & a_n \\ * & * & \cdots & * & 0 \\ \vdots & \vdots & \ddots & \vdots & \vdots \\ e.s.b_1 & e.s.b_2 & \cdots & e.s.b_{n-1} & r \end{pmatrix}$ where $e = \pm 1$ has determinant $dr \pm a_n s.e$. By choosing e properly, we can make the determinant of this matrix to be 1. □

Lemma 5.4.11. *Let $x_1, x_2, \ldots, x_n$ be generators of an abelian group G written additively. Let $a_1, a_2, \ldots, a_n$ be integers with gcd 1. Then $a_1x_1 + a_2x_2 + \cdots + a_nx_n$ may be chosen as one of the generators for the group G.*

Proof. Let $a_1, a_2, \ldots, a_n$ be integers with gcd 1 and $A = (a_{ij})$ be a matrix as in lemma 5.4.8. Since $A^{-1} = \det(A) \cdot \operatorname{adj}(A)$ and $\operatorname{adj}(A)$ is a matrix with integer entries, it follows that $A^{-1} = (b_{ij})$ is also an integer matrix. Let X be the column vector $\begin{pmatrix} x_1 \\ x_2 \\ \vdots \\ x_n \end{pmatrix}$ and $Y = AX$. If $Y = \begin{pmatrix} y_1 \\ y_2 \\ \vdots \\ y_n \end{pmatrix}$ then we can write $y_i = \sum_{j=1}^n a_{ij}x_j$. In particular $y_1 = \sum_{j=1}^n a_{1j}x_j = \sum_{j=1}^n a_jx_j$.

Further since $X = A^{-1}Y$, we can also write $x_i = \sum_{j=1}^n b_{ij}y_j$ for $1 \leq i \leq n$.

Let x be an element of the group G. Then there exist integers α_i such that $x = \sum_{i=1}^n \alpha_i x_i$. Now we substitute the expression for x_i in terms of y_j in

$x = \sum_{i=1}^{n} \alpha_i x_i$ and get

$$\begin{aligned} x &= \sum_{i=1}^{n} \alpha_i x_i \\ &= \sum_{i=1}^{n} \alpha_i (\sum_{j=1}^{n} b_{ij} y_j) \\ &= \sum_{j=1}^{n} (\sum_{i=1}^{n} \alpha_i b_{ij}) y_j. \end{aligned}$$

This proves that the set $\{y_1 = \sum_{j=1}^{n} a_j x_j, y_2, \cdots, y_n\}$ generates G. □

Theorem 5.4.12 (Basis Theorem). *Let G be a finitely generated abelian group. Then G is a direct product of cyclic groups.*

Proof. Let $\mathcal{S} := \{S \subseteq G : S \text{ is finite and } \langle S \rangle = G\}$ and $k = \min\{|S| : S \in \mathcal{S}\}$.

If $k = 1$, then G is cyclic and the proof is complete.

If $k \geq 2$, let $S = \{x_1, x_2, \ldots, x_k\}$ in $\mathcal{S}_k$ be a set of generators of G with the following properties:

1. The elements x_i are labeled such that $o(x_i) \geq o(x_{i+1})$ for $1 \leq i \leq k-1$ and

2. if $T = \{y_1, y_2, \ldots y_k\}$ is another generating set with k-elements then $o(x_k) \leq o(y_i)$ for $1 \leq i \leq k$.

Let $S_1 = S \setminus \{x_k\}$ and H is the group generated by S_1. Then by induction hypothesis, the subgroup H is a direct product of cyclic groups.

Let $K = \langle x_k \rangle$ be the cyclic group generated by x_k and if $o(x_k)$ is finite, let $o(x_k)$ be m_k. We will now prove that $H \cap K = \{0\}$. Suppose not. Let $0 \neq y$ be an element in $H \cap K$. Then there exist integers $r_k < m_k$ and r_i such that $r_k x_k = r_1 x_1 + r_2 x_2 + \cdots + r_{k-1} x_{k-1}$. This means that $r_1 x_1 + r_2 x_2 + \cdots + r_{k-1} x_{k-1} - r_k x_k = 0$.

- Suppose $d := \gcd(r_1, r_2, \cdots, r_k) = 1$. Then by Lemma 5.4.11 we can choose $r_1 x_1 + r_2 x_2 + \cdots + r_{k-1} x_{k-1} - r_k x_k$ as one of the elements of a set of generators. This shows that G has a generating set F with $k-1$ elements, a contradiction.

- If $d \neq 1$, let $s_i = \frac{r_i}{d}$. Then $\gcd(s_1, s_2, \cdots, s_k) = 1$ and hence we can choose a set of generators with $y_1 := s_1 x_1 + s_2 x_2 + \cdots + s_k x_k$ as one of the elements of the generating set. Further $d(s_1 x_1 + s_2 x_2 + \cdots + s_k x_k) = r_1 x_1 + r_2 x_2 + \cdots + r_k x_k = 0$. This is again a contradiction to our assumption.

  If $d = m_k$, then $r_k = s_k m_k$ and $r_k x_k = s_k(m_k x_k) = 0$. Therefore $d \neq m_k$ and the proof is complete.

□

Let G be a finitely generated abelian group.

Let $S = \{x_1, x_2, \ldots, x_r, x_{r+1}, \ldots x_{r+s}\}$ be a minimal set of generators such that $o(x_i)$ is infinite for $1 \leq i \leq r$ and $o(x_{r+j})$ is finite for $1 \leq j \leq s$. Then G is isomorphic to the direct product $\mathbb{Z}^r \times G_1 \times G_2 \times \cdots \times G_s$ where $\mathbb{Z}^r := \overbrace{\mathbb{Z} \times \mathbb{Z} \times \cdots \times \mathbb{Z}}^{r-\text{times}}$ is the r-fold product and $G_j := \langle x_{r+j} \rangle$ for $1 \leq j \leq s$. Let $o(x_{r+j}) = n_j$.

Definition 5.4.13. The integer r is called the rank of the group G.

The integers $n_1, n_2, \ldots, n_s$ are called invariant factors of G and this decomposition of G is called the invariant factor decomposition of G.

Let G be a finite abelian group of order n, say. Then G is finitely generated and by Theorem 5.4.12, we can write it as a product of cyclic groups, say, $G_1 \times G_2 \times \cdots \times G_k$ where $G_i := \langle x_i \rangle$ and $n_{i+1} | n_i$ for $1 \leq i \leq k-1$. The order of the group G is $n = n_1 n_2 \cdots n_s$.

Theorem 5.4.12 also tells us that, for every sequence of integers $n_1, n_2, \ldots, n_s$ such that $n_j \geq 2$, $n_{j+1} | n_j$ and $n = n_1 n_2 \cdots n_s$, there is a unique abelian group of order n.

Let G be an abelian group of order $n \geq 2$ and $n_1, n_2, \ldots, n_s$ be its invariant factors. If p is a prime number that divides $n = n_1 n_2 \cdots n_s$, then p divides n_i for some i. Since n_i divides n_{i-1}, it follows that p divides n_1, the largest invariant factor of G. This shows that every prime divisor of n divides the largest invariant factor. In particular if $n = p_1 p_2 \cdots p_s$, the product of distinct primes, then $n = n_1$. As a consequence it follows that there is only one abelian group of order n, if $n = p_1 p_2 \cdots p_s$.

We will now work out an example if $n = p_1^{\alpha_1} p_2^{\alpha_2} \cdots p_t^{\alpha_t}$ with at least one $\alpha_i \geq 2$.

Example 5.4.14. Let $n = 3^2 \times 5^2 \times 7 = 1575$. The largest invariant factor is either $n = n_1 = 3^2 \times 5^2 \times 7 = 1575$, $n_1 = 3 \times 5^2 \times 7 = 525$, $n_1 = 3^2 \times 5 \times 7 = 315$ or $n_1 = 3 \times 5 \times 7 = 105$. Then $n_2 = 1$, 3, 5 or $3 \times 5 = 15$ and any abelian group G of order 1575 must be isomorphic to either $\mathbb{Z}_{1575}$, $\mathbb{Z}_{525} \times \mathbb{Z}_3$, $\mathbb{Z}_{315} \times \mathbb{Z}_5$ or $\mathbb{Z}_{105} \times \mathbb{Z}_{15}$.

If $n = p_1^{\alpha_1} p_2^{\alpha_2} \cdots p_t^{\alpha_t}$ is the prime factorization of n, then invariant factors for abelian groups of order n can be determined by finding all abelian groups of order $p_i^{\alpha_i}$ for $1 \leq i \leq t$. We will do this at present.

Let G be a finite abelian group of order p^α. Suppose $(\alpha_1, \alpha_2, \ldots, \alpha_k)$ be a partition of α. That is $\alpha = \alpha_1 + \alpha_2 + \cdots + \alpha_k$ such that $\alpha_i \geq \alpha_{i+1} \geq 1$ for $1 \leq i \leq k-1$. Let $G = \mathbb{Z}_{p^{\alpha_1}} \times \mathbb{Z}_{p^{\alpha_2}} \times \cdots \times \mathbb{Z}_{p^{\alpha_k}}$.

We claim that for each partition $(\alpha_1, \alpha_2, \ldots, \alpha_k)$ of α there is a unique abelian group of order p^α that is isomorphic to $\mathbb{Z}_{p^{\alpha_1}} \times \mathbb{Z}_{p^{\alpha_2}} \times \cdots \times \mathbb{Z}_{p^{\alpha_k}}$.

Lemma 5.4.15. *Let p be a prime number and α be a natural number. Let $\alpha = \alpha_1 + \alpha_2 + \cdots + \alpha_k = \beta_1 + \beta_2 + \cdots + \beta_l$ where α_i, β_j are natural numbers such that $\alpha_i \geq \alpha_{i+1}$ for $1 \leq i \leq k-1$ and $\beta_j \geq \beta_{j+1}$ for $1 \leq j \leq l-1$. If the abelian groups $\mathbb{Z}_{p^{\alpha_1}} \times \mathbb{Z}_{p^{\alpha_2}} \times \cdots \times \mathbb{Z}_{p^{\alpha_k}}$ and $\mathbb{Z}_{p^{\beta_1}} \times \mathbb{Z}_{p^{\beta_2}} \times \cdots \times \mathbb{Z}_{p^{\beta_l}}$ are isomorphic, then $k = l$ and $\alpha_i = \beta_i$.*

Proof. We will prove the result by induction on α. If $\alpha = 1$ then there is a unique cyclic subgroup $\mathbb{Z}_p$ of order p and we are done.

Let $\alpha \geq 2$ and assume the result to be true for all $\beta < \alpha$.

Let $\alpha = \alpha_1 + \alpha_2 + \cdots + \alpha_k = \beta_1 + \beta_2 + \cdots + \beta_l$ where α_i, β_j are natural numbers such that $\alpha_i \geq \alpha_{i+1}$ for $1 \leq i \leq k-1$ and $\beta_j \geq \beta_{j+1}$ for $1 \leq j \leq l-1$. Let $G_1 := \mathbb{Z}_{p^{\alpha_1}} \times \mathbb{Z}_{p^{\alpha_2}} \times \cdots \times \mathbb{Z}_{p^{\alpha_k}}$ and $G_2 := \mathbb{Z}_{p^{\beta_1}} \times \mathbb{Z}_{p^{\beta_2}} \times \cdots \times \mathbb{Z}_{p^{\beta_l}}$. Let $\varphi : G_1 \to G_2$ be an isomorphism of these two groups.

Let x be an element of order p^{α_1} in G_1. Then $\varphi(x)$ is an element of order p^{α_1} and it is an element of maximal order. Therefore $p^{\alpha_1} \geq p^{\beta_1}$. The maximal order of an element in G_2 is p^{β_1}. Therefore $\alpha_1 = \beta_1$.

Let $H_1 = G_1/\mathbb{Z}_p^{\alpha_1}$ and $H_2 = G_2/\mathbb{Z}_p^{\alpha_1}$. Then H_1 and H_2 are isomorphic of order $p^{\alpha-\alpha_1}$. Hence by induction hypothesis $k-1 = l-1$ and $\alpha_i = \beta_i$ for $2 \leq i \leq k$. □

We will now state the following theorem.

Theorem 5.4.16. *Let G be an abelian group of order n and $n = p_1^{\alpha_1} p_2^{\alpha_2} \cdots p_t^{\alpha_t}$ be the unique factorization of n. Then*

1. *G is isomorphic to $G_1 \times G_2 \times \cdots \times G_t$ where $|G_i| = p_i^{\alpha_i}$ and each*

2. *G_i is isomorphic to $\mathbb{Z}_{p_i^{\alpha_{i1}}} \times \mathbb{Z}_{p_i^{\alpha_{i2}}} \times \cdots \times \mathbb{Z}_{p_i^{\alpha_{it_i}}}$ where $(\alpha_{i1}, \alpha_{i2}, \ldots, \alpha_{it_i})$ is a partition of α_i.*

Further this decomposition is unique.

Definition 5.4.17. The decomposition of G as in Theorem 5.4.16 is called the elementary divisor decomposition of G.

We will not prove this result. We illustrate the result with two examples.

Example 5.4.18. 1. Let $n = 8 = 2^3$. Then the partitions of 3 and the distinct isomorphism classes of abelian groups are given in the following table.

partition	Group
3	$\mathbb{Z}_8$
$(1,2)$	$\mathbb{Z}_4 \times \mathbb{Z}_2$
$(1,1,1)$	$\mathbb{Z}_2 \times \mathbb{Z}_2 \times \mathbb{Z}_2$

2. Let $n = 108 = 2^2 \times 3^3$. We can read the isomorphism classes of abelian groups from the following table.

Partitions of 2 and 3	Group
2, 3	$\mathbb{Z}_4 \times \mathbb{Z}_{27}$
$(1,1)$, 3	$\mathbb{Z}_2 \times \mathbb{Z}_2 \times \mathbb{Z}_{27}$
2, $(2,1)$	$\mathbb{Z}_4 \times \mathbb{Z}_9 \times \mathbb{Z}_3$
$(1,1)$, $(2,1)$	$\mathbb{Z}_2 \times \mathbb{Z}_2 \times \mathbb{Z}_9 \times \mathbb{Z}_3$
2, $(1,1,1)$	$\mathbb{Z}_4 \times \mathbb{Z}_3 \times \mathbb{Z}_3 \times \mathbb{Z}_3$
$(1,1)$, $(1,1,1)$	$\mathbb{Z}_2 \times \mathbb{Z}_2 \times \mathbb{Z}_3 \times \mathbb{Z}_3 \times \mathbb{Z}_3$

5.5 Finite subgroups of $E(n)$

This section is devoted to classifying finite subgroups of $E(n)$ for $n = 2, 3$.

The first crucial fact is that any finite subgroup of $E(n)$ is conjugate to a finite subgroup of $\mathrm{O}(n, \mathbb{R})$. This is proved in the following theorem.

Theorem 5.5.1. *Let G be a finite subgroup of $E(n)$. Then*

1. *there exists a point p_0 in $\mathbb{R}^n$ such that $gp_0 = p_0$ for all $g \in G$ and*
2. *G is conjugate to a subgroup of $O(n, \mathbb{R})$.*

Proof. We choose a point $p \in \mathbb{R}^n$ and let $p_0 := \frac{1}{|G|} \sum_{g \in G} g(p)$. We will now prove that the group G fixes the point p_0.

Let $h \in G$. We write $h = (A, v)$ where $A \in \mathrm{O}(n, \mathbb{R})$ and $v \in \mathbb{R}^n$. Then

$$\begin{aligned} h(p_0) &= h\left(\frac{1}{|G|}\sum_{g\in G} g(p)\right) \\ &= \frac{1}{|G|}\sum_{g\in G} A(g(p)) + v \\ &= \frac{1}{|G|}\sum_{g\in G} (A(g(p)) + v) \\ &= \frac{1}{|G|}\sum_{g\in G} h(g(p)) \\ &= \frac{1}{|G|}\sum_{g\in G} g(p) \\ &= p_0. \end{aligned}$$

This proves the first part of the theorem. For the second part of the statement, define $\tau : \mathbb{R}^n \to \mathbb{R}^n$ by $\tau(x) := x + p_0$.

If g is in G, then $\tau^{-1} g \tau(0) = 0$. Hence $\tau^{-1} g \tau$ is in $\mathrm{O}(n, \mathbb{R})$ for all $g \in G$. This proves that every finite subgroup of $E(n)$ is conjugate to a finite subgroup of $\mathrm{O}(n, \mathbb{R})$. □

Remark 5.5.2. If G is a nontrivial finite subgroup of $E(n)$, the point p_0 is unique.

Corollary 5.5.3. *If G is a finite subgroup of $E(n)$, then G does not contain translations of $\mathbb{R}^n$.*

5.5.1 Finite Subgroups of $E(2)$

In view of Theorem 5.5.1, if we find all finite subgroups of $O(2, \mathbb{R})$ we know all finite subgroups of $E(2)$.

Let G be a finite subgroup of $O(2, \mathbb{R})$ and $|G| \geq 2$.

Let $H := \{A \in G : \det(A) = 1\}$. Then H is a finite subgroup of $\mathrm{SO}(2, \mathbb{R})$. By Theorem 2.2.47, the subgroup H is cyclic. Let $|H| = n$.

Assume $n = 1$. Then we claim that the group G is of order 2.

Let $\rho \in G \setminus H$. Then $\det(\rho) = -1$ and ρ is a reflection about an axis and $\rho^2 = Id$. (See [10].) If there is an element A in $G \setminus H$, then $A\rho$ is of determinant 1 and hence $A\rho = Id$. This means that $A = \rho^{-1} = \rho$. Therefore $G = \{Id, \rho\}$.

We will now assume that $n \geq 2$.

If $G = H$, then G is cyclic and it is the group of rotations of a regular $|G|$-sided polygon.

Let us now assume that $G \neq H$. Then there exists a $\rho \in G \setminus H$. As earlier ρ is a reflection about an axis and $\rho^2 = Id$. In this case there exists a line $\ell_v = \mathbb{R} \cdot v$ and a line $\ell_u := \mathbb{R} \cdot u$, orthogonal to each other and such that $\rho(w) = w$ if w is in ℓ_v and $\rho(w) = -w$ if w is in ℓ_u. (For a proof see [10]). Further if A is in $G \setminus H$, then ρA is a rotation and hence it is in H. This proves that A is in ρH, the coset of H. Therefore the subgroup H is of index 2 in G and H is normal in G.

Let r_θ be a generator of H. Then $\rho r_\theta \rho = {r_\theta}^i$ for some $i < |H|$. Since $\rho^2 = Id$, it follows that $r_\theta = \rho^2 r_\theta \rho^2 = {r_\theta}^{i^2}$. Hence we can write ${r_\theta}^{i^2-1} = Id$. Now by Lagrange's theorem n divides $i^2 - 1$. Therefore $i = 1$ or $i = n - 1$.

Assume $i = 1$. This means that ρ and r_θ commute and therefore $\rho r_\theta(w) = r_\theta \rho(w) = r_\theta(w)$ for all $w \in \ell_v$. From this we conclude that $r_\theta(w)$ is also in ℓ_v, if w lies on the line ℓ_v. A nontrivial rotation fixes only the origin and if it fixes a line, then it must be a rotation by an angle π. Therefore $r_\theta = Id$ or $-Id$. This proves that $G = \{\pm Id, \pm\rho\}$ is isomorphic to $\mathbb{Z}_2 \times \mathbb{Z}_2$.

Now assume that $i = n - 1$. Then ${r_\theta}^i = {r_\theta}^{n-1} = {r_\theta}^{-1}$. Therefore $\rho r_\theta = {r_\theta}^{-1}\rho$ and the group G can be described as

$$G = \{\rho^i {r_\theta}^j : 1 \leq i \leq 2, 1 \leq j \leq n, {r_\theta}^n = \rho^2 = Id \ \& \ \rho r_\theta = {r_\theta}^{-1}\rho\}.$$

In this case the group G is the group of symmetries of a regular n-sided polygon.

We summarize this in the following theorem.

Theorem 5.5.4. *Let G be a finite subgroup of $O(2, \mathbb{R})$. Then one of the following holds.*

1. *G is the cyclic group of rotational symmetries of a regular n-sided polygon.*

2. *$G = D_n$, the dihedral group (of order $2n$) of all symmetries of a regular n-sided polygon.*

Theorem 5.5.5. *Let G be a finite group of $\mathbb{E}^2$. Then G is conjugate to one of the groups listed in Theorem 5.5.4.*

5.5.2 Finite Subgroups of $E(3)$

Let us first recall the Geometric version of Euler's theorem.

Theorem 5.5.6 (Euler's Theorem). *Let g be an element of $SO(3, \mathbb{R})$. Then there exists a non-zero vector $v \in \mathbb{R}^3$ such that $gv = v$ and $g(v^\perp) = v^\perp$.*

Proof. For a proof we refer to [11]. □

The theorem tells us that every element $g(\neq e)$ of $\mathrm{SO}(3,\mathbb{R})$ is a rotation about an axis: Let $g(\neq e)$ be an element of G. Then by the theorem, there exists a unit vector $v \in \mathbb{R}^3$ such that $gv =$ and $g(v^{\perp}) = v^{\perp}$. Therefore we can write the matrix of g as $g = \begin{pmatrix} 1 & * \\ * & A \end{pmatrix}$ with respect to an orthonormal basis $\{v, u, w\}$ of $\mathbb{R}^3$. Since g is in $\mathrm{SO}(3,\mathbb{R})$, it follows that A is in $\mathrm{SO}(2,\mathbb{R})$. Thus g is a rotation about the axis $\mathbb{R}v$.

Now we state the main theorem of this section.

Theorem 5.5.7. *Let G be a finite subgroup of $SO(3,\mathbb{R})$. Then G is one of the following.*

1. *A cyclic group of rotations about a line ℓ.*

2. *Dihedral group of symmetries of a regular n-sided polygon lying in a plane.*

3. *The tetrahedral group of symmetries of order 12 of a regular tetrahedron.*

4. *The octahedral group of order 24 of rotations of a cube or of a regular octahedron.*

5. *The icosahedral group of order 60 of rotations of a dodecahedron or a icosahedron.*

Proof of this theorem is beyond the scope of this book. For a proof we refer to [2] and [1]. For more details on Groups and Symmetry the reader may refer to [2].

5.6 Solvable Groups

In this section we will very briefly discuss a special class of groups called *Solvable Groups.*

Let G be a group. Given two elements a and b in G, the product $aba^{-1}b^{-1}$ is called the commutator of the elements a and b and it is denoted by $[a,b]$. We denote by $[G,G]$ the subgroup generated by all commutators $[a,b]$ of elements a, b of G.

Let us denote by $C := \{[a,b] : a, b \in G\}$. Then $[G,G]$ is the subgroup generated by the set C in G. The commutator subgroup $[G,G]$ is also denoted by $G^{(1)}$ and for $k \geq 2$, we denote by $G^{(k)}$, the commutator subgroup of $G^{(k-1)}$. That is $G^{(k)} = [G^{(k-1)}, G^{(k-1)}]$.

Definition 5.6.1. A group G is called a solvable group if there exists an integer $k \geq 1$ such that $G^{(k)} = e$.

Example 5.6.2. 1. Let G be an abelian group. Then $[a,b] = e$ for any two elements a, b in G. Hence $[G,G] = (e)$. Hence G is solvable.

2. The group S_3 is solvable.

Let us first list all the elements of G. If τ denotes an element of order 2 and σ denotes an element of order 3, then

$$G = \{e, \tau, \sigma, \sigma^2, \tau\sigma, \tau\sigma^2 : \tau\sigma = \sigma^{-1}\tau\}.$$

Now we use the relation $\tau\sigma = \sigma^{-1}\tau$ to conclude that $[a, b]$ is either σ or σ^2 for any two elements $a, b \in G$. Therefore $G^{(1)} = [G, G]$ is the cyclic subgroup of order 3 in S_3. Since $G^{(1)}$ is abelian, it follows that $G^{(2)} = (e)$. Hence G is solvable.

3. Let $G = D_n$, the dihedral group of order $2n$. Then G is solvable.

   The proof is same as the earlier example.
   Let τ denote a reflection and σ a rotation of degree $\frac{2\pi}{n}$ of a regular n-sided polygon. Then $\tau^2 = Id = \sigma^n$ and

   $$G = \{e, \tau, \sigma, \sigma^2, \ldots, \sigma^{n-1}, \tau\sigma, \tau\sigma^2 \ldots, \tau\sigma^{n-1} : \tau\sigma = \sigma^{-1}\tau\}.$$

   As in the earlier example, we use the relation $\tau\sigma = \sigma^{-1}\tau$ to conclude that, for any two elements $a, b \in G$, the commutator $[a, b]$ is σ^j for some $j \leq n$. Therefore $G^{(1)} = [G, G]$ is the subgroup of rotations order n in D_n and it is cyclic. Since $G^{(1)}$ is abelian, it follows that $G^{(2)} = (e)$. Hence G is solvable.

4. The group S_4 is solvable.

   Let us first observe that for any two elements σ and τ in S_4, the commutator $[\sigma, \tau] = \sigma\tau\sigma^{-1}\tau^{-1}$ is an even permutation. Hence the commutator subgroup $G^{(1)} = [G, G]$ is contained in A_4, the subgroup of even permutations.
   Let $\tau = (12)$ and $\sigma = (123)$. Then $(1,2)(1,2,3)(1,2)(321) = (123)$. This shows that (123) is in $G^{(1)} = [G, G]$. Since all the three cycles are conjugate in S_4 and $G^{(1)}$ is normal in G it follows that all the three cycles are in $G^{(1)}$. We have already shown that A_4 is generated by three cycles. Hence $A_4 \subseteq G^{(1)}$. On the other hand $G^{(1)}$ is already contained in A_4. This proves that $G^{(1)} = A_4$. (*Let us observe that the same arguments show that* $[S_n, S_n] = A_n$ *for any* $n \geq 3$).
   We have shown that A_4 does not have subgroups of order 12 or 8. Hence $G^{(2)}$ must be a group of order 4. However a group of order 4 is abelian and therefore $G^{(3)} = [G^{(2)}, G^{(2)}] = (e)$. This proves that S_4 is solvable.

5. The group S_n is not solvable for $n \geq 5$.

   We have shown that, for $n \geq 5$ the group A_n is simple. That is A_n does not have any proper normal subgroup. This means that $[A_n, A_n] = A_n$ for $n \geq 5$. Hence S_n is not solvable for $n \geq 5$.

Theorem 5.6.3. *Let G be a solvable group and H be a subgroup of G. Then H is solvable.*

Proof. Let H be a subgroup of a solvable group G. We need to show that there exists $l \geq 1$ such that $H^{(l)} = (e)$.

Since G is solvable there exists $k \geq 1$ be such that $G^{(k)} = (e)$. On the other hand, for every subgroup H of a group G, the subgroup $H^{(j)}$ is contained in the subgroup $G^{(j)}$ for every $j \geq 1$. Since $G^{(k)} = (e)$, it follows that $H^{(k)} = (e)$. Hence H is solvable. □

Theorem 5.6.4. *Let G and K be two groups and $f : G \to K$ be a homomorphism. If G is solvable then $f(G)$ is solvable.*

Conversely if $f(G)$ is solvable and $ker(f)$ solvable, then G is solvable.

Proof. Let a, b be two elements in G. Then

$$[f(a), f(b)] = f(a)f(b)f(a)^{-1}f(b)^{-1} = f(aba^{-1}b^{-1}) = f([a, b]).$$

This shows that $f(G^{(j)}) = [f(G)]^{(j)}$ for every $j \geq 1$. As a consequence it follows that $f(G)$ is solvable if G is solvable.

Conversely let us now assume that $f(G)$ and $H := \ker(f)$ are solvable. Let $k, l \geq 1$ be such that $[f(G)]^{(l)} = (e)$ and $H^{(k)} = (e)$. Since $f(G^{(l)}) = [f(G)]^{(l)} = (e)$, it follows that $G^{(l)} = H$. Therefore $H^{(1)} = [H, H] = [G^{(l)}, G^{(l)}] = G^{(l+1)}$ and $H^{(t)} = G^{(l+t)}$ for $t \geq 1$. This proves that $G^{(l+k)} = H^{(k)} = (e)$ and G is solvable. □

Corollary 5.6.5. *Let G be a group. If H is a normal subgroup of G such that H and G/H are solvable, then G is solvable.*

Proof. Let $\pi : G \to G/H$ be the quotient map. Now we apply Theorem 5.6.4 to conclude that G is solvable. □

Bibliography

[1] Michael Artin, *Algebra*, Prentice-Hall India, New Delhi, 1994.

[2] M. A. Armstrong, *Groups and Symmetry*, Undergraduate Texts in Mathematics, Springer, 1997.

[3] Oscar A. Campoli, *A Principal Ideal Domain That Is Not a Euclidean Domain*, The American Mathematical Monthly, Vol. 95, No. 9 (1988), pp. 868-871.

[4] David M Burton, *Elementary Number Theory*, McGraw-Hill International Edition, 2011.

[5] D. Dummit and R. Foote, *Abstract Algebra*, Wiley, 2004.

[6] J.A. Gallian, *Contemporary Abstract Algebra*, Narosa Publishing House, New Delhi, 1999.

[7] I. N. Herstein, Topics in Algebra, Wiley-India Edition, 2012.

[8] Thomas W. Hungerford, *A Counterexample in Galois Theory*, The American Mathematical Monthly, Vol.97, No.1 (1990), pp.54-57.

[9] Gareth A. Jones and J. Mary Jones, *Elementary Number Theory*, Springer Undergraduate Mathematics Series, Springer, 2011.

[10] S. Kumaresan, *Linear Algebra– A Geometric Approach*, Prentice-Hall India, 2000.

[11] S. Kumaresan and G. Santhanam, *An Expedition to Geometry*, TRIM Series 30, Hindustan Book Agency, New Delhi, 2005.

[12] J. P. Serre, *A Course in Arithmetic*, Graduate Texts in Mathematics 7, Spinger, 1973.

Index